Study and Solutions Guide

to Accompany

College Algebra: Concepts and Models

Second Edition

Larson/Hostetler/Hodgkins

Dianna L. Zook

Indiana University, Purdue University at Fort Wayne

D. C. Heath and Company

Lexington, Massachusetts   Toronto

Address editorial correspondence to:
D. C. Heath and Company
125 Spring Street
Lexington, MA 02173

Published simultaneously in Canada.

Printed in the United States of America.

International Standard Book Number: 0–669–41632–0

10 9 8 7 6 5 4 3 2 1

TO THE STUDENT

This *Study and Solutions Guide* is a supplement to the textbook *College Algebra: Concepts and Models, Second Edition* by Roland E. Larson, Robert P. Hostetler, and Anne V. Hodgkins. Solutions to selected exercises in the text are given with all essential algebraic steps included.

As a mathematics instructor, I often have students come to me with questions about the assigned homework. When I ask to see their work, the reply often is "I didn't know where to start." The purpose of the *Study Guide* is to provide brief summaries of the topics covered in the textbook and enough detailed solutions to problems so that you will be able to work the remaining exercises.

I would like to thank Meridian Creative Group for helping in the production of this guide. Also, I would like to thank my husband, Edward Schlindwein, for his support during the many months in which I worked on this project.

Good luck with your study of algebra.

Dianna L. Zook
Indiana University
Purdue University at Fort Wayne
Fort Wayne, Indiana 46805

CONTENTS

Chapter 7 Matrices and Determinants

Chapter 8 Sequences and Probability

Appendix A Conic Sections

CHAPTER P
Review of Fundamental Concepts of Algebra

P.1 Real Numbers: Order and Absolute Value

- You should know the following sets.
  (a) The set of real numbers includes the rational numbers and the irrational numbers.
  (b) The set of rational numbers includes all real numbers that can be written as the ratio p/q of two integers, where $q \neq 0$.
  (c) The set of irrational numbers includes all real numbers which are not rational.
  (d) The set of integers: $\{\ldots, -3, -2, -1, 0, 1, 2, 3, \ldots\}$
  (e) The set of natural numbers: $\{1, 2, 3, 4, \ldots\}$
  (f) The set of whole number: $\{0, 1, 2, 3, 4, \ldots\}$
- The real number line is used to represent the real numbers.
- Know the inequality symbols.
  (a) $a < b$ means a is less than b.
  (b) $a \leq b$ means a is less than or equal to b.
  (c) $a > b$ means a is greater than b.
  (d) $a \geq b$ means a is greater than or equal to b.
- You should know that

$$|a| = \begin{cases} a, & \text{if } a \geq 0 \\ -a, & \text{if } a < 0 \end{cases}$$

- Know the properties of absolute value.
  (a) $|a| \geq 0$
  (b) $|-a| = |a|$
  (c) $|ab| = |a||b|$
  (d) $\left|\frac{a}{b}\right| = \frac{|a|}{|b|}$
- The distance between a and b on the real line is $|b - a| = |a - b|$.

SOLUTIONS TO SELECTED EXERCISES

3. Determine which numbers in the given set are (a) natural numbers, (b) integers, (c) rational numbers, and (d) irrational numbers.

$$\left\{12,\ -13,\ 1,\ \sqrt{4},\ \sqrt{6},\ \tfrac{3}{2}\right\}$$

Solution

Note: $\sqrt{4} = 2$

(a) Natural numbers: $\{12,\ 1,\ 2\}$

(b) Integers: $\{12,\ -13,\ 1,\ 2\}$

(c) Rational numbers: $\left\{12,\ -13,\ 1,\ 2,\ \tfrac{3}{2}\right\}$

(d) Irrational numbers: $\{\sqrt{6}\}$

9. Plot -4 and -8 on the real number line and place the appropriate inequality sign (< or >) between them.

Solution

$-4 > -8$

11. Plot $\frac{5}{6}$ and $\frac{2}{3}$ on the real number line and place the appropriate inequality sign (< or >) between them.

Solution

$\frac{5}{6} > \frac{2}{3}$

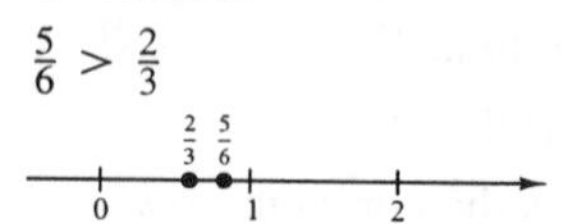

17. Describe the subset of real numbers that is represented by $x \geq 4$ and sketch the subset on the real number line.

Solution

All real numbers that are greater than or equal to 4.

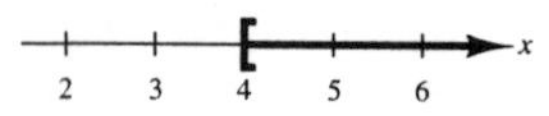

21. Describe the subset of real numbers that is represented by $-1 \leq x < 0$ and sketch the subset on the real number line.

Solution

All real numbers between -1 and 0, including -1 but not including 0.

25. Use inequality notation to describe "the person's age, A, is at least 30."

Solution

$A \geq 30$

27. Use ineqality notation to describe "the annual rate of inflation, r, is expected to be at least 3.5%, but no more than 6%."

Solution

$3.5\% \leq r \leq 6\%$

$0.035 \leq r \leq 0.06$

31. Write the expression $|3 - \pi|$ without using absolute value signs.

Solution

Since $3 < \pi$ we have

$|3 - \pi| = -(3 - \pi) = \pi - 3$

35. Write the expression $-3| - 3|$ without using absolute value signs.

Solution

$-3| - 3| = -3[-(-3)] = -3[3] = -9$

39. Place the correct symbol ($<$, $>$, or $=$) between -5 and $-|5|$.

Solution

$-5 = -|5|$ since $-5 = -5$

45. Find the distance between $a = -\frac{5}{2}$ and $b = 0$. (See figure in the textbook.)

Solution

Distance $= \left|0 - \left(-\frac{5}{2}\right)\right| = \frac{5}{2}$

49. Find the distance between $a = 126$ and $b = 75$.

Solution

Distance $= |126 - 75| = |51| = 51$

55. Use absolute value notation to describe the expression: the distance between z and $\frac{3}{2}$ is greater than 1.

Solution

$\left|z - \frac{3}{2}\right| > 1$

57. Use absolute value notation to describe the expression: y is at least six units from 0.

Solution

$|y - 0| \geq 6 \Rightarrow |y| \geq 6$

59. Use a calculator to order the following numbers (from smallest to largest).

$\frac{7{,}071}{5{,}000}, \frac{584}{413}, \sqrt{2}, \frac{47}{33}, \frac{127}{90}$

Solution

$\frac{127}{90} \approx 1.41111,\ \frac{584}{413} \approx 1.41404,\ \frac{7071}{5000} \approx 1.41420,$

$\sqrt{2} \approx 1.41421,\ \frac{47}{33} \approx 1.42424$

61. Use a calculator to find the decimal form of $\frac{5}{8}$. If it is a nonterminating decimal, write the repeating pattern.

Solution

$\frac{5}{8} = 0.625$

67. *Budget Variance* The accounting department of a company is checking to see whether the actual expenses of $9,972.59 for utilities differ from the budgeted expense of $9,400.00 by more than $500 or 5%. Determine whether the actual expense passes the "budget variance test."

Solution

	Budgeted Expense, b	*Actual Expense, a*	$\lvert a - b\rvert$	$0.05b$
Utilities	$9,400.00	$9,972.59	$572.59	$470.00

The actual expense for utilities does not pass either test.

73. *Median Incomes* Use the bar graph in the textbook that shows the median income for American households that purchase equipment for skiing, tennis, golf, fishing, camping, and hunting. (*Source:* National Sporting Good Association 1989 Survey.) Given a household income of $37,300, find the amount that the income differs from the median income for the sport of tennis.

Solution

	Median Income, y	*Household Income, x*	$\lvert y - x\rvert$
Tennis	$52,984	$37,300	$15,684

P.2 The Basic Rules of Algebra

- You should be able to identify the terms in an algebraic expression.
- You should know and be able to use the basic rules of algebra.
- (a) Subtraction: $a - b = a + (-b)$
  (b) Division: If $b \neq 0$, $a(1/b) = a/b$
- Commutative Property
  (a) Addition: $a + b = b + a$
  (b) Multiplication: $a \cdot b = b \cdot a$
- Associative Property
  (a) Addition: $(a + b) + c = a + (b + c)$
  (b) Multiplication: $(ab)c = a(bc)$
- Identity Property
  (a) Addition: 0 is the identity; $a + 0 = 0 + a = a$.
  (b) Multiplication: 1 is the identity; $a \cdot 1 = 1 \cdot a = a$.
- Inverse Property
  (a) Addition: $-a$ is the inverse of a; $a + (-a) = (-a) + a = 0$.
  (b) Multiplication: $1/a$ is the inverse of a, $a \neq 0$; $a(1/a) = (1/a)a = 1$.
- Distributive Property
  (a) Left: $a(b + c) = ab + ac$
  (b) Right: $(a + b)c = ac + bc$
- Properties of Negatives
  (a) $(-1)a = -a$
  (b) $-(-a) = a$
  (c) $(-a)b = a(-b) = -(ab)$
  (d) $(-a)(-b) = ab$
  (e) $-(a + b) = (-a) + (-b) = -a - b$
- Properties of Zero
  (a) $a \pm 0 = a$
  (b) $a \cdot 0 = 0$
  (c) $0 \div a = 0/a = 0, a \neq 0$
  (d) If $ab = 0$, then $a = 0$ or $b = 0$.
  (e) $a/0$ is undefined.

- Properties of Fractions ($b \neq 0,\ d \neq 0$)
  - (a) Equivalent Fractions: $a/b = c/d$ if and only if $ad = bc$.
  - (b) Rule of Signs: $-a/b = a/-b = -(a/b)$ and $-a/(-b) = a/b$
  - (c) Equivalent Fractions: $a/b = ac/bc,\ c \neq 0$
  - (d) Addition and Subtraction
    1. Like Denominators: $(a/b) \pm (c/b) = (a \pm c)/b$
    2. Unlike Denomiators: $(a/b) \pm (c/d) = (ad \pm bc)/bd$
  - (e) Multiplication: $(a/b) \cdot (c/d) = ac/bd$
  - (f) Division: $(a/b) \div (c/d) = (a/b) \cdot (d/c) = ad/bc$ if $c \neq 0$.
- Properties of Equality
  - (a) Reflexive: $a = a$
  - (b) Symmetric: If $a = b$, then $b = a$.
  - (c) Transitive: If $a = b$ and $b = c$, then $a = c$.
  - (d) Substitution:

    If $a = b$, a can be replaced by b in any statement involving a or b.
    1. If $a = b$, then $a + c = b + c$.
    2. If $a = b$, then $ac = bc$.
- Cancellation Laws
  - (a) If $a + c = b + c$, then $a = b$.
  - (b) If $ac = bc$, then $a = b,\ c \neq 0$.
- Know how to round decimal numbers
  - (a) If the decision digit is 5 or greater, round up.
  - (b) If the decision digit is 4 or less, round down.

SOLUTIONS TO SELECTED EXERCISES

3. Identify the terms of $x^2 - 4x + 8$.

Solution

Terms: $x^2,\ -4x,\ 8$

9. Evaluate $x^2 - 3x + 4$ for the values of $x = -2$ and $x = 2$.

Solution

(a) $(-2)^2 - 3(-2) + 4 = 4 + 6 + 4 = 14$

(b) $(2)^2 - 3(2) + 4 = 4 - 6 + 4 = 2$

13. Evaluate $\dfrac{x+1}{x-1}$ for the values of $x = 1$ and $x = -1$. (If not possible, state the reason.)

Solution

(a) $\dfrac{1+1}{1-1} = \dfrac{2}{0}$ is undefined. You cannot divide by zero.

(b) $\dfrac{-1+1}{-1-1} = \dfrac{0}{-2} = 0$

17. Identify the rule(s) of algebra illustrated by the equation $-15 + 15 = 0$.

Solution

Inverse (addition)

19. Identify the rule(s) of algebra illustrated by the equation $2(x+3) = 2x+6$.

Solution

Left distributive

23. Identify the rule(s) of algebra illustrated by the equation $h + 0 = h$.

Solution

Property of Zero, identify (addition)

27. Identify the rule(s) of algebra illustrated by the equation $6 + (7+8) = (6+7) + 8$.

Solution

Associative (addition)

31. Write the prime factorization of 36.

Solution

$36 = 4 \cdot 9 = 2 \cdot 2 \cdot 3 \cdot 3 = 2^2 \cdot 3^2$

35. Perform the indicated operations.

$$(8 - 17) + 3$$

Solution

$(8 - 17) + 3 = (-9) + 3 = -6$

39. Perform the indicated operations.

$$(4-7)(-2)$$

Solution

$(4-7)(-2) = (-3)(-2) = 6$

45. Perform the indicated operations.

$$\frac{6}{7} - \frac{4}{7}$$

Solution

$$\frac{6}{7} - \frac{4}{7} = \frac{6-4}{7} = \frac{2}{7}$$

49. Perform the indicated operations. (Write fractional answers in reduced form.)

$$\frac{4}{5} \cdot \frac{1}{2} \cdot \frac{3}{4}$$

Solution

$$\frac{4}{5} \cdot \frac{1}{2} \cdot \frac{3}{4} = \frac{\not{4} \cdot 1 \cdot 3}{5 \cdot 2 \cdot \not{4}} = \frac{3}{10}$$

53. Perform the indicated operations. (Write fractional answers in reduced form.)

$$12 \div \tfrac{1}{4}$$

Solution

$12 \div \frac{1}{4} = 12 \cdot \frac{4}{1} = 48$

57. Use a calculator to evaluate the expression $\frac{1}{8} + \frac{1}{7}$. (Round your answer to two decimal places.)

Solution

$\frac{1}{8} + \frac{1}{7} \approx 0.27$

61. Use a calculator to evaluate the following expression. (Round your answer to two decimal places.)

$$\frac{11.46 - 5.37}{3.91}$$

Solution

$$\frac{11.46 - 5.37}{3.91} = \frac{6.09}{3.91} \approx 1.56$$

65. Use a calculator to solve 33% of 57.

Solution

$0.33(57) = 18.81$

71. Find the percentage that corresponds to the unlabeled portion of the pie chart. (See figure in the textbook.)

Solution

$1 - (0.143 + 0.286 + 0.226 + 0.179) = 0.166 = 16.6\%$

73. *Vehicle Sales* The pie chart (see figure in textbook) shows the market share for American vehicle sales (cars and trucks) for 1990. (*Source: USA Today.*) What percent of the market did General Motors have? If 15 million vehicles were sold during 1990, find the number of vehicles sold by General Motors, Ford, Chrysler, Toyota, Honda, and Nissan. (Round your answers to the nearest tenth of a million vehicles.)

Solution

General Motors: $1 - (0.100 + 0.044 + 0.062 + 0.076 + 0.120 + 0.240) = 0.358 = 35.8\%$

General Motors: $0.358(15) \approx 5.4$ million vehicles

Ford: $0.240(15) = 3.6$ million vehicles

Chrysler: $0.120(15) = 1.8$ million vehicles

Toyota: $0.076(15) \approx 1.1$ million vehicles

Honda: $0.062(15) \approx 0.9$ million vehicles

Nissan: $0.044(15) \approx 0.7$ million vehicles

Other: $0.10(15) = 1.5$ million vehicles

77. *Calculator Keystrokes* Write the keystrokes used to evaluate $5(18 - 2^3) \div 10$.

Solution

Scientific Calculator: 5 [×] [(] 18 [−] 2 [y^x] 3 [)] [÷] 10 [=]

Graphing Calculator: 5 [(] 18 [−] 2 [∧] 3 [)] [÷] 10 [ENTER]

79. *College Enrollment* The percent of students 24 years old or older at a college is 44.7%. If the college enrollment for the 1992–1993 academic year was 13,385 students, how many students were under 24 years old?

Solution

$$\begin{aligned}(\text{Total students}) - (\text{Students 24 and older}) &= 13{,}385 - 0.447(13{,}385)\\ &= 13{,}385 - 5983 \quad \text{(Round to the nearest student.)}\\ &= 7402 \quad \text{Students under 24 years old}\end{aligned}$$

P.3 Integer Exponents

- You should know the properties of exponents.
  - (a) $a^1 = a$
  - (b) $a^0 = 1,\ a \neq 0$
  - (c) $a^m a^n = a^{m+n}$
  - (d) $a^m/a^n = a^{m-n},\ a \neq 0$
  - (e) $a^{-n} = 1/a^n,\ a \neq 0$
  - (f) $(a^m)^n = a^{m \cdot n}$
  - (g) $(ab)^n = a^n b^n$
  - (h) $(a/b)^n = a^n/b^n,\ b \neq 0$
  - (i) $(a/b)^{-n} = (b/a)^n,\ a \neq 0,\ b \neq 0$
  - (j) $|a^2| = |a|^2 = a^2$
- You should be able to write numbers in scientific notation, $c \times 10^n$, where $1 \leq c < 10$ and n is an integer.
- You should be able to use your calculator to evaluate expression involving exponents.
- You should know the formulas for the balance in an account.
  - (a) Simple interest: $A = P(1 + rt)$
  - (b) Compound interest: $A = P\left(1 + \dfrac{r}{n}\right)^{nt}$

SOLUTIONS TO SELECTED EXERCISES

3. Evaluate $\dfrac{5^5}{5^2}$.

Solution

$$\frac{5^5}{5^2} = 5^{5-2} = 5^3 = 125$$

7. Evaluate -3^4.

Solution

$$-3^4 = -(3)^4 = -81$$

11. Evaluate $5^{-1} + 2^{-1}$.

Solution

$$5^{-1} + 2^{-1} = \tfrac{1}{5} + \tfrac{1}{2} = \tfrac{2}{10} + \tfrac{5}{10} = \tfrac{7}{10}$$

17. Evaluate $\left(-\dfrac{3}{5}\right)^3\left(\dfrac{5}{3}\right)^2$.

Solution

$$\left(-\frac{3}{5}\right)^3\left(\frac{5}{3}\right)^2 = (-1)^3\left(\frac{3^3}{5^3}\right)\left(\frac{5^2}{3^2}\right)$$
$$= (-1)\left(\frac{3}{5}\right) = -\frac{3}{5}$$

19. Evaluate $6^5 \cdot 6^{-3}$.

Solution

$$6^5 \cdot 6^{-3} = 6^{5+(-3)} = 6^2 = 36$$

25. Evaluate $4x^{-3}$ when $x = 2$.

Solution

$$4x^{-3} = 4(2)^{-3} = \frac{4}{2^3} = \frac{4}{8} = \frac{1}{2}.$$

27. Evaluate $6x^0 - (6x)^0$ when $x = 10$.

Solution

$$6x^0 - (6x)^0 = 6(10)^0 - (60)^0$$
$$= 6(1) - 1 = 5.$$

31. Simplify $5x^4(x^2)$.

Solution

$$5x^4(x^2) = 5x^{4+2} = 5x^6$$

35. Simplify $6y^2(2y^4)^2$.

Solution

$$6y^2(2y^4)^2 = 6y^2 \cdot 2^2y^8 = 6 \cdot 4y^{2+8} = 24y^{10}$$

39. Simplify $\frac{7x^2}{x^3}$.

Solution

$$\frac{7x^2}{x^3} = 7x^{2-3} = 7x^{-1} = \frac{7}{x}$$

43. Simplify $(x+5)^0$, when $x \neq -5$.

Solution

$(x+5)^0 = 1,\ x \neq -5$

47. Simplify $(-2x^2)^3(4x^3)^{-1}$.

Solution

$$(-2x^2)^3(4x^3)^{-1} = \frac{-8x^6}{4x^3} = -2x^3$$

51. Simplify $\left(\frac{3z^2}{x}\right)^{-2}$.

Solution

$$\left(\frac{3z^2}{x}\right)^{-2} = \left(\frac{x}{3z^2}\right)^2 = \frac{x^2}{9z^4}$$

53. Simplify $3^n \cdot 3^{2n}$.

Solution

$3^n \cdot 3^{2n} = 3^{n+2n} = 3^{3n}$

57. Write the number in scientific notation.

Land Area of Earth: 57,500,000 square miles

Solution

$57,500,000 = 5.75 \times 10^7$ square miles

61. Write the number in scientific notation.

Relative Density of Hydrogen: 0.0000899

Solution

$0.0000899 = 8.99 \times 10^{-5}$

65. Write the number in decimal form.

U.S. Daily Coca-Cola Consumption: 5.24×10^8 servings

Solution

$5.24 \times 10^8 = 524,000,000$ servings

69. Write the number in decimal form.

Charge of Electron: 4.8×10^{-10} electrostatic units

Solution

$4.8 \times 10^{-10} = 0.00000000048$ electrostatic units

75. Use a calculator to evaluate the following expressions. (Round your answer to three decimal places.)

(a) $0.000345(8,900,000,000)$ (b) $\frac{67,000,000 + 93,000,000}{0.0052}$

Solution

(a)

3.45 [EE] 4 [+/−] [×] 8.90 [EE] 9 [=] 3,070,500 Scientific

3.45 [EE] −4 [×] 8.90 [EE] 9 [ENTER] Graphing

(b)

6.7 [EE] 7 [+] 9.3 [EE] 7 [=] [÷] 5.2 [EE] 3 [+/−] [=] 3.077×10^{10} Scientific

6.7 [EE] 7 [+] 9.3 [EE] 7 [ENTER] [÷] 5.2 [EE] [(−)] 3 [ENTER] Graphing

77. Use a calculator to evaluate the following expressions. (Round your answer to three decimal places.)

(a) $(9.3 \times 10^6)^3(6.1 \times 10^{-4})$ (b) $\dfrac{(2.414 \times 10^4)^6}{(1.68 \times 10^5)^5}$

Solution

(a)

[(] 9.3 [EE] 6 [)] [y^x] 3 [×] 6.1 [EE] 4 [+/−] [=] 4.907×10^{17} Scientific

[(] 9.3 [EE] 6 [)] [∧] 3 [×] 6.1 [EE] [−] 4 [ENTER] Graphing

(b)

[(] 2.414 [EE] 4 [)] [y^x] 6 [÷] [(] 1.68 [EE] 5 [)] [y^x] 5 [=] 1.479 Scientific

[(] 2.414 [EE] 4 [)] [∧] 6 [÷] [(] 1.68 [EE] 5 [)] [∧] 5 [ENTER] Graphing

81. *Balance in an Account* Ten thousand dollars is deposited in an account with an annual percentage rate of 6.5% for 10 years. What is the balance in the account if the interest is compounded (a) quarterly and (b) monthly?

Solution

(a) $A = 10,000\left(1 + \dfrac{0.065}{4}\right)^{4(10)} \approx \$19,055.59$

(b) $A = 10,000\left(1 + \dfrac{0.065}{12}\right)^{12(10)} \approx \$19,121.84$

87. *Calculator Keystrokes* Write the arithmetic expression that corresponds to the keystrokes.

[(] 5.1 [−] 3.6 [)] [y^x] 5 [=] Scientific

[(] 5.1 [−] 3.6 [)] [∧] 5 [ENTER] Graphing

Solution

$(5.1 - 3.6)^5$

P.4 Radicals and Rational Exponents

- You should know the properties of radicals.
  - (a) $\sqrt[n]{a^m} = (\sqrt[n]{a})^m$
  - (b) $\sqrt[n]{a} \cdot \sqrt[n]{b} = \sqrt[n]{ab}$
  - (c) $\dfrac{\sqrt[n]{a}}{\sqrt[n]{b}} = \sqrt[n]{\dfrac{a}{b}}$
  - (d) $\sqrt[m]{\sqrt[n]{a}} = \sqrt[mn]{a}$
  - (e) $(\sqrt[n]{a})^n = a$
  - (f) For n even, $\sqrt[n]{a^n} = |a|$

    For n odd, $\sqrt[n]{a^n} = a$
  - (g) $a^{1/n} = \sqrt[n]{a}$
  - (h) $a^{m/n} = (\sqrt[n]{a})^m = \sqrt[n]{a^m}$
- You should be able to simplify radicals.
  - (a) All possible factors have been removed from the radical.
  - (b) All fractions have radical-free denominators.
  - (c) The index for the radical has been reduced as far as possible.
- You should be able to rationalize denominators.
- Realize that the properties of exponents listed in Section P.3 also apply to rational exponents.
- You should be able to use your calculator to evaluate radicals.

SOLUTIONS TO SELECTED EXERCISES

3. Find the radical form of $32^{1/5} = 2$.

Solution

Radical form: $\sqrt[5]{32} = 2$

7. Find the rational exponent form of $\sqrt[3]{-216} = -6$.

Solution

Rational exponent form: $(-216)^{1/3} = -6$

11. Find the rational exponent form of $\sqrt[4]{81^3} = 27$.

Solution

Rational exponent form: $(81^3)^{1/4} = 81^{3/4} = 27$

15. Evaluate $\sqrt[3]{8}$.

Solution

$\sqrt[3]{8} = \sqrt[3]{2^3} = 2$

19. Evaluate $-\sqrt[3]{-27}$.

Solution

$-\sqrt[3]{-27} = -\sqrt[3]{(-3)^3} = -(-3) = 3$

23. Evaluate $\left(\sqrt[3]{-125}\right)^3$.

Solution

$(\sqrt[3]{-125})^3 = ((-125)^{1/3})^3 = -125$

29. Evaluate $32^{-3/5}$.

Solution

$$32^{-3/5} = \frac{1}{32^{3/5}} = \frac{1}{(\sqrt[5]{32})^3} = \frac{1}{8}$$

33. Evaluate $\left(-\dfrac{1}{64}\right)^{-1/3}$.

Solution

$$\left(-\frac{1}{64}\right)^{-1/3} = (-64)^{1/3}$$
$$= \sqrt[3]{-64}$$
$$= -4$$

37. Simplify the radical $\sqrt[3]{16x^5}$.

Solution

$\sqrt[3]{16x^5} = \sqrt[3]{8x^3 \cdot 2x^2} = 2x\sqrt[3]{2x^2}$

39. Simplify the radical $\sqrt{75x^2y^{-4}}$.

Solution

$$\sqrt{75x^2y^{-4}} = \sqrt{3 \cdot 25x^2y^{-4}}$$
$$= |5xy^{-2}| \cdot \sqrt{3}$$
$$= \frac{5|x|\sqrt{3}}{y^2}$$

43. Rewrite $8/\sqrt[3]{2}$ by rationalizing the denominator. Simplify your answer.

Solution

$$\frac{8}{\sqrt[3]{2}} = \frac{8}{\sqrt[3]{2}} \cdot \frac{\sqrt[3]{4}}{\sqrt[3]{4}} = \frac{8\sqrt[3]{4}}{\sqrt[3]{8}} = \frac{8\sqrt[3]{4}}{2} = 4\sqrt[3]{4}$$

47. Rewrite $3/(\sqrt{5}+\sqrt{6})$ by rationalizing the denominator. Simplify your answer.

Solution

$$\frac{3}{\sqrt{5}+\sqrt{6}} \cdot \frac{\sqrt{5}-\sqrt{6}}{\sqrt{5}-\sqrt{6}} = \frac{3(\sqrt{5}-\sqrt{6})}{5-6}$$
$$= -3(\sqrt{5}-\sqrt{6})$$
$$= 3(\sqrt{6}-\sqrt{5})$$

51. Simplify $2\sqrt{50} + 12\sqrt{8}$.

Solution

$$2\sqrt{50} + 12\sqrt{8} = 2\sqrt{25 \cdot 2} + 12\sqrt{4 \cdot 2}$$
$$= 2 \cdot 5\sqrt{2} + 12 \cdot 2\sqrt{2}$$
$$= 10\sqrt{2} + 24\sqrt{2}$$
$$= 34\sqrt{2}$$

53. Simplify $2\sqrt{4y} - 2\sqrt{9y}$.

Solution

$$2\sqrt{4y} - 2\sqrt{9y} = 2 \cdot 2\sqrt{y} - 2 \cdot 3\sqrt{y}$$
$$= 4\sqrt{y} - 6\sqrt{y}$$
$$= -2\sqrt{y}$$

57. Simplify $5^{1/2} \cdot 5^{3/2}$.

Solution

$$5^{1/2}5^{3/2} = 5^{1/2+3/2}$$
$$= 5^{4/2}$$
$$= 5^2 = 25$$

61. Simplify $\dfrac{x^2}{x^{1/2}}$.

Solution

$$\frac{x^2}{x^{1/2}} = x^{2-1/2}$$
$$= x^{4/2-1/2}$$
$$= x^{3/2},\ x \neq 0$$

67. Use rational exponents to reduce the index of the radical $\sqrt[6]{(x+1)^4}$.

Solution

$$\sqrt[6]{(x+1)^4} = (x+1)^{4/6}$$
$$= (x+1)^{2/3}$$
$$= \sqrt[3]{(x+1)^2}$$

73. Use a calculator to approximate $0.26^{-0.8}$. Round your answer to three decimal places.

Solution

0.26 [y^x] 0.8 [+/−] [=] 2.938	Scientific
0.26 [∧] −0.8 [ENTER]	Graphing

75. Use a calculator to approximate $(3 - \sqrt{5})/2$. Round your answer to three decimal places.

Solution

[(] 3 [−] 5 [√] [)] [÷] 2 [=] 0.382 Scientific

[(] 3 [−] [√] 5 [)] [÷] 2 [ENTER] Graphing

79. Fill in the blank with <, =, or >.

$5 \_\_\_\_ \sqrt{3^2 + 2^2}$

Solution

$\sqrt{3^2 + 2^2} = \sqrt{9+4} = \sqrt{13} \approx 3.606$

Therefore, $5 > \sqrt{3^2 + 2^2}$.

83. *Dimensions of a Cube* Find the dimensions of a cube that has a volume of 13,824 cubic inches. (See figure in the textbook.)

Solution

$x^3 = 13,824$

$x = \sqrt[3]{13,824} = 24$ inches

24 in × 24 in × 24 in

87. *Period of a Pendulum* The period T in seconds of a pendulum is given by $T = 2\pi\sqrt{\frac{L}{32}}$ where L is the length of the pendulum in feet. Find the period of a pendulum whose length is 2 feet.

Solution

$T = 2\pi\sqrt{\frac{2}{32}} = 2\pi\sqrt{\frac{1}{16}} = 2\pi\left(\frac{1}{4}\right) = \frac{1}{2}\pi = \frac{\pi}{2}$ or approximately 1.57 seconds.

91. *Calculator Experiment* Enter any positive real number in your calculator and repeatedly take the square root. What real number does the display appear to be approaching?

Solution

$x^{(1/2) \cdot (1/2) \cdot (1/2) \cdot (1/2) \cdots} \approx 1$

Mid-Chapter Quiz for Chapter P

1. Place the correct symbol (<, >, or =) between $-|-7|$ and $|-7|$.

Solution

$-|-7| = -[(-7)] = -[7] = -7$

$|-7| = -(-7) = 7$

Thus, $-|-7| < |-7|$.

2. Place the correct symbol (<, >, or =) between $-(-3)$ and $|-3|$.

Solution

$-(-3) = 3$

$|-3| = -(-3) = 3$

Thus, $-(-3) = |-3|$.

3. Use inequality notation to describe "x is positive or x is equal to zero."

Solution

$x \geq 0$

4. Use inequality notation to describe "the apartment occupancy rate r will be at least 96.5% during the coming year."

Solution

Since the largest r can ever be is 100%, we have

$96.5\% \leq r \leq 100\%$

$0.965 \leq r \leq 1$

5. Describe the subset of real numbers that is represented by the inequality $0 \leq x < 3$ and sketch the subset on the real number line.

Solution

$0 \leq x < 3$ means x is greater than or equal to zero *and* x is less than 3.

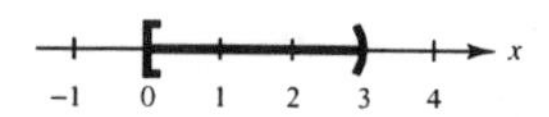

6. Identify the terms of the algebraic expression.

$$3x^2 - 7x + 2.$$

Solution

$3x^2 - 7x + 2 = 3x^2 + (-7x) + 2$

Terms: $3x^2, -7x, 2$

7. Perform the indicated operations.

$$-5 - (-7)$$

Solution

$-5 - (-7) = -5 + 7 = 2$

8. Perform the indicated operations.

$$\frac{28 - 4}{(-6)}$$

Solution

$$\frac{28 - 4}{(-6)} = \frac{24}{(-6)} = -4$$

9. Perform the indicated operations.

$$\frac{2}{3} \cdot \frac{5}{4} \cdot \frac{3}{7}$$

Solution

$$\frac{2}{3} \cdot \frac{5}{4} \cdot \frac{3}{7} = \frac{\cancel{2} \cdot 5 \cdot \cancel{3}}{\cancel{3} \cdot \cancel{4} \cdot 7} = \frac{5}{14}$$

2

10. Perform the indicated operations.

$$\frac{11}{15} \div \frac{3}{5}$$

Solution

$$\frac{11}{15} \div \frac{3}{5} = \frac{11}{15} \cdot \frac{5}{3} = \frac{11 \cdot \cancel{5}}{\cancel{15} \cdot 3} = \frac{11}{9}$$

3

11. Simplify.

$$(-x)^3(2x^4)$$

Solution

$(-x)^3(2x^4) = -x^3(2x^4) = -2x^7$

12. Simplify.

$$\frac{5y^7}{15y^3}$$

Solution

$$\frac{5y^7}{15y^3} = \frac{\cancel{5}y^4\cancel{y^3}}{3 \cdot \cancel{5}\cancel{y^3}} = \frac{y^4}{3}$$

13. Simplify.

$$(3t^{-2})(6t^5)$$

Solution

$(3t^{-2})(6t^5) = 18t^{-2+5} = 18t^3$

14. Simplify.

$$\left(\frac{x^{-2}y^2}{3}\right)^{-3}$$

Solution

$$\left(\frac{x^{-2}y^2}{3}\right)^{-3} = \left(\frac{y^2}{3x^2}\right)^{-3} = \left(\frac{3x^2}{y^2}\right)^3 = \frac{27x^6}{y^6}$$

15. One thousand dollars is deposited in an account with an annual rate of 8% compounded monthly. Find the balance in the account in 10 years.

Solution

$P = 1000, r = 0.08, n = 12, t = 10$

$$A = P\left(1 + \frac{r}{n}\right)^{nt} = 1000\left(1 + \frac{0.08}{12}\right)^{(12)(10)} \approx \$2,219.64$$

16. Evaluate $\dfrac{-\sqrt[4]{81}}{3}$.

Solution

$$\frac{-\sqrt[4]{81}}{3} = \frac{-\sqrt[4]{3^4}}{3} = \frac{-3}{3} = -1$$

17. Evaluate $\left(\sqrt[3]{-64}\right)^3$.

Solution

$$\left(\sqrt[3]{-64}\right)^3 = -64$$

18. Simplify.

$$3^{1/2} \cdot 3^{3/2}$$

Solution

$$3^{1/2} \cdot 3^{3/2} = 3^{1/2+3/2} = 3^{4/2} = 3^2 = 9$$

19. Simplify.

$$\sqrt[3]{81} - 4\sqrt[3]{3}$$

Solution

$$\sqrt[3]{81} - 4\sqrt[3]{3} = \sqrt[3]{27 \cdot 3} - 4\sqrt[3]{3} = 3\sqrt[3]{3} - 4\sqrt[3]{3} = -\sqrt[3]{3}$$

20. Find the dimensions of a cube that has a volume of 15,625 cubic centimeters.

Solution

$x^3 = 15,625$

$x = \sqrt[3]{15,625} = 25$

Dimensions: 25 cm × 25 cm × 25cm

P.5 Polynomials and Special Products

- Given a polynomial in x, $a_nx^n + a_{n-1}x^{n-1} + \ldots + a_1x + a_0$, where $a_n \neq 0$, you should be able to identify the following:
  (a) Degree: n
  (b) Terms: $a_nx^n, a_{n-1}x^{n-1}, \ldots, a_1x, a_0$
  (c) Coefficients: $a_n, a_{n-1}, \ldots, a_1, a_0$
  (d) Leading Coefficient: a_n
  (e) Constant term: a_0
- You should be able to add and subtract polynomials.
- You should be able to multiply polynomials by either
  (a) The Distributive Properties or
  (b) The Vertical Method.
- You should know the special binomial products.
  (a) $(ax+b)(cx+d) = acx^2 + adx + bcx + bd$ FOIL
  $= acx^2 + (ad+bc)x + bd$
  (b) $(u \pm v)^2 = u^2 \pm 2uv + v^2$
  (c) $(u+v)(u-v) = u^2 - v^2$
  (d) $(u \pm v)^3 = u^3 \pm 3u^2v + 3uv^2 \pm v^3$

SOLUTIONS TO SELECTED EXERCISES

3. Find the degree and leading coefficient of the polynomial $x^5 - 1$.

Solution

Standard form: $x^5 - 1$

Degree: 5

Leading coefficient: 1

11. Determine whether $y^2 - y^4 + y^3$ is a plynomial. If it is, write the polynomial in standard form and state its degree.

Solution

$y^2 - y^4 + y^3$ *is* a polynomial.

Standard form: $-y^4 + y^3 + y^2$

Degree: 4

9. Determine whether $(3x+4)/x$ is a polynomial. If it is, write the polynomial in standard form and state its degree.

Solution

$\frac{3x+4}{x} = 3 + \frac{4}{x}$ is *not* a polynomial.

15. Evaluate $x^2 - 2x + 3$ for the indicated values of x.

(a) $x = -2$ (b) $x = -1$
(c) $x = 0$ (d) $x = 1$

Solution

(a) $(-2)^2 - 2(-2) + 3 = 11$

(b) $(-1)^2 - 2(-1) + 3 = 6$

(c) $(0)^2 - 2(0) + 3 = 3$

(d) $(1)^2 - 2(1) + 3 = 2$

19. Perform the indicated operations and write the resulting polynomial in standard form.

$$(6x + 5) - (8x + 15)$$

Solution

$(6x + 5) - (8x + 15) = 6x + 5 - 8x - 15 = (6x - 8x) + (5 - 15) = -2x - 10$

23. Perform the indicated operations and write the resulting polynomial in standard form.

$$(15x^2 - 6) - (-8x^3 - 14x^2 - 17)$$

Solution

$$\begin{aligned}(15x^2 - 6) - (-8x^3 - 14x^2 - 17) &= 15x^2 - 6 + 8x^3 + 14x^2 + 17 \\ &= 8x^3 + (15x^2 + 14x^2) + (-6 + 17) \\ &= 8x^3 + 29x^2 + 11\end{aligned}$$

27. Perform the indicated operations and write the resulting polynomial in standard form.

$$3x(x^2 - 2x + 1)$$

Solution

$3x(x^2 - 2x + 1) = 3x(x^2) + 3x(-2x) + 3x(1) = 3x^3 - 6x^2 + 3x$

31. Perform the indicated operations and write the resulting polynomial in standard form.

$$(-2x)(-3x)(5x + 2)$$

Solution

$(-2x)(-3x)(5x + 2) = 6x^2(5x + 2) = 6x^2(5x) + 6x^2(2) = 30x^3 + 12x^2$

35. Find the product of $(3x - 5)(2x + 1)$.

Solution

$(3x - 5)(2x + 1) = 6x^2 + 3x - 10x - 5 = 6x^2 - 7x - 5$

39. Find the product of $(2x - 5y)^2$.

Solution

$(2x - 5y)^2 = 4x^2 - 2(5y)(2x) + 25y^2 = 4x^2 - 20xy + 25y^2$

41. Find the product of $[(x - 3) + y]^2$.

Solution

$$\begin{aligned}[(x - 3) + y]^2 &= (x - 3)^2 + 2y(x - 3) + y^2 \\ &= x^2 - 6x + 9 + 2xy - 6y + y^2 \\ &= x^2 + 2xy + y^2 - 6x - 6y + 9\end{aligned}$$

45. Find the product of $(x + 2y)(x - 2y)$.

Solution

$(x + 2y)(x - 2y) = x^2 - 4y^2$

47. Find the product of $(m - 3 + n)(m - 3 - n)$.

Solution

$(m - 3 + n)(m - 3 - n) = [(m - 3) + n][(m - 3) - n] = (m - 3)^2 - n^2 = m^2 - n^2 - 6m + 9$

51. Find the product of $(x+1)^3$.

Solution

$(x+1)^3 = x^3 + 3x^2(1) + 3x(1^2) + 1^3 = x^3 + 3x^2 + 3x + 1$

55. Find the product of $(\sqrt{x}+\sqrt{y})(\sqrt{x}-\sqrt{y})$.

Solution

$(\sqrt{x}+\sqrt{y})(\sqrt{x}-\sqrt{y}) = (\sqrt{x})^2 - (\sqrt{y})^2 = x - y$

61. Find the product of $(x^2 - x + 1)(x^2 + x + 1)$.

Solution

$$\begin{array}{l} x^2 - x + 1 \\ x^2 + x + 1 \\ \hline x^4 - x^3 + x^2 \\ \quad x^3 - x^2 + x \\ \qquad x^2 - x + 1 \\ \hline x^4 + 0x^3 + x^2 + 0x + 1 \end{array}$$

Thus, $(x^2 - x + 1)(x^2 + x + 1) = x^4 + x^2 + 1$.

63. Find the product of $5x(x+1) - 3x(x+1)$.

Solution

$5x(x+1) - 3x(x+1) = (x+1)(5x-3x) = (x+1)(2x) = 2x^2 + 2x$

67. *Compound Interest* After two years an investment of \$500 compounded annually at an interest rate r will yield an amount of $500(1+r)^2$. Write this polynomial in standard form.

Solution

$500(1+r)^2 = 500(r+1)^2 = 500(r^2 + 2r + 1) = 500r^2 + 1000r + 500$

69. *Geometry* An open box is made by cutting squares out of the corners of a piece of metal that is 18 inches by 26 inches. (See figure in the textbook.) If the edge of each cut-out square is x inches, what is the volume of the box? Find the volume when $x = 1$, $x = 2$, and $x = 3$.

Solution

$$\begin{aligned} V = l \cdot w \cdot h &= (26-2x)(18-2x)(x) \\ &= 468x - 88x^2 + 4x^3 \\ &= 4x^3 - 88x^2 + 468x \end{aligned}$$

When $x = 1: V = 4(1)^3 - 88(1)^2 + 468(1) = 384$ cubic inches

When $x = 2: V = 4(2)^3 - 88(2)^2 + 468(2) = 616$ cubic inches

When $x = 3: V = 4(3)^3 - 88(3)^2 + 468(3) = 720$ cubic inches

73. *Geometry* Find a polynomial that represents the total number of square feet for the following floor plan. (See figure in the textbook.)

Solution

$A = (x + 6 + 24)(x + 12) = (x + 30)(x + 12) = x^2 + 42x + 360$

P.6 Factoring

- You should be able to factor out all common factors, the first step in factoring.
- You should be able to factor the following special polynomial forms.

  (a) $u^2 - v^2 = (u + v)(u - v)$

  (b) $u^2 \pm 2uv + v^2 = (u \pm v)^2$

  (c) $mx^2 + nx + r = (ax + b)(cx + d)$, where $m = ac,\ r = bd,\ n = ad + bc$

  Note: Not all trinomials can be factored (using real coefficients).

  (d) $u^3 \pm v^3 = (u \pm v)(u^2 \mp uv + v^2)$
- You should be able to factor by grouping.

SOLUTIONS TO SELECTED EXERCISES

3. Factor out the common factor.

$2x^3 - 6x.$

Solution

$2x^3 - 6x = 2x(x^2 - 3)$

9. Factor $16y^2 - 9$.

Solution

$16y^2 - 9 = (4y)^2 - (3)^2 = (4y + 3)(4y - 3)$

11. Factor $(x - 1)^2 - 4$.

Solution

$(x - 1)^2 - 4 = [(x - 1) + 2][(x - 1) - 2] = (x + 1)(x - 3)$

15. Factor the perfect square trinomial $4t^2 + 4t + 1$.

Solution

$4t^2 + 4t + 1 = (2t)^2 + 2(2t)(1) + 1^2 = (2t + 1)^2$

17. Factor the perfect square trinomial $25y^2 - 10y + 1$.

Solution

$25y^2 - 10y + 1 = (5y)^2 - 2(5y)(1) + 1^2 = (5y - 1)^2$

21. Factor the trinomial $s^2 - 5s + 6$.

Solution

$s^2 - 5s + 6 = (s - 2)(s - 3)$

25. Factor the trinomial $x^2 - 30x + 200$.

Solution

$x^2 - 30x + 200 = (x - 20)(x - 10)$

31. Factor the trinomial $5x^2 + 26x + 5$.

Solution

$5x^2 + 26x + 5 = (5x + 1)(x + 5)$

35. Factor $y^3 + 64$.

Solution

$$\begin{aligned} y^3 + 64 &= y^3 + 4^3 \\ &= (y + 4)(y^2 - 4y + 16) \end{aligned}$$

37. Factor $8t^3 - 1$.

Solution

$$\begin{aligned} 8t^3 - 1 &= (2t)^3 - 1^3 \\ &= (2t - 1)(4t^2 + 2t + 1) \end{aligned}$$

41. Factor $2x^3 - x^2 - 6x + 3$ by grouping.

Solution

$2x^3 - x^2 - 6x + 3 = x^2(2x - 1) - 3(2x - 1) = (2x - 1)(x^2 - 3)$

47. Completely factor $x^3 - 9x$.

Solution

$x^3 - 9x = x(x^2 - 9) = x(x + 3)(x - 3)$

53. Completely factor $1 - 4x + 4x^2$.

Solution

$1 - 4x + 4x^2 = 1 - 2(2x) + (2x)^2 = (1 - 2x)^2$

55. Completely factor $-2x^2 - 4x + 2x^3$

Solution

$$\begin{aligned} -2x^2 - 4x + 2x^3 &= 2x^3 - 2x^2 - 4x \\ &= 2x(x^2 - x - 2) \\ &= 2x(x - 2)(x + 1) \end{aligned}$$

61. Completely factor $x^4 - 4x^3 + x^2 - 4x$.

Solution

$$\begin{aligned} x^4 - 4x^3 + x^2 - 4x &= x(x^3 - 4x^2 + x - 4) \\ &= x[x^2(x - 4) + (x - 4)] \\ &= x(x - 4)(x^2 + 1) \end{aligned}$$

65. Completely factor $(x^2 + 1)^2 - 4x^2$.

Solution

$$\begin{aligned} (x^2 + 1)^2 - 4x^2 &= [(x^2 + 1) + 2x][(x^2 + 1) - 2x] \\ &= (x^2 + 2x + 1)(x^2 - 2x + 1) \\ &= (x + 1)^2(x - 1)^2 \end{aligned}$$

69. Completely factor $4x(2x - 1) + (2x - 1)^2$.

Solution

$4x(2x - 1) + (2x - 1)^2 = (2x - 1)(4x + (2x - 1)) = (2x - 1)(6x - 1)$

71. Factor $x^3 + 6x^2 + 12x + 8$ by using the following formula.

$$(x + a)^3 = x^3 + 3x^2a + 3xa^2 + a^3$$

Solution

$x^3 + 6x^2 + 12x + 8 = x^3 + 3x^2(2) + 3x(2)^2 + (2)^3 = (x + 2)^3$

73. Make a "geometric factoring model" to represent $3x^2 + 7x + 2 = (3x + 1)(x + 2)$. For instance, a factoring model for $2x^2 + 5x + 2 = (2x + 1)(x + 2)$ is shown in the figure in the textbook.

Solution

$3x^2 + 7x + 2 = (3x + 1)(x + 2)$

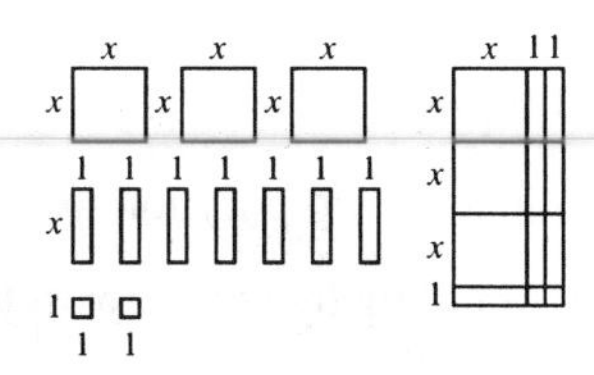

77. Match the "geometric factoring model" with the formula $a^2 - b^2 = (a + b)(a - b)$.

Solution

$a^2 - b^2 = (a + b)(a - b)$ Matches model (b).

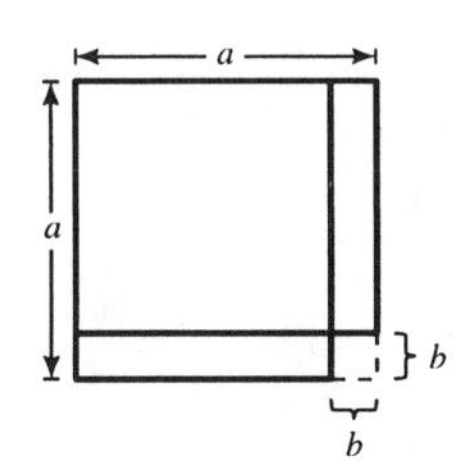

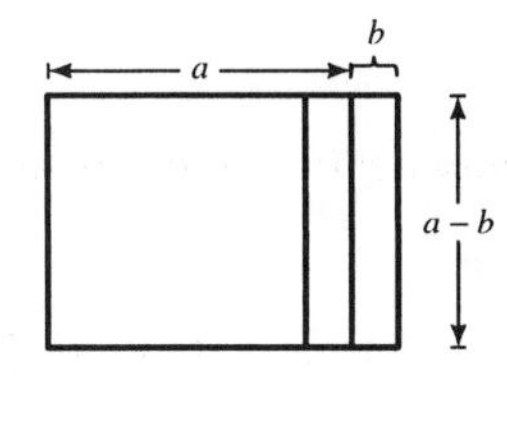

81. *Dimensions of a Room* The room shown in the figure in the textbook has a floor space of $2x^2 + 3x + 1$ square feet. If the width of the room is $(x + 1)$ feet, what is the length?

Solution

$2x^2 + 3x + 1 = (x + 1)(2x + 1)$

The length is $(2x + 1)$ feet.

P.7 Fractional Expressions and Probability

- You should be able to find the domain of an algebraic expression.
- You should know that a rational expression is the quotient of two polynomials.
- You should be able to simplify rational expressions by reducing them to lowest terms. This may involve factoring both the numerator and the denominator.
- You should be able to add, subtract, multiply, and divide rational expressions.
- You should be able to simplify compound fractions.
- If an event E has $n(E)$ equally likely outcomes and its sample space has $n(S)$ equally likely outcomes, then the probability of event E is

$$P(E) = \frac{n(E)}{n(S)}.$$

- You should know the following properties of the probability of an event.
  (a) $0 \leq P(E) \leq 1$
  (b) If $P(E) = 0$, then E is an impossible event.
  (c) If $P(E) = 1$, then E is a certain event.

SOLUTIONS TO SELECTED EXERCISES

3. Find the domain of $4x^3 + 5x + 3,\ x \geq 0$.

Solution

The domain of the polynomial $4x^3 + 5x + 3$ is the set of all nonnegative real numbers because it is specifically restricted to that set.

7. Find the domain of

$$\frac{x-1}{x^2 - 4x}.$$

Solution

The domain of $\dfrac{x-1}{x(x-4)}$ is the set of all real numbers except $x = 0$ and $x = 4$.

13. Find the missing factor in the numerator so that the two fractions will
be equivalent.

$$\frac{x+1}{x} = \frac{(x+1)(\quad)}{x(x-2)}$$

Solution

$$\frac{x+1}{x} = \frac{(x+1)(x-2)}{x(x-2)},\ x \neq 2$$

17. Write the following rational expression in reduced form.

$$\frac{15x^2}{10x}$$

Solution

$$\frac{15x^2}{10x} = \frac{5x(3x)}{5x(2)} = \frac{3x}{2},\ x \neq 0$$

21. Write the following rational expression in reduced form.

$$\frac{x-5}{10-2x}$$

Solution

$$\frac{x-5}{10-2x} = \frac{x-5}{-2(x-5)} = -\frac{1}{2},\ x \neq 5$$

25. Write the following rational expression in reduced form.

$$\frac{y^2 - 7y + 12}{y^2 + 3y - 18}$$

Solution

$$\frac{y^2 - 7y + 12}{y^2 + 3y - 18} = \frac{(y-3)(y-4)}{(y+6)(y-3)} = \frac{y-4}{y+6},\ y \neq -6,\ 3$$

27. Write the following rational expression in reduced form.

$$\frac{2 - x + 2x^2 - x^3}{x - 2}$$

Solution

$$\frac{2 - x + 2x^2 - x^3}{x - 2} = \frac{(2 - x) + x^2(2 - x)}{-(2 - x)} = \frac{(2 - x)(1 + x^2)}{-(2 - x)} = -(1 + x^2),\ x \neq 2$$

33. Perform the indicated operations and simplify.

$$\frac{(x - 9)(x + 7)}{x + 1} \cdot \frac{x}{9 - x}$$

Solution

$$\frac{(x - 9)(x + 7)}{x + 1} \cdot \frac{x}{9 - x} = \frac{x(x - 9)(x + 7)}{(x + 1)(-1)(x - 9)} = -\frac{x(x + 7)}{x + 1},\ x \neq -1,\ 9$$

37. Perform the indicated operations and simplify.

$$\frac{t^2 - t - 6}{t^2 + 6t + 9} \cdot \frac{t + 3}{t^2 - 4}$$

Solution

$$\frac{t^2 - t - 6}{t^2 + 6t + 9} \cdot \frac{t + 3}{t^2 - 4} = \frac{(t - 3)(t + 2)(t + 3)}{(t + 3)^2(t + 2)(t - 2)} = \frac{t - 3}{(t + 3)(t - 2)},\ t \neq -3,\ \pm 2$$

43. Perform the indicated operations and simplify.

$$\frac{\left(\dfrac{x^2}{(x + 1)^2}\right)}{\left(\dfrac{x}{(x + 2)^3}\right)}$$

Solution

$$\frac{\left(\dfrac{x^2}{(x + 1)^2}\right)}{\left(\dfrac{x}{(x + 1)^3}\right)} = \frac{x^2}{(x + 1)^2} \cdot \frac{(x + 1)^3}{x} = x(x + 1),\ x \neq -1,\ 0$$

47. Perform the indicated operations and simplify.

$$6 - \frac{5}{x + 3}$$

Solution

$$6 - \frac{5}{x + 3} = \frac{6(x + 3)}{(x + 3)} - \frac{5}{x + 3} = \frac{6(x + 3) - 5}{x + 3} = \frac{6x + 13}{x + 3},\ x \neq -3$$

51. Perform the indicated operations and simplify.

$$\frac{2}{x^2 - 4} - \frac{1}{x^2 - 3x + 2}$$

Solution

$$\frac{2}{x^2 - 4} - \frac{1}{x^2 - 3x + 2} = \frac{2}{(x + 2)(x - 2)} - \frac{1}{(x - 1)(x - 2)}$$

$$= \frac{2(x - 1) - (x + 2)}{(x + 2)(x - 2)(x - 1)} = \frac{x - 4}{(x + 2)(x - 2)(x - 1)},\ x \neq 1,\ \pm 2$$

55. Simplify the following compound fraction.

$$\frac{\left(\frac{x}{2}-1\right)}{(x-2)}$$

Solution

$$\frac{\left(\frac{x}{2}-1\right)}{(x-2)} = \frac{\left(\frac{x}{2}-\frac{2}{2}\right)}{\left(\frac{x-2}{1}\right)} = \frac{x-2}{2}\cdot\frac{1}{x-2} = \frac{1}{2},\ x \neq 2$$

59. Simplify the following compound fraction.

$$\frac{\left(\frac{1}{(x+2)^2}-\frac{1}{x^2}\right)}{2}$$

Solution

$$\frac{\left(\frac{1}{(x+2)^2}-\frac{1}{x^2}\right)}{2} = \frac{1}{2}\left[\frac{x^2-(x+2)^2}{x^2(x+2)^2}\right] = \frac{1}{2}\left[\frac{x^2-(x^2+4x+4)}{x^2(x+2)^2}\right]$$

$$= \frac{1}{2}\left[\frac{-4x-4}{x^2(x+2)^2}\right] = \frac{-4(x+1)}{2x^2(x+2)^2} = \frac{-2(x+1)}{x^2(x+2)^2},\ x \neq 0,\ -2$$

61. Simplify the following compound fraction.

$$\frac{\left(\sqrt{x}-\frac{1}{2\sqrt{x}}\right)}{\sqrt{x}}$$

Solution

$$\frac{\left(\sqrt{x}-\frac{1}{2\sqrt{x}}\right)}{\sqrt{x}} = \frac{\left(\sqrt{x}-\frac{1}{2\sqrt{x}}\right)}{\sqrt{x}}\cdot\frac{2\sqrt{x}}{2\sqrt{x}} = \frac{2x-1}{2x}$$

65. *Income Tax Audit* In 1990 the Internal Revenue Service audited 388,000 of the 43,694,000 income tax returns that were filed with Form 1040A. If you filed a Form 1040A return during that year, what is the probability that your return was audited? (*Source: Internal Revenue Service*)

Solution

$$P(E) = \frac{388{,}000}{43{,}694{,}000} = \frac{388}{43{,}694} = \frac{194}{21{,}847} \approx 0.009$$

69. *Monthly Payment* Use the following formula, which gives the approximate annual percentage rate r of a monthly installment loan:

$$r = \frac{\left(\frac{24(NM - P)}{N}\right)}{\left(P + \frac{NM}{12}\right)}$$

where N is the total number of payments, M is the monthly payment, and P is the amount financed.

(a) Approximate the annual percentage rate r for a four-year car loan of \$15,000 that has monthly payment of \$400.

(b) Simplify the expression for the annual percentage rate r, and then rework part (a).

Solution

(a) $$r = \frac{\left(\frac{24[48(400) - 15,000]}{48}\right)}{\left(15,000 + \frac{48(400)}{12}\right)} \approx 0.1265 = 12.65\%$$

(b) $$r = \frac{\left(\frac{24(NM - P)}{N}\right)}{\left(P + \frac{NM}{12}\right)} = \frac{24(NM - P)}{N} \cdot \frac{12}{12P + NM} = \frac{288(NM - P)}{N(12P + NM)}$$

$$r = \frac{288[48(400) - 15,000]}{48[12(15,000) + 48(400)]} \approx 0.1265 = 12.65\%$$

73. *Probability* Consider an experiment in which a marble is tossed into a box whose dimensions are $2x + 1$ inches by x inches. Find the probability that the marble will come to rest in the shaded portion of the box. (See figure in the textbook.)

Solution

$$P(E) = 1 - \frac{\pi\left(\frac{x}{2}\right)^2}{x(2x + 1)} = 1 - \frac{\pi x^2}{4} \cdot \frac{1}{x(2x + 1)}$$

$$= 1 - \frac{\pi x}{4(2x + 1)} = \frac{4(2x + 1) - \pi x}{4(2x + 1)} = \frac{(8 - \pi)x + 4}{4(2x + 1)}$$

Review Exercises for Chapter P

SOLUTIONS TO SELECTED EXERCISES

1. Determine which numbers in the set are (a) natural numbers, (b) integers, (c) rational numbers, and (d) irrational numbers.

$$\left\{11,\ -14,\ -\tfrac{8}{9},\ \tfrac{5}{2},\ \sqrt{6},\ 0.4\right\}$$

Solution

(a) Natural numbers: $\{11\}$

(b) Integers: $\{11,\ -14\}$

(c) Rational numbers: $\left\{11,\ -14,\ -\tfrac{8}{9},\ \tfrac{5}{2},\ 0.4\right\}$

(d) Irrational numbers: $\{\sqrt{6}\}$

7. Use inequality notation to describe "x is nonnegative."

Solution

$x \geq 0$

13. Find the distance between a and b when $a = 48$, and $b = 45$.

Solution

$d(a, b) = |45 - 48| = 3$

15. Use absolute value notation to describe "the distance between x and 7 is at least 4."

Solution

$|x - 7| \geq 4$

19. Evaluate $-4x^2 - 6x$ for the values (a) $x = -1$ and (b) $x = 0$.

Solution

(a) $-4(-1)^2 - 6(-1) = 2$

(b) $-4(0)^2 - 6(0) = 0$

25. Perform the indicated operations. (Write fractional answers in reduced form.)

$\tfrac{1}{2} + \tfrac{1}{3} - \tfrac{1}{6}$

Solution

$\tfrac{1}{2} + \tfrac{1}{3} - \tfrac{1}{6} = \tfrac{3}{6} + \tfrac{2}{6} - \tfrac{1}{6} = \tfrac{4}{6} = \tfrac{2}{3}$

29. Use a calculator to evaluate $4(\tfrac{1}{6} - \tfrac{1}{7})$. (Round your answer to two decimal places.)

Solution

$4(\tfrac{1}{6} - \tfrac{1}{7}) = 4(\tfrac{7}{42} - \tfrac{6}{42}) = \tfrac{4}{1} \cdot \tfrac{1}{42} = \tfrac{2}{21} \approx 0.10$

4 [×] [(] 1 [÷] 6 [−] 1 [÷] 7 [)] [=] Scientific

4 [(] 1 [÷] 6 [−] 1 [÷] 7 [)] [ENTER] Graphing

33. Simplify $\dfrac{(4x)^2}{2x}$.

Solution

$$\frac{(4x)^2}{2x} = \frac{16x^2}{2x} = 8x,\ x \neq 0$$

35. Write the number in scientific notation.

Daily U.S. Consumption of Dunkin' Donuts: 2,740,000

Solution

$2,740,000 = 2.74 \times 10^6$

39. Use a calculator to evaluate the following expressions. (Round your answer to three decimal places.)

(a) $1,800(1 + 0.08)^{24}$ Scientific

(b) $0.0024(7,658,400)$ Graphing

Solution

(a) $1800(1 + 0.08)^{24} \approx 11,414.125$

1800 [×] 1.08 [y^x] 24 [=] Scientific

1800 [×] 1.08 [∧] 24 [ENTER] Graphing

(b) $0.0024(7,658,400) = 18,380.160$

0.0024 [×] 7,658,400 [=] Scientific

0.0024 [×] 7,658,400 [ENTER] Graphing

43. Write the rational exponent form of $\sqrt{16} = 4$.

Solution

Radical form: $\sqrt{16} = 4$

Rational exponent form: $16^{1/2} = 4$

47. Simplify $\sqrt{4x^4}$ by removing all possible factors.

Solution

$\sqrt{4x^4} = 2x^2$

49. Rewrite the following expression by rationalizing the denominator. Simplify your answer.

$$\frac{1}{2 - \sqrt{3}}$$

Solution

$$\frac{1}{2-\sqrt{3}} = \frac{1}{2-\sqrt{3}} \cdot \frac{2+\sqrt{3}}{2+\sqrt{3}} = \frac{2+\sqrt{3}}{4-3} = \frac{2+\sqrt{3}}{1} = 2 + \sqrt{3}$$

55. Use rational exponents to reduce the index of $\sqrt[4]{5^2}$.

Solution

$\sqrt[4]{5^2} = (5^2)^{1/4} = 5^{1/2} = \sqrt{5}$

61. Perform the indicated operations and write the resulting polynomial in standard form.

$x(x - 2) - 2(3x + 7)$

Solution

$x(x - 2) - 2(3x + 7) = x^2 - 2x - 6x - 14 = x^2 - 8x - 14$

65. Perform the indicated operations and write the resulting polynomial in standard form.

$$(x-1)(x^2+2)$$

Solution

$$(x-1)(x^2+2) = x^3 - x^2 + 2x - 2$$

71. Completely factor $x^3 - 4x^2 - 2x + 8$.

Solution

$$\begin{aligned} x^3 - 4x^2 - 2x + 8 &= x^2(x-4) - 2(x-4) \\ &= (x-4)(x^2-2) \\ &= (x-4)(x+\sqrt{2})(x-\sqrt{2}) \end{aligned}$$

73. Find the domain of

$$\frac{2x+1}{x-3}.$$

Solution

The domain of $\frac{2x+1}{x-3}$ is the set of all real numbers except $x = 3$.

77. Write the following rational expression in reduced form.

$$\frac{x^2-4}{2x+4}$$

Solution

$$\frac{x^2-4}{2x+4} = \frac{(x+2)(x-2)}{2(x+2)} = \frac{x-2}{2},\ x \neq -2$$

81. Perform the indicated operations and simplify.

$$\frac{2x-1}{x+1} \cdot \frac{x^2-1}{2x^2-7x+3}$$

Solution

$$\frac{2x-1}{x+1} \cdot \frac{x^2-1}{2x^2-7x+3} = \frac{2x-1}{x+1} \cdot \frac{(x+1)(x-1)}{(2x-1)(x-3)} = \frac{x-1}{x-3},\ x \neq -1,\ \frac{1}{2},\ 3$$

83. Perform the indicated operations and simplify.

$$\frac{x}{x-1} + \frac{2x}{x-2}$$

Solution

$$\begin{aligned} \frac{x}{x-1} + \frac{2x}{x-2} &= \frac{x(x-2) + 2x(x-1)}{(x-1)(x-2)} = \frac{x^2 - 2x + 2x^2 - 2x}{(x-1)(x-2)} \\ &= \frac{3x^2 - 4x}{(x-1)(x-2)} = \frac{x(3x-4)}{(x-1)(x-2)},\ x \neq 1,\ 2 \end{aligned}$$

85. Simplify the following compound fraction.

$$\frac{\left(\frac{1}{x} - \frac{1}{y}\right)}{(x^2 - y^2)}$$

Solution

$$\begin{aligned} \frac{\left(\frac{1}{x} - \frac{1}{y}\right)}{(x^2 - y^2)} &= \frac{\left(\frac{y-x}{xy}\right)}{\left(\frac{(x+y)(x-y)}{1}\right)} = \frac{-(x-y)}{xy} \cdot \frac{1}{(x+y)(x-y)} \\ &= \frac{-1}{xy(x+y)} = -\frac{1}{xy(x+y)},\ x \neq 0,\ y \neq 0,\ x \neq \pm y \end{aligned}$$

87. *Dialing a Phone* Suppose you are dialing a phone number. After dialing the first six digits correctly, your finger slips and you are not sure which number (from 0 to 9) you dialed as the seventh digit. What is the probability that you dialed the correct phone number?

Solution

$P(E) = \frac{1}{10}$

Test for Chapter P

1. Evaluate $-3x^2 - 5x$ when $x = -3$.

Solution

$-3(-3)^2 - 5(-3) = -3(9) + 15 = -27 + 15 = -12$

2. Evaluate $-\frac{(-x)^3}{4}$ when $x = 2$.

Solution

$-\frac{(-2)^3}{4} = -\frac{(-8)}{4} = -(-2) = 2$

3. Simplify $8(-2x^2)^3$.

Solution

$8(-2x^2)^3 = 8(-8x^6) = -64x^6$

4. Simplify $3\sqrt{x} - 7\sqrt{x}$.

Solution

$3\sqrt{x} - 7\sqrt{x} = (3 - 7)\sqrt{x} = -4\sqrt{x}$

5. Simplify $5^{1/4} \cdot 5^{7/4}$.

Solution

$$\begin{aligned} 5^{1/4}5^{7/4} &= 5^{1/4+7/4} \\ &= 5^{8/4} \\ &= 5^2 = 25 \end{aligned}$$

6. Simplify $\sqrt{128} - \sqrt{72}$.

Solution

$$\begin{aligned} \sqrt{128} - \sqrt{72} &= \sqrt{64 \cdot 2} - \sqrt{36 \cdot 2} \\ &= 8\sqrt{2} - 6\sqrt{2} \\ &= 2\sqrt{2} \end{aligned}$$

7. Simplify $\sqrt{12x^3}$ by removing all possible factors from the radical.

Solution

$\sqrt{12x^3} = \sqrt{4x^2(3x)} = 2|x|\sqrt{3x}$

8. Rewrite the following expression by rationalizing the denominator and simplifying your answer.

$$\frac{2}{5 - \sqrt{7}}$$

Solution

$$\frac{2}{5 - \sqrt{7}} = \frac{2}{5 - \sqrt{7}} \cdot \frac{5 + \sqrt{7}}{5 + \sqrt{7}} = \frac{2(5 + \sqrt{7})}{25 - 7} = \frac{2(5 + \sqrt{7})}{18} = \frac{5 + \sqrt{7}}{9}$$

9. Perform the indicated operations and write the resulting polynomial in standard form.

$$(3x+5)^2$$

Solution

$(3x+5)^2 = (3x)^2 + 2(3x)(5) + (5)^2 = 9x^2 + 30x + 25$

10. Perform the indicated operations and write the resulting polynomial in standard form.

$$3x(x+5) - 2x(4x-7)$$

Solution

$3x(x+5) - 2x(4x-7) = 3x^2 + 15x - 8x^2 + 14x = -5x^2 + 29x$

11. Completely factor $5x^2 - 80$.

Solution

$5x^2 - 80 = 5(x^2 - 16) = 5(x+4)(x-4)$

12. Completely factor $x^3 - 6x^2 - 3x + 18$

Solution

$x^3 - 6x^2 - 3x + 18 = x^2(x-6) - 3(x-6) = (x-6)(x^2-3)$

13. Write the following rational expression in reduced form.

$$\frac{x^2-16}{3x+12}$$

Solution

$$\frac{x^2-16}{3x+12} = \frac{(x+4)(x-4)}{3(x+4)} = \frac{x-4}{3}, \quad x \neq -4$$

14. Perform the indicated operations and simplify:

$$\frac{3x-5}{x+3} \cdot \frac{x^2+7x+12}{9x^2-25}$$

Solution

$$\frac{3x-5}{x+3} \cdot \frac{x^2+7x+12}{9x^2-25} = \frac{3x-5}{x+3} \cdot \frac{(x+3)(x+4)}{(3x+5)(3x-5)} = \frac{x+4}{3x+5}, \quad x \neq -3, \pm\frac{5}{3}$$

15. Perform the indicated operations and simplify:

$$\frac{x}{x-3} + \frac{3x}{x-4}$$

Solution

$$\frac{x}{x-3} + \frac{3x}{x-4} = \frac{x(x-4) + 3x(x-3)}{(x-3)(x-4)} = \frac{x^2 - 4x + 3x^2 - 9x}{(x-3)(x-4)}$$

$$= \frac{4x^2 - 13x}{(x-3)(x-4)} = \frac{x(4x-13)}{(x-3)(x-4)}, \quad x \neq 3, 4$$

16. Perform the indicated operations and simplify:

$$\frac{3}{x+5} - \frac{4}{x-2}$$

Solution

$$\frac{3}{x+5} - \frac{4}{x-2} = \frac{3(x-2) - 4(x+5)}{(x+5)(x-2)} = \frac{3x - 6 - 4x - 20}{(x+5)(x-2)} = \frac{-x-26}{(x+5)(x-2)}, \quad x \neq -5, 2$$

17. Complete the table shown in the textbook, given $3,000 is deposited in an account with an annual percentage rate of 8%, compounded monthly. What can you conclude from the table?

Solution

$$B = 3000\left(1 + \frac{0.08}{12}\right)^{12t}$$

Number of Years	5	10	15	20	25
Balance	$4,469.54	$6,658.92	$9,920.76	$14,780.41	$22,020.53

The money grows at a faster rate the longer it is left in the account.

18. Simplify $\dfrac{\left(\frac{1}{x} + \frac{1}{y}\right)}{\left(\frac{1}{x} - \frac{1}{y}\right)}$.

Solution

$$\frac{\left(\frac{1}{x} + \frac{1}{y}\right)}{\left(\frac{1}{x} - \frac{1}{y}\right)} = \frac{\frac{y+x}{xy}}{\frac{y-x}{xy}} = \frac{y+x}{xy} \cdot \frac{xy}{y-x} = \frac{y+x}{y-x}$$

19. You have just bought a new microwave oven. The manufacturer made 500,000 microwaves of your model and of those, 10,000 had a defective timer. What is the probability that your new microwave oven has a defective timer?

Solution

$p = \frac{10{,}000}{500{,}000} = \frac{1}{50} = 0.02$

20. 42 of the students in your class took a test. Your instructor lost two of the tests. What is the probability that your test got lost?

Solution

$p = \frac{2}{42} = \frac{1}{21} \approx 0.048$

Practice Test for Chapter P

1. Evaluate $-2x^3 + 7x - 4$ when $x = -2$.

2. Simplify $\frac{x}{z} - \frac{z}{y}$.

3. The distance between x and 7 is no more than 4. Use absolute value notation to describe this expression.

4. Evaluate $10(-x)^3$ for $x = 5$.

5. Simplify $(-4x^3)(2x^{-5})\left(\frac{1}{16}x\right)$.

6. Change 0.0000412 to scientific notation.

7. Evaluate $125^{2/3}$.

8. Simplify $\sqrt[4]{64x^7y^9}$.

9. Rationalize the denominator and simplify $6/\sqrt{12}$.

10. Simplify $3\sqrt{80} - 7\sqrt{500}$.

11. Simplify $(8x^4 - 9x^2 + 2x - 1) - (3x^3 + 5x + 4)$.

12. Multiply $(x - 3)(x^2 + x - 7)$.

13. Multiply $[(x - 2) - y]^2$.

14. Factor $16x^4 - 1$.

15. Factor $6x^2 + 5x - 4$.

16. Factor $x^3 - 64$.

17. Combine and simplify:

$$-\frac{3}{x} + \frac{x}{x^2 + 2}$$

18. Combine and simplify:

$$\frac{x - 3}{4x} \div \frac{x^2 - 9}{x^2}$$

19. Simplify:

$$\frac{1 - (1/x)}{1 - \dfrac{1}{1 - (1/x)}}$$

20. Find the balance if \$5,000 is deposited in an account for 20 years with an annual percentage rate of 7.5% compounded quarterly.

CHAPTER ONE

Algebraic Equations and Inequalities

1.1 Linear Equations

- You should know how to solve linear equations. $ax + b = 0, \ a \neq 0$
- An identity is an equation whose solution consists of every real number in its domain.
- To solve an equation you can:
  (a) Add or subtract the same quantity from both sides.
  (b) Multiply or divide both sides by the same nonzero quantity.

  To solve an equation that can be simplified to a linear equation:
  (a) Remove all symbols of grouping and all fractions.
  (b) Combine like terms.
  (c) Solve by algebra.
  (d) Check the answer.
- A "solution" that does not satisfy the original equation is called an extraneous solution.

SOLUTIONS TO SELECTED EXERCISES

1. Determine whether $2(x - 1) = 2x - 2$ is an identity or a conditional equation.

Solution

The equation $2(x - 1) = 2x - 2$ is an *identity* because (by the Distributive Property) it is true for every real value of x.

5. Determine whether $2(x + 1) = 2x + 1$ is an identity or a conditional equation.

Solution

The equation $2(x + 1) = 2x + 1$ is *conditional* since there are *no* real number values of x for which the equation is true.

9. Determine whether the values of x are solutions for $3x^2 + 2x - 5 = 2x^2 - 2$.

(a) $x = -3$ (b) $x = 1$ (c) $x = 4$ (d) $x = -5$

Solution

(a) $3(-3)^2 + 2(-3) - 5 \stackrel{?}{=} 2(-3)^2 - 2$

$$16 = 16$$

$x = -3$ *is* a solution.

(b) $3(1)^2 + 2(1) - 5 \stackrel{?}{=} 2(1)^2 - 2$

$$0 = 0$$

$x = 1$ *is* a solution.

(c) $3(4)^2 + 2(4) - 5 \stackrel{?}{=} 2(4)^2 - 2$

$$51 \neq 30$$

$x = 4$ *is not* a solution.

(d) $3(-5)^2 + 2(-5) - 5 \stackrel{?}{=} 2(-5)^2 - 2$

$$60 \neq 48$$

$x = -5$ *is not* a solution.

11. Determine whether the values of x are solutions for $(5/2x) - (4/x) = 3$.

(a) $x = -\dfrac{1}{2}$ (b) $x = 4$ (c) $x = 0$ (d) $x = \frac{1}{4}$

Solution

(a) $\dfrac{5}{2(-1/2)} - \dfrac{4}{(-1/2)} \stackrel{?}{=} 3$

$$3 = 3$$

$x = -\dfrac{1}{2}$ *is* a solution.

(b) $\dfrac{5}{2(4)} - \dfrac{4}{4} \stackrel{?}{=} 3$

$$-\frac{3}{8} \neq 3$$

$x = 4$ *is not* a solution.

(c) $\dfrac{5}{2(0)} - \dfrac{4}{0}$ is undefined.

$x = 0$ *is not* a solution.

(d) $\dfrac{5}{2(1/4)} - \dfrac{4}{1/4} \stackrel{?}{=} 3$

$$-6 \neq 3$$

$x = \dfrac{1}{4}$ *is not* a solution.

17. Solve $7 - 2x = 15$ and check your answer.

Solution

$$7 - 2x = 15$$
$$7 - 2x - 7 = 15 - 7$$
$$-2x = 8$$
$$x = -4$$

Check

$$7 - 2(-4) \stackrel{?}{=} 15$$
$$7 + 8 \stackrel{?}{=} 15$$
$$15 = 15 \checkmark$$

23. Solve $6[x - (2x + 3)] = 8 - 5x$ and check your answer.

Solution

$$6[x - (2x + 3)] = 8 - 5x$$
$$6[-x - 3] = 8 - 5x$$
$$-6x - 18 = 8 - 5x$$
$$-x = 26$$
$$x = -26$$

Check

$$6[-26 - (2(-26) + 3)] \stackrel{?}{=} 8 - 5(-26)$$
$$6[-26 - (-52 + 3)] \stackrel{?}{=} 8 + 130$$
$$6[-26 - (-49)] \stackrel{?}{=} 138$$
$$6[-26 + 49] \stackrel{?}{=} 138$$
$$6(23) \stackrel{?}{=} 138$$
$$138 = 138 \checkmark$$

27. Solve $\frac{3}{2}(z + 5) - \frac{1}{4}(z + 24) = 0$ and check your answer.

Solution

$$\tfrac{3}{2}(z + 5) - \tfrac{1}{4}(z + 24) = 0$$
$$4(\tfrac{3}{2})(z + 5) - 4(\tfrac{1}{4})(z + 24) = 4(0)$$
$$6(z + 5) - (z + 24) = 0$$
$$6z + 30 - z - 24 = 0$$
$$5z = -6$$
$$z = -\tfrac{6}{5}$$

Check

$$\tfrac{3}{2}\left(-\tfrac{6}{5} + 5\right) - \tfrac{1}{4}\left(-\tfrac{6}{5} + 24\right) \stackrel{?}{=} 0$$
$$\tfrac{3}{2}\left(-\tfrac{6}{5} + \tfrac{25}{5}\right) - \tfrac{1}{4}\left(-\tfrac{6}{5} + \tfrac{120}{5}\right) \stackrel{?}{=} 0$$
$$\tfrac{3}{2}\left(\tfrac{19}{5}\right) - \tfrac{1}{4}\left(\tfrac{114}{5}\right) \stackrel{?}{=} 0$$
$$\tfrac{57}{10} - \tfrac{57}{10} \stackrel{?}{=} 0$$
$$0 = 0 \checkmark$$

31. Solve $x + 8 = 2(x - 2) - x$ and check your answer.

Solution

$$x + 8 = 2(x - 2) - x$$
$$x + 8 = 2x - 4 - x$$
$$x + 8 = x - 4$$
$$8 = -4$$

Contradiction: no solution

35. Solve $\frac{5x-4}{5x+4} = \frac{2}{3}$ and check your answer.

Solution

$$\frac{5x-4}{5x+4} = \frac{2}{3}$$
$$3(5x-4) = 2(5x+4)$$
$$15x - 12 = 10x + 8$$
$$5x = 20$$
$$x = 4$$

Check

$$\frac{5(4)-4}{5(4)+4} \stackrel{?}{=} \frac{2}{3}$$
$$\frac{16}{24} \stackrel{?}{=} \frac{2}{3}$$
$$\frac{2}{3} = \frac{2}{3} \checkmark$$

41. Solve $\frac{x}{x+4} + \frac{4}{x+4} + 2 = 0$ and check your answer.

Solution

$$\frac{x}{x+4} + \frac{4}{x+4} + 2 = 0$$
$$\frac{x+4}{x+4} + 2 = 0$$
$$1 + 2 = 0$$
$$3 = 0$$

Contradiction: no solution

45. Solve $\frac{3}{x(x-3)} + \frac{4}{x} = \frac{1}{x-3}$ and check your answer.

Solution

$$\frac{3}{x(x-3)} + \frac{4}{x} = \frac{1}{x-3}$$
$$3 + 4(x-3) = x$$
$$3 + 4x - 12 = x$$
$$3x = 9$$
$$x = 3$$

A check reveals that $x = 3$ is an extraneous solution since it makes the denominator zero, so there is no solution.

47. Solve $(x+2)^2 + 5 = (x+3)^2$ and check your answer.

Solution

$$(x+2)^2 + 5 = (x+3)^2$$
$$x^2 + 4x + 4 + 5 = x^2 + 6x + 9$$
$$4x + 9 = 6x + 9$$
$$-2x = 0$$
$$x = 0$$

Check

$$(0+2)^2 + 5 \stackrel{?}{=} (0+3)^2$$
$$4 + 5 \stackrel{?}{=} 9$$
$$9 = 9 \checkmark$$

51. Solve $(2x+1)^2 = 4(x^2 + x + 1)$ and check your answer.

Solution

$$(2x+1)^2 = 4(x^2 + x + 1)$$
$$4x^2 + 4x + 1 = 4x^2 + 4x + 4$$
$$1 = 4$$

Contradiction: no solution

55. Use a calculator to solve the following equation for x.

$$\frac{x}{0.6321} + \frac{x}{0.0692} = 1,000$$

Solution

$$\frac{x}{0.6321} + \frac{x}{0.0692} = 1000$$
$$0.0692x + 0.6321x = 1000(0.6321)(0.0692)$$
$$0.7013x = 43.74132$$
$$x = \frac{43.74132}{0.7013} \approx 62.372$$

63. Evaluate the following expression in two ways. (a) Calculate entirely on your calculator by storing intermediate results, and then round the answer to two decimal places. (b) Round both the numerator and the denominator to two decimal places before dividing, and then round the final answer to two decimal places. Does the second method introduce an additional round-off error?

$$\frac{333 + \dfrac{1.98}{0.74}}{4 + \dfrac{6.25}{3.15}}$$

Solution

(a) 56.09 (b) $\dfrac{335.68}{5.98} \approx 56.13$

The second method introduced an additional round-off error.

67. *Human Height* The relationship between the length of an adult's thigh bone and the height of the adult can be approximated by the linear equation

$y = 0.432x - 10.44$ Female

$y = 0.449x - 12.15$ Male

where y is the length of the thigh bone in inches and x is the height of the adult in inches. An anthropologist discovers a thigh bone that belonged to an adult human female. The bone is 16 inches long. How tall would you estimate the female to have been?

Solution

$$16 = 0.432x - 10.44$$

$$26.44 = 0.432x$$

$$\frac{26.44}{0.432} = x$$

$$x \approx 61.20 \text{ inches}$$

71. *Probability* The equation $p + p' = 1$ states that the sum of the probability that an event *will* occur p and the probability that it *will not* occur p' is 1. If the probability that an event will occur is 0.54, what is the probability that the event will not occur?

Solution

$$0.54 + p' = 1$$

$$p' = 0.46$$

1.2 Linear Equations and Modeling

- Be able to set up mathematical models to solve problems.
- Be able to translate key words and phrases.
  (a) Consecutive: Next, subsequent
  (b) Addition: Sum, plus, greater, increased by, more than, exceeds, total of
  (c) Subtraction: Difference, minus, less than, decreased by, subtracted from, reduced by, the remainder
  (d) Multiplication: Product, multiplied by, twice, times, percent of
  (e) Division: Quotient, divided by, ratio, per
  (f) Equality: Equals, equal to, is, are, was, will be, represents
- You should know the following formulas:
  (a) Perimeter
    1. Square: $P = 4s$
    2. Rectangle: $P = 2L + 2W$
    3. Circle: $C = 2\pi r$
  (b) Area
    1. Square: $A = s^2$
    2. Rectangle: $A = LW$
    3. Circle: $A = \pi r^2$
    4. Triangle: $A = (1/2)bh$
  (c) Volume
    1. Cube: $V = s^3$
    2. Rectangular solid: $V = LWH$
    3. Cylinder: $V = \pi r^2 h$
    4. Sphere: $V = (4/3)\pi r^3$
  (d) Simple Interest: $I = Prt$
  (e) Distance: $D = r \cdot t$
  (f) Temperature: $F = (9/5)C + 32$
- You should be able to solve word problems. Study the examples in the text carefully.

SOLUTIONS TO SELECTED EXERCISES

3. Write the algebraic expression for the following verbal expression.

Distance Traveled The distance traveled in t hours by a car traveling at 50 miles per hour.

Solution

Model: (Distance) = (rate) × (time)

Labels: Distance = d, rate = 50 mph, time = t

Expression: $d = 50t$

7. Write the algebraic expression for the following verbal expression.

Geometry The perimeter of a rectangle whose width is x and whose length is twice the width.

Solution

Model: Perimeter = 2 (width) +2 (length)

Labels: Perimeter = P, width = x, length = 2 (width) = $2x$

Expression: $P = 2x + 2(2x) = 6x$

11. Write a mathematical model for "the sum of two consecutive natural numbers is 525." Find the two numbers and solve the problem.

Solution

Model: sum = (first number) + (second number)

Labels: sum = 525, first number = n, second number = $n + 1$

Equation: $525 = n + (n + 1)$

$$n = 262$$

Answer: first number = $n = 262$, second number = $n + 1 = 263$

17. *Weekly Paycheck* Your weekly paycheck is 15% *more* than your co-worker's. Your two paychecks total \$645. Find the amount of each paycheck.

Solution

(Total of paychecks) = (co-worker's paycheck) + (your paycheck)

Labels: Total of paychecks = \$645, co-worker's paycheck = x,

your paycheck = $x + 15\%$ of $x = x + 0.15x$

Equation: $645 = x + (x + 0.15x) = 2.15x$

$$x = \frac{645}{2.15} = 300$$

Answer: co-worker's paycheck = x = \$300,

your paycheck = $x + 0.15x$ = \$345

23. *Then and Now* The prices for a pound of ground beef are \$0.15 in 1940 and \$1.44 in 1990. Find the percentage increase for the price of ground beef. (*Source: USA Today.*)

Solution

Model: (1990 lb. of ground beef price) = (percentage increase)(1940 lb. of ground beef price) + (1940 lb. of ground beef price)

Labels: 1990 lb. of ground beef price = \$1.44, percentage increase = p, 1940 lb. of ground beef price = \$0.15

Equation: $1.44 = 0.15p + 0.15$

$$\frac{1.29}{0.15} = p$$

$$p = 8.6 = 860\%$$

Answer: percentage increase = $p = 860\%$

27. *Retirement Nest Eggs* Based on a survey of 740 companies, the most common types of investments by corporate retirement plans are shown in the bar graph in the textbook. (*Source:* Wyatt Company.) How many respondents in the survey used each type of investment? (Most respondents used more than one type of investment.) (See figure in the textbook.)

Solution

Model: (Number of respondents who used a type of investment) = (percent)(740)

Labels: Guaranteed investment = R_1, percent = 73%;

Money fund = R_2, percent = 45%;

Diversified stock = R_3, percent = 42%;

Aggressive stock = R_4, percent = 41%;

Company stock = R_5, percent = 40%;

Equations: $R_1 = 0.73(740) \approx 540$

$$R_2 = 0.45(740) = 333$$

$$R_3 = 0.42(740) \approx 311$$

$$R_4 = 0.41(740) \approx 303$$

$$R_5 = 0.40(740) = 296$$

Answer: Guaranteed investment: 540 respondents

Money fund: 333 respondents

Diversified stock: 311 respondents

Aggressive stock: 303 respondents

Company stock: 296 respondents

31. *Geometry* A room is 1.5 times as long as it is wide, and its perimeter is 75 feet. (See figure in the textbook.) Find the dimensions of the room.

Solution

Model: perimeter = 2(width) +2(length)

Labels: perimeter = 75 (feet), width (in feet) = x, length (in feet) = 1.5, (width) = $1.5x$

Equation: $75 = 2x + 2(1.5x)$

$$x = 15$$

Answer: width = x = 15 feet, length = $1.5x$ = 22.5 feet

35. *Loan Payments* A family has annual loan payments totaling $13,077.75, or 58.6% of its annual income. What is the family's income?

Solution

(Annual loan payments) = 58.6% × (annual income)

$$\$13,077.75 = 58.6\% \times x$$

$$13,077.75 = 0.586x$$

$$x = \frac{13,077.75}{0.586} \approx \$22,316.98$$

39. *Discount Rate* The price of a television set has been discounted by \$150. The sale price is \$245. What percent is the discount of the original list price?

Solution

$$\text{(Sale price)} = \text{(original price)} - \text{(discount percent)(original price)}$$

$$245 = (245 + 150) - p(245 + 150)$$

$$245 = 395 - 395p$$

$$395p = 150$$

$$p = \tfrac{150}{395} \approx 0.3797 \approx 38\%$$

43. *Travel Time* Two cars start at a given point and travel in the same direction at average speeds of 40 miles per hour and 55 miles per hour. How much time must elapse before the two cars are five miles apart?

Solution

$$\text{Distance} = \text{rate} \times \text{time}$$

$$d_1 = 40 \text{ mph} \times t$$

$$d_2 = 55 \text{ mph} \times t$$

$$\text{(Distance between cars)} = \text{(second distance)} - \text{(first distance)}$$

$$5 = d_2 - d_1$$

$$5 = 55t - 40t = 15t$$

$$t = \tfrac{1}{3} \text{ hour}$$

47. *Radio Waves* Radio waves travel at the same speed as light, 3.0×10^8 meters per second. Find the time required for a radio wave to travel from mission control in Houston to NASA astronauts on the surface of the moon 3.86×10^8 meters away.

Solution

$$\text{Time} = \frac{\text{distance}}{\text{rate}}$$

$$t = \frac{3.86 \times 10^8 \text{ meters}}{3.0 \times 10^8 \text{ meters per second}}$$

$$t \approx 1.29 \text{ seconds}$$

51. *Projected Expenses* From January through May, a company's expenses have totaled \$234,980. If the monthly expenses continue at this rate, what will the total expenses for the year be?

Solution

$$\text{Total expenses} = 12 \text{ (Monthly expenses)}$$

$$T = 12\left(\frac{234{,}980}{5}\right) = \$563{,}952$$

55. *Comparing Investment Returns* You invested \$12,000 in a fund paying $9\frac{1}{2}\%$ simple interest and \$8,000 in a fund with a variable interest rate. At the end of the year, you received notification that the total interest for both funds was \$2,054.40. Find the equivalent simple interest rate on the variable rate fund.

Solution

$$\text{Interest} = (\text{interest rate}) \times \text{principal}$$

$$i_1 = 9.5\% \times \$12,000$$

$$i_2 = r \times \$8,000$$

$$\text{Total interest} = (\text{interest in first account}) + (\text{interest in second account})$$

$$\$2054.40 = i_1 + i_2$$

$$2054.40 = 0.095(12000) + 8000r$$

$$r = \tfrac{914.40}{8000} = 0.1143 = 11.43\%$$

59. *Length of a Tank* The diameter of a cylindrical propane gas tank is 4 feet. (See figure in the textbook.) The total volume of the tank is 603.2 cubic feet. Find the length of the tank.

Solution

$$\text{Volume} = \pi r^2 h,$$

$$\text{Diameter} = 2r = 4 \Rightarrow r = 2$$

$$603.2 = \pi(2)^2 h$$

$$h = \frac{603.2}{4\pi} \approx 48 \text{ feet long}$$

63. *New York City Marathon* Find the average speed of the record-holding runners in the New York City Marathon. The men's record time is 2 hours, 8 minutes. The length of the course is 26 miles and 385 yards. (Note that 1 mile = 5,280 feet = 1,760 yards.)

Solution

$$\text{Rate} = \frac{\text{distance}}{\text{time}} = \frac{\left(26 + \frac{385}{1760}\right) \text{ miles}}{\left(2 + \frac{8}{60}\right) \text{ hours}} \approx 12.29 \text{ miles per hour.}$$

65. Solve the following variable for h.

Area of a Triangle: $A = \frac{1}{2}bh$

Solution

$$A = \frac{1}{2}bh$$

$$2A = bh$$

$$\frac{2A}{b} = h$$

69. Solve the following variable for C.

Markup: $S = C + RC$

Solution

$$S = C + RC$$

$$S = C(1 + R)$$

$$\frac{S}{1 + R} = C$$

73. Solve the following variable for b.

Area of a Trapezoid: $A = \frac{1}{2}(a + b)h$

Solution

$$A = \frac{1}{2}(a+b)h$$

$$\frac{2A}{h} = a + b$$

$$\frac{2A}{h} - a = b$$

$$\frac{2A - ah}{h} = b$$

1.3 Quadratic Equations

- You should be able to solve quadratic equations:
  - (a) by factoring
  - (b) by extracting square roots

SOLUTIONS TO SELECTED EXERCISES

3. Write $x^2 = 25x$ in standard form, $ax^2 + bx + c = 0$.

Solution

$x^2 = 25x$

$x^2 - 25x = 0$

$a = 1, b = -25, c = 0$

9. Write the following quadratic equation in standard form, $ax^2 + bx + c = 0$.

$$\frac{3x^2 - 10}{5} = 12x$$

Solution

$$\frac{3x^2 - 10}{5} = 12x$$

$3x^2 - 10 = 60x$

$3x^2 - 60x - 10 = 0$

$a = 3, b = -60, c = -10$

13. Solve $x^2 - 2x - 8 = 0$ by factoring.

Solution

$x^2 - 2x - 8 = 0$

$(x - 4)(x + 2) = 0$

$x - 4 = 0$ or $x + 2 = 0$

$x = 4$ or $x = -2$

17. Solve $3 + 5x - 2x^2 = 0$ by factoring.

Solution

$3 + 5x - 2x^2 = 0$

$(3 - x)(1 + 2x) = 0$

$3 - x = 0$ or $1 + 2x = 0$

$x = 3$ or $x = -\frac{1}{2}$

21. Solve $-x^2 - 7x = 10$ by factoring.

Solution

$$-x^2 - 7x = 10$$
$$0 = x^2 + 7x + 10$$
$$0 = (x+5)(x+2)$$
$$x + 5 = 0 \quad \text{or} \quad x + 2 = 0$$
$$x = -5 \quad \text{or} \quad x = -2$$

25. Solve $x^2 = 7$ by extracting square roots. List both the exact answer *and* a decimal answer that has been rounded to two decimal places.

Solution

$$x^2 = 7$$
$$x = \pm\sqrt{7}$$
$$\approx \pm 2.65$$

29. Solve $(x - 12)^2 = 18$ by extracting square roots. List both the exact answer *and* a decimal answer that has been rounded to two decimal places.

Solution

$$(x - 12)^2 = 18$$
$$x - 12 = \pm 3\sqrt{2}$$
$$x = 12 \pm 3\sqrt{2}$$
$$x \approx 16.24 \text{ or,}$$
$$x \approx 7.76$$

35. Solve $3x^2 + 2(x^2 - 4) = 15$ by extracting square roots. List both the exact answer *and* a decimal answer that has been rounded to two decimal places.

Solution

$$3x^2 + 2(x^2 - 4) = 15$$
$$3x^2 + 2x^2 - 8 = 15$$
$$5x^2 = 23$$
$$x^2 = \frac{23}{5}$$
$$x = \pm\sqrt{\frac{23}{5}} = \pm\frac{\sqrt{115}}{5}$$
$$x \approx \pm 2.14$$

39. Solve $x^2 - 2x + 1 = 0$ by any convenient method.

Solution

$$x^2 - 2x + 1 = 0$$
$$(x - 1)^2 = 0$$
$$x - 1 = 0$$
$$x = 1$$

45. Solve $(x + 3)^2 = 81$ by any convenient method.

Solution

$$(x + 3)^3 = 81$$
$$x + 3 = \pm 9$$
$$x + 3 = 9 \quad \text{or} \quad x + 3 = -9$$
$$x = 6 \quad \text{or} \quad x = -12$$

49. Solve $50 + 5x = 3x^2$ by any convenient method.

Solution

$$50 + 5x = 3x^2$$
$$3x^2 - 5x - 50 = 0$$
$$(3x + 10)(x - 5) = 0$$
$$3x + 10 = 0 \quad \text{or} \quad x - 5 = 0$$
$$x = -\tfrac{10}{3} \quad \text{or} \quad x = 5$$

53. Solve $50x^2 - 60x + 10 = 0$ by any convenient method.

Solution

$$50x^2 - 60x + 10 = 0$$
$$10(5x^2 - 6x + 1) = 0$$
$$10(5x - 1)(x - 1) = 0$$
$$5x - 1 = 0 \quad \text{or} \quad x - 1 = 0$$
$$x = \tfrac{1}{5} \quad \text{or} \quad x = 1$$

57. Solve $(x+1)^2 = x^2$ by any convenient method.

Solution

$$(x+1)^2 = x^2$$
$$(x+1)^2 - x^2 = 0$$
$$[(x+1)+x][(x+1)-x] = 0$$
$$2x+1 = 0$$
$$x = -\tfrac{1}{2}$$

61. *Geometry* A triangular sign has a height that is equal to its base. The area of the sign is four square feet. Find the base and height of the sign.

Solution

$$\text{Area} = \tfrac{1}{2}(\text{base})(\text{height})$$
$$4 = \tfrac{1}{2}(b)(b)$$
$$8 = b^2$$
$$\pm\sqrt{8} = b$$
$$b = 2\sqrt{2}\ (-2\sqrt{2} \text{ is extraneous})$$

Answer: base $=$ height $= 2\sqrt{2} \approx 2.83$ feet

65. *Olympic Diver* The high-dive platform in the Olympics is 10 meters above the water. A diver wants to perform a double flip into the water. How long will the diver be in the air?

Solution

$$S = -16t^2 + V_o t + S_o$$
$$V_0 = 0 \text{ and } S_0 = 10 \text{ meters} \approx 32.808 \text{ feet}$$

Let $S = 0$.

$$0 = -16t^2 + 32.808$$
$$16t^2 = 32.808$$
$$t^2 = 2.0505$$
$$t \approx 1.43 \text{ seconds}$$

69. *Flying Distance* The cities of Chicago, Atlanta, and Buffalo approximately form the vertices of a right triangle. The distance from Atlanta to Buffalo is about 650 miles and the distance from Atlanta to Chicago is about 565 miles. Use the map in the textbook to approximate the flying distance from Atlanta to Buffalo *by way of* Chicago.

Solution

$$a^2 + b^2 = c^2$$
$$a^2 + 565^2 = 650^2$$
$$a^2 + 319,225 = 422,500$$
$$a^2 = 103,275$$
$$a \approx 321.36 \text{ miles}$$

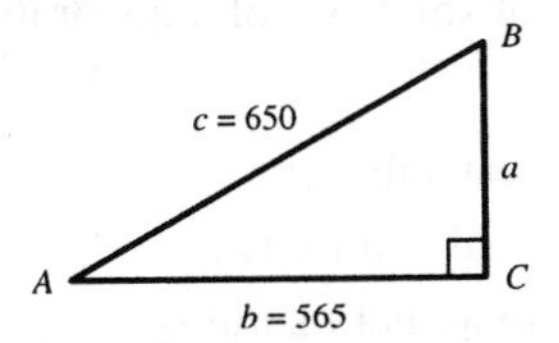

The flying distance from Atlanta to Buffalo by way of Chicago $= a + b = 886.36 \approx 886$ miles.

73. *Total Revenue* The demand equation for a certain product is $p = 20 - 0.0002x$, where p is the price per unit and x is the number of units sold. The total revenue for selling x units is given by

$$\text{Revenue} = xp - x(20 = 0.0002x).$$

How many units must be sold to produce a revenue of $500,000?

Solution

$$\begin{aligned}
\text{Revenue} &= xp = x(20 - 0.0002x) \\
500,000 &= 20x - 0.0002x^2 \\
0.0002x^2 - 20x + 500,000 &= 0 \\
x^2 - 100,000x + 2,500,000,000 &= 0 \\
(x - 50,000)^2 &= 0 \\
x &= 50,000 \text{ units}
\end{aligned}$$

79. *Airline Flight* Suppose that you are going on a round-trip airline flight. You are told that the probability that both flights will be late is $p^2 = 0.2016$. What is the probability p that you will be late on the return trip?

Solution

$p^2 = 0.2016$

$p = \sqrt{0.2016} \approx 0.4490$

1.4 The Quadratic Formula

- You should know the Quadratic Formula: For $ax^2 + bx + c = 0,\ a \neq 0$,
$$x = \frac{-b \pm \sqrt{b^2 - 4ac}}{2a}.$$
- You should be able to determine the types of solutions of a quadratic equation by checking the discriminant $b^2 - 4ac$.
  (a) If $b^2 - 4ac > 0$, there are two distinct real solution.
  (b) If $b^2 - 4ac = 0$, there is one repeating real solution.
  (c) If $b^2 - 4ac < 0$, there are two distinct imaginary solutions.

  You should be able to use your calculator to solve quadratic equations.
- You should be able to solve word problems involving quadratic equations. Study the examples in the text carefully.

SOLUTIONS TO SELECTED EXERCISES

1. Use the discriminant to determine the number of real solutions of $4x^2 - 4x + 1 = 0$.

Solution

$4x^2 - 4x + 1 = 0$

$b^2 - 4ac = (-4)^2 - 4(4)(1) = 0$

One real solution

5. Use the discriminant to determine the number of real solutions of $2x^2 - 5x + 5 = 0$.

Solution

$2x^2 - 5x + 5 = 0$

$b^2 - 4ac = (-5)^2 - 4(2)(5) = -15 < 0$

No real solutions

9. Use the Quadratic Formula to solve $2x^2 + x - 1 = 0$.

Solution

$a = 2, b = 1, c = -1$

$$x = \frac{-b \pm \sqrt{b^2 - 4ac}}{2a}$$

$$= \frac{-1 \pm \sqrt{1^2 - 4(2)(-1)}}{2(2)}$$

$$= \frac{-1 \pm \sqrt{9}}{4}$$

$$= \frac{-1 \pm 3}{4} = \frac{1}{2}, -1$$

13. Use the Quadratic Formula to solve $2 + 2x - x^2 = 0$.

Solution

$-x^2 + 2x + 2 = 0$

$a = -1, b = 2, c = 2$

$$x = \frac{-b \pm \sqrt{b^2 - 4ac}}{2a}$$

$$= \frac{-2 \pm \sqrt{2^2 - 4(-1)(2)}}{2(-1)}$$

$$= \frac{-2 \pm \sqrt{12}}{-2}$$

$$= \frac{-2 \pm 2\sqrt{3}}{-2} = 1 \pm \sqrt{3}$$

17. Use the Quadratic Formula to solve $x^2 + 8x - 4 = 0$.

Solution

$a = 1, b = 8, c = -4$

$$x = \frac{-b \pm \sqrt{b^2 - 4ac}}{2a}$$

$$= \frac{-8 \pm \sqrt{8^2 - 4(1)(-4)}}{2(1)}$$

$$= \frac{-8 \pm \sqrt{80}}{2}$$

$$= \frac{-8 \pm 4\sqrt{5}}{2} = -4 \pm 2\sqrt{5}$$

21. Use the Quadratic Formula to solve $36x^2 + 24x = 7$.

Solution

$36x^2 + 24x - 7 = 0$

$a = 36, b = 24, c = -7$

$$x = \frac{-b \pm \sqrt{b^2 - 4ac}}{2a}$$

$$= \frac{-24 \pm \sqrt{24^2 - 4(36)(-7)}}{2(36)}$$

$$= \frac{-24 \pm 12\sqrt{11}}{72} = -\frac{1}{3} \pm \frac{\sqrt{11}}{6}$$

25. Use the Quadratic Formula to solve $28x - 49x^2 = 4$.

Solution

$$28x - 49x^2 = 4$$

$$-49x^2 + 28x - 4 = 0$$

$$a = -49,\ b = 28,\ c = -4$$

$$x = \frac{-b \pm \sqrt{b^2 - 4ac}}{2a}$$

$$= \frac{-28 \pm \sqrt{28^2 - 4(-49)(-4)}}{2(-49)}$$

$$= \frac{-28 \pm 0}{-98} = \frac{2}{7}$$

31. Use a calculator to solve $5.1x^2 - 1.7x - 3.2 = 0$. Round your answers to three decimal places.

Solution

$$x = \frac{1.7 \pm \sqrt{(-1.7)^2 - 4(5.1)(-3.2)}}{2(5.1)}$$

$$x \approx 0.976,\ -0.643$$

35. Use a calculator to solve $422x^2 - 506x - 347 = 0$. Round your answers to three decimal places.

Solution

$$x = \frac{506 \pm \sqrt{(-506)^2 - 4(422)(-347)}}{2(422)}$$

$$x \approx 1.687,\ -0.488$$

39. Solve $4x^2 - 15 = 25$ by any convenient method.

Solution

$$4x^2 - 15 = 25$$

$$4x^2 = 40$$

$$x^2 = 10$$

$$x = \pm\sqrt{10}$$

45. Solve $100x^2 - 400 = 0$ by any convenient method.

Solution

$$100x^2 - 400 = 0$$

$$100x^2 = 400$$

$$x^2 = 4$$

$$x = \pm 2$$

49. *Writing Real-life Problems* One number is one more than another number. The sum of their squares is 113. Find the numbers. Solve the number problem and write a real-life problem that could be represented by this verbal model.

Solution

Let $n =$ first integer and $n + 1 =$ next integer.

$$n^2 + (n+1)^2 = 113$$

$$n^2 + n^2 + 2n + 1 = 113$$

$$2n^2 + 2n - 112 = 0$$

$$2(n^2 + n - 56) = 0$$

$$2(n+8)(n-7) = 0$$

$$n = -8 \text{ or } n = 7$$

Since we want a positive integer, $n = 7$ and $n + 1 = 8$. The hypotenuse of a right triangle is $\sqrt{113}$ feet. One leg of the triangle is one foot more than the other leg. Find the lengths of the legs.

51. *Cost Equation* Use the cost equation $C = 0.125x^2 + 20x + 5,000$ to find the number of units x that a manufacturer can produce when $C = \$14,000$. (Round your answer to the nearest positive integer.)

Solution

$$0.125x^2 + 20x - 9000 = 0$$

$$x = \frac{-20 \pm \sqrt{20^2 - 4(0.125)(-9000)}}{2(0.125)} = \frac{-20 \pm 70}{0.25}$$

$$x = 200 \text{ units}$$

57. *Geometry* An open box is to be made from a square piece of material by cutting 2-inch squares from each corner and turning up the sides. (See figure in the textbook.) The volume of the finished box is to be 200 cubic inches. Find the size of the original piece of material.

Solution

$$\text{Volume} = 2x^2 = 200$$

$$x^2 = 100$$

$$x = 10$$

The original piece of material was $x + 4 = 14$ square inches.

61. *Flying Distance* A small commuter airline flies to three cities whose locations form the vertices of a right triangle. (See figure in the textbook.) The total flight distance (from City A to City B to City C and back to City A) is 1,400 miles. It is 600 miles between the two cities that are farthest apart. Find the other two distances between cities.

Solution

Distance between A and $C = x$

Distance between C and $B = y$

Total distance $= x + y + 600 = 1400$

$y = 800 - x$

$x^2 + (800 - x)^2 = 600^2$

$2x^2 - 1600x + 280{,}000 = 0$

$$x = \frac{1600 \pm \sqrt{(-1600)^2 - 4(2)(280{,}000)}}{2(2)} = \frac{1600 \pm 400\sqrt{2}}{4} = 400 \pm 100\sqrt{2}$$

The other two distances are $400 - 100\sqrt{2} \approx 259$ miles and $400 + 100\sqrt{2} \approx 541$ miles.

65. *Computers in the Classroom* The number of personal computers (in millions) used in grades K through 12 in the United States from 1980 to 1988 can be modeled by

$$\text{Number} = 0.016t^2 + 0.1t - 0.04$$

where $t = 0$ represents 1980. From this model, when will three million computers be used in grades K through 12? *(Source: Future Computing/Data Inc.)*

Solution

Number $= 3$ (in millions)

$$3 = 0.016t^2 + 0.1t - 0.04$$

$$0 = 0.016t^2 + 0.1t - 3.04$$

$$t = \frac{-0.1 \pm \sqrt{(0.1)^2 - 4(0.016)(-3.04)}}{2(0.016)}$$

Use the positive value of $t \approx 11$, which represents the year 1991. $(1980 + 11)$

67. *Flying Speed* Two planes leave simultaneously from the same airport, one flying due east and the other due south. (See figure in the textbook.) The eastbound plane is flying 50 miles per hour faster than the southbound plane. After three hours the planes are 2440 miles apart. Find the speed of each plane.

Solution

$d_E = (3 \text{ hours})(r + 50 \text{ mph})$

$d_S = (3 \text{ hours})(r \text{ mph})$

$d_E^2 + d_S^2 = 2440^2$

$9(r + 50)^2 + 9r^2 = 2440^2$

$18r^2 + 900r - 5{,}931{,}100 = 0$

$$r = \frac{-900 \pm \sqrt{900^2 - 4(18)(-5{,}931{,}100)}}{2(18)} = \frac{-900 \pm 60\sqrt{118{,}847}}{36}$$

Using the positive value for r, we have the southbound plane moving at $r \approx 550$ mph and the eastbound plane moving at $r + 50 \approx 600$ mph.

Mid-Chapter Quiz for Chapter 1

1. Solve $2(x+3)-5(2x-3)=5$ and check the solution.

Solution

$$\begin{aligned}2(x+3)-5(2x-3)&=5\\2x+6-10x+15&=5\\-8x+21&=5\\-8x&=-16\\x&=2\end{aligned}$$

Check

$$\begin{aligned}2(2+3)-5(2(2)-3)&\stackrel{?}{=}5\\2(5)-5(1)&\stackrel{?}{=}5\\10-5&\stackrel{?}{=}5\\5&=5\ \checkmark\end{aligned}$$

2. Solve $\dfrac{3x+3}{5x-2}=\dfrac{3}{4}$ and check the solution.

Solution

$$\begin{aligned}\frac{3x+3}{5x-2}&=\frac{3}{4}\\4(3x+3)&=3(5x-2)\\12x+12&=15x-6\\-3x&=-18\\x&=6\end{aligned}$$

Check

$$\begin{aligned}\frac{3(6)+3}{5(6)-2}&\stackrel{?}{=}\frac{3}{4}\\\frac{21}{28}&\stackrel{?}{=}\frac{3}{4}\\\frac{3}{4}&=\frac{3}{4}\ \checkmark\end{aligned}$$

3. Solve $\dfrac{2}{x(x-1)}+\dfrac{1}{x}=\dfrac{1}{x-4}$ and check the solution.

Solution

$$\begin{aligned}x(x-1)(x-4)\left[\frac{2}{x(x-1)}+\frac{1}{x}\right]&=x(x-1)(x-4)\left(\frac{1}{x-4}\right)\\2(x-4)+(x-1)(x-4)&=x(x-1)\\2x-8+x^2-5x+4&=x^2-x\\x^2-3x-4&=x^2-x\\-2x&=4\\x&=-2\end{aligned}$$

Check

$$\begin{aligned}\frac{2}{(-2)(-2-1)}+\frac{1}{(-2)}&\stackrel{?}{=}\frac{1}{-2-4}\\\frac{1}{3}-\frac{1}{2}&\stackrel{?}{=}-\frac{1}{6}\\\frac{2}{6}-\frac{3}{6}&\stackrel{?}{=}-\frac{1}{6}\\-\frac{1}{6}&=-\frac{1}{6}\ \checkmark\end{aligned}$$

4. Solve $(x+3)^2 - x^2 = 6(x+2)$ and check the solution.

Solution

$$\begin{aligned}(x+3)^2 - x^2 &= 6(x+2)\\ x^2 + 6x + 9 - x^2 &= 6x + 12\\ 6x + 9 &= 6x + 12\\ 9 &\neq 12\end{aligned}$$

Contradiction: No solution

5. Solve $\frac{x}{3.057} + \frac{x}{4.392} = 100$ and round to three places.

Solution

$$\begin{aligned}(3.057)(4.392)\left[\frac{x}{3.057} + \frac{x}{4.392}\right] &= (3.057)(4.392)(100)\\ 4.392x + 3.057x &= 1342.6344\\ 7.449x &= 1342.6344\\ x &= \frac{1342.6344}{7.449}\\ x &\approx 180.244\end{aligned}$$

6. Solve $0.378x + 0.757(500 - x) = 215$ and round to three places.

Solution

$$\begin{aligned}0.378x + 0.757(500 - x) &= 215\\ 0.378x + 378.5 - 0.757x &= 215\\ -0.379x &= -163.5\\ x &= \frac{163.5}{0.379}\\ x &\approx 431.398\end{aligned}$$

7. Write an algebraic expression for "the total cost of producing x units for which the fixed costs are \$1,400 and the cost per unit is \$36."

Solution

$$\begin{aligned}\text{Total cost} &= (\text{cost per unit})(\text{number of units}) + (\text{fixed costs})\\ &= 36x + 1400\end{aligned}$$

8. Write an algebraic expression for "the amount of acid in x gallons of a 25% acid solution."

Solution

$$\begin{aligned}\text{Amount of acid} &= (\text{percent of acid})(\text{number of gallons})\\ &= (25\%)(x)\\ &= 0.25x\end{aligned}$$

9. *Production Limit* A company has fixed costs of \$20,000 per month and variable costs of \$7.50 per unit manufactured. The company has \$80,000 available to cover monthly costs. How many units can the company manufacture?

Solution

$$7.50x + 20{,}000 = 80{,}000$$
$$7.50x = 60{,}000$$
$$x = 8000 \text{ units}$$

10. Solve $3x^2 + 13x = 10$ by factoring.

Solution

$$3x^2 + 13x - 10 = 0$$
$$(3x - 2)(x + 5) = 0$$
$$3x - 2 = 0 \quad \text{or } x + 5 = 0$$
$$x = \frac{2}{3} \quad \text{or } x = -5$$

11. Solve $3x^2 = 15$ by extracting roots.

Solution

$$3x^2 = 15$$
$$x^2 = 5$$
$$x = \pm\sqrt{5} \approx \pm 2.24$$

12. Solve $(x + 3)^2 = 17$ by extracting roots.

Solution

$$(x + 3)^2 = 17$$
$$x + 3 = \pm\sqrt{17}$$
$$x = -3 \pm \sqrt{17}$$
$$x \approx 1.12 \quad \text{or } x \approx -7.12$$

13. Solve $3x^2 + 7x - 2 = 0$ by the Quadratic Formula.

Solution

$$3x^2 + 7x - 2 = 0$$
$$a = 3, b = 7, c = -2$$
$$x = \frac{-7 \pm \sqrt{(7)^2 - 4(3)(-2)}}{2(3)} = \frac{-7 \pm \sqrt{49 + 24}}{6} = \frac{-7 \pm \sqrt{73}}{6}$$

14. Solve $2x + x^2 = 5$ by the Quadratic Formula.

Solution

$$x^2 + 2x - 5 = 0$$
$$a = 1, b = 2, c = -5$$
$$x = \frac{-2 \pm \sqrt{(2)^2 - 4(1)(-5)}}{2(1)} = \frac{-2 \pm \sqrt{4 + 20}}{2} = \frac{-2 \pm \sqrt{24}}{2}$$
$$= \frac{-2 \pm 2\sqrt{6}}{2}$$
$$= -1 \pm \sqrt{6}$$

15. Solve $3x^2 - 4.50x - 0.32 = 0$ by the Quadratic Formula.

Solution

$3x^2 - 4.50x - 0.32 = 0$

$a = 3, b = -4.50, c = -0.32$

$$x = \frac{-(-4.50) \pm \sqrt{(-4.50)^2 - 4(3)(-0.32)}}{2(3)}$$

$$= \frac{4.50 \pm \sqrt{24.09}}{6}$$

$x \approx 1.568$ or $x \approx -0.068$

16. Use the determinant to determine the number of real solutions for $x^2 + 3x - 5 = 0$.

Solution

$x^2 + 3x - 5 = 0$

$b^2 - 4ac = (3)^2 - 4(1)(-5) = 29 > 0$

Two distinct real solutions

17. Use the determinant to determine the number of real solutions for $2x^2 - 4x + 9 = 0$.

Solution

$2x^2 - 4x + 9 = 0$

$b^2 - 4ac = (-4)^2 - 4(2)(9) = -56 < 0$

No real solutions

18. Use the determinant to determine the number of real solutions for $x^2 + 3 = 0$.

Solution

$x^2 + 3 = 0$

$b^2 - 4ac = (0)^2 - 4(1)(3) = -12 < 0$

No real solutions

19. A fire retardant is dropped from a height of 250 feet. With no initial velocity, what is the minimum time it would take to hit the ground? Explain your reasoning.

Solution

$S = -16t + Vot + So$

Since $Vo = 0$ and $So = 250$, we have

$S = -16t^2 + 250.$

When it hits the ground, the height, s, is zero.

$$0 = -16t^2 + 250$$

$$16t^2 = 250$$

$$t^2 = \frac{250}{16}$$

$$t = \sqrt{\frac{250}{16}}$$ Since t represents time, it must be nonnegative.

$$= \frac{5\sqrt{10}}{4} \approx 3.95 \text{ seconds}$$

20. An open box 5 inches deep and 180 cubic inches in volume is to be constructed. Find the dimensions of the square base.

Solution

Volume = (length)(width)(height)

$$180 = (x)(x)(5)$$

$$180 = 5x^2$$

$$36 = x^2$$

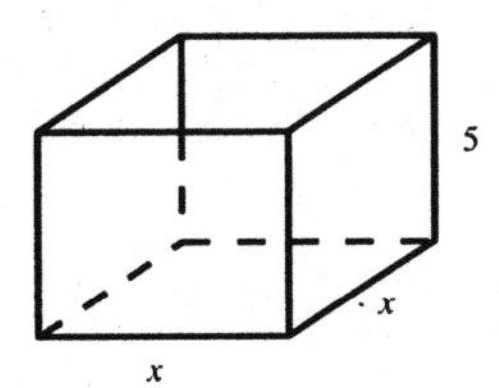

Since x must be nonnegative, we have $x = 6$.

Dimensions of square base: 6 inches × 6 inches

1.5 Other Types of Equations

- You should be able to solve certain types of nonlinear or nonquadratic equations.
- For equations involving radicals or fractional powers, raise both sides to the same power.
- For equations that are of the quadratic type, $au^2 + bu + c = 0,\ a \neq 0$, use either factoring or the quadratic equation.
- For equations involving absolute value, remember that the expression inside the absolute value can be positive or negative.
- Always check for extraneous solution.

SOLUTIONS TO SELECTED EXERCISES

5. Find the real solutions of $x^4 - 81 = 0$.

Solution

$x^4 - 81 = 0$

$(x^2 + 9)(x + 3)(x - 3) = 0$

$x^2 + 9 = 0 \quad \Rightarrow \quad$ No real solution.

$x + 3 = 0 \quad \Rightarrow \quad x = -3$

$x - 3 = 0 \quad \Rightarrow \quad x = 3$

9. Find the real solutions of $x^3 - 3x^2 - x + 3 = 0$.

Solution

$x^3 - 3x^2 - x + 3 = 0$

$x^2(x - 3) - (x - 3) = 0$

$(x - 3)(x^2 - 1) = 0$

$(x - 3)(x + 1)(x - 1) = 0$

$x - 3 = 0 \quad \Rightarrow \quad x = 3$

$x + 1 = 0 \quad \Rightarrow \quad x = -1$

$x - 1 = 0 \quad \Rightarrow \quad x = 1$

13. Find the real solutions of $x^4 - 10x^2 + 9 = 0$.

Solution

$x^4 - 10x^2 + 9 = 0$

$(x^2 - 1)(x^2 - 9) = 0$

$(x + 1)(x - 1)(x + 3)(x - 3) = 0$

$x + 1 = 0 \quad \Rightarrow \quad x = -1$

$x - 1 = 0 \quad \Rightarrow \quad x = 1$

$x + 3 = 0 \quad \Rightarrow \quad x = -3$

$x - 3 = 0 \quad \Rightarrow \quad x = 3$

17. Find the real solutions of $4x^4 - 65x^2 + 16 = 0$.

Solution

$4x^4 - 65x^2 + 16 = 0$

$(4x^2 - 1)(x^2 - 16) = 0$

$(2x + 1)(2x - 1)(x + 4)(x - 4) = 0$

$2x + 1 = 0 \quad \Rightarrow \quad x = -\frac{1}{2}$

$2x - 1 = 0 \quad \Rightarrow \quad x = \frac{1}{2}$

$x + 4 = 0 \quad \Rightarrow \quad x = -4$

$x - 4 = 0 \quad \Rightarrow \quad x = 4$

21. Find the real solutions of $\sqrt{2x} - 10 = 0$.

Solution

$$\sqrt{2x} - 10 = 0$$
$$\sqrt{2x} = 10$$
$$2x = 100$$
$$x = 50$$

25. Find the real solutions of $\sqrt[3]{2x+5} + 3 = 0$.

Solution

$$\sqrt[3]{2x+5} + 3 = 0$$
$$\sqrt[3]{2x+5} = -3$$
$$2x + 5 = -27$$
$$2x = -32$$
$$x = -16$$

29. Find the real solutions of $x = \sqrt{11x - 30}$.

Solution

$$x = \sqrt{11x - 30}$$
$$x^2 = 11x - 30$$
$$x^2 - 11x + 30 = 0$$
$$(x-5)(x-6) = 0$$
$$x - 5 = 0 \quad \Rightarrow \quad x = 5$$
$$x - 6 = 0 \quad \Rightarrow \quad x = 6$$

33. Find the real solutions of $\sqrt{x+1} - 3x = 1$.

Solution

$$\sqrt{x+1} - 3x = 1$$
$$\sqrt{x+1} = 3x + 1$$
$$x + 1 = 9x^2 + 6x + 1$$
$$0 = 9x^2 + 5x$$
$$0 = x(9x+5)$$
$$x = 0$$
$$9x + 5 = 0 \quad \Rightarrow \quad x = -\tfrac{5}{9},$$ This solution is extraneous.

37. Find the real solutions of $(x+3)^{3/2} = 8$.

Solution

$$(x+3)^{3/2} = 8$$
$$x + 3 = 8^{2/3}$$
$$x + 3 = 4$$
$$x = 1$$

41. Find the real solutions of $\frac{1}{x} - \frac{1}{x+1} = 3$.

Solution

$$\frac{1}{x} - \frac{1}{x+1} = 3$$
$$x + 1 - x = 3x(x+1),\ x \neq 0, -1$$
$$-3x^2 - 3x + 1 = 0$$
$$x = \frac{-b \pm \sqrt{b^2 - 4ac}}{2a}$$
$$= \frac{-(-3) \pm \sqrt{(-3)^2 - 4(-3)(1)}}{2(-3)}$$
$$= \frac{3 \pm \sqrt{21}}{-6} = \frac{-3 \pm \sqrt{21}}{6}$$

47. Find the real solutions of $\frac{1}{x} = \frac{4}{x-1} + 1$.

Solution

$$\frac{1}{x} = \frac{4}{x-1} + 1$$

$$x - 1 = 4x + x(x-1), \ x \neq 0, \ 1$$

$$x - 1 = x^2 + 3x$$

$$x^2 + 2x + 1 = 0$$

$$(x+1)^2 = 0$$

$$x + 1 = 0 \quad \Rightarrow \quad x = -1$$

51. Find the real solutions of $|x + 1| = 2$.

Solution

$$x + 1 = 2 \quad \Rightarrow \quad x = 1$$

$$-(x + 1) = 2 \quad \Rightarrow \quad x = -3$$

55. Find the real solutions of $|x| = x^2 + x - 3$.

Solution

$$x = x^2 + x - 3 \qquad \text{OR} \qquad -x = x^2 + x - 3$$

$$x^2 - 3 = 0 \qquad\qquad x^2 + 2x - 3 = 0$$

$$x = \pm\sqrt{3} \qquad\qquad (x-1)(x+3) = 0$$

$$x - 1 = 0 \quad \Rightarrow \quad x = 1$$

$$x + 3 = 0 \quad \Rightarrow \quad x = -3$$

Only $x = \sqrt{3}$ and $x = -3$ are solutions.

59. Use a calculator to find the real solutions $3.2x^4 - 1.5x^2 - 2.1 = 0$. Round your answers to three decimal places.

Solution

$$x^2 = \frac{1.5 \pm \sqrt{1.5^2 - 4(3.2)(-2.1)}}{2(3.2)}$$

$$x = \pm\sqrt{\frac{1.5 + \sqrt{29.13}}{6.4}} \approx \pm 1.038$$

63. *Sharing the Cost* A college charters a bus for \$1,700 to take a group of students to the World's Fair. When six more students join the group, the cost per student drops by \$7.50. How many students were in the original group?

Solution

$$\text{Number of students} = x$$

$$\text{Cost per student} = f$$

$$fx = 1700 \Rightarrow f = \frac{1700}{x}$$

$$(f - 7.50)(x + 6) = 1700$$

$$\left(\frac{1700}{x} - 7.50\right)(x + 6) = 1700$$

$$(3400 - 15x)(x + 6) = 3400x$$

$$-15x^2 - 90x + 20,400 = 0$$

$$x = \frac{90 \pm \sqrt{(-90)^2 - 4(-15)(20,400)}}{2(-15)} = \frac{90 \pm 1110}{-30}$$

The original number was $x = 34$.

67. *Borrowing Money* Suppose you borrow \$100 from a friend and agree to pay the money back, plus \$10 in interest, after six months. Assuming that the interest is compounded monthly, what annual percentage rate are you paying?

Solution

$$A = P\left(1 + \frac{r}{n}\right)^{nt}$$

$$110 = 100\left(1 + \frac{r}{12}\right)^{(12)(1/2)}$$

$$1.1 = \left(1 + \frac{r}{12}\right)^6$$

$$1.1^{1/6} = 1 + \frac{r}{12}$$

$$[1.1^{1/6} - 1](12) = r$$

$$r \approx 0.192 = 19.2\%$$

71. *Life Expectancy* The life expectancy table (for ages 58-75) used by the U.S. National Center for Health Statistics is modeled by

$$y = \sqrt{0.6632x^2 - 110.55x + 4680.24}$$

where x represents a person's current age and y represents the average number of years the person is expected to live. (See figure in the textbook.) If a person's life expectancy is estimated to be 20 years, how old is the person? (*Source: U.S. National Center for Health Statistics*)

Solution

$$20 = \sqrt{0.6632x^2 - 110.55x + 4680.24}$$

$$400 = 0.6632x^2 - 110.55x + 4680.24$$

$$0 = 0.6632x^2 - 110.55x + 4280.24$$

$$x = \frac{110.55 \pm \sqrt{110.55^2 - 4(0.6632)(4280.24)}}{2(0.6632)}$$

$$x \approx 105.54 \text{ or } x \approx 61.15$$

Using the value in the domain ($58 \leq x \leq 75$), we have $x \approx 61$ years old. (Compare this answer with the graph in the textbook.)

75. *Sailboat Stay* Two stays for the mast on a sailboat are attached to the boat at two points. (See figure in the textbook.) One point is 10 feet from the mast and the other point is 15 feet. The total length of the two stays is 35 feet. How high on the mast are the stays attached?

Solution

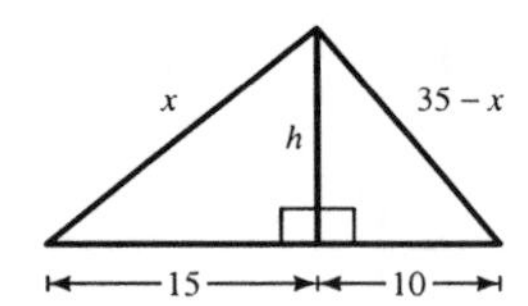

By the Pythagorean Theorem

$$15^2 + h^2 = x^2 \Rightarrow h^2 = x^2 - 15^2$$

$$10^2 + h^2 = (35 - x)^2 \Rightarrow h^2 = (35 - x)^2 - 10^2$$

Thus, $x^2 - 15^2 = (35 - x)^2 - 10^2$

$$x^2 - 225 = 1225 - 70x + x^2 - 100$$

$$x^2 - 225 = x^2 - 70x + 1125$$

$$70x = 1350$$

$$x = \frac{1350}{70} = \frac{135}{7} \approx 19.286 \text{ feet}$$

$$h^2 = \left(\frac{135}{7}\right)^2 - 15^2$$

$$h = \sqrt{\left(\frac{135}{7}\right)^2 - 225} \approx 12.12 \text{ feet}$$

1.6 Linear Inequalities

- You should be able to solve linear inequalities.
- You should know the properties of inequalities.
  (a) Transitive: $a < b$ and $b < c$ implies $a < c$.
  (b) Addition: $a < b$ and $c < d$ implies $a + c < b + d$.
  (c) Adding or Subtracting a Constant: $a \pm c < b \pm c$ if $a < b$.
  (d) Multiplying or Dividing a Constant: For $a < b$,
  If $c > 0$, then $ac < bc$ and $\frac{a}{c} < \frac{b}{c}$.
  If $c < 0$, then $ac > bc$ and $\frac{a}{c} > \frac{b}{c}$.
- You should be able to solve absolute value inequalities.
  (a) $|x| < a$ if and only if $-a < x < a$.
  (b) $|x| > a$ if and only if $x < -a$ or $x > a$.

SOLUTIONS TO SELECTED EXERCISES

1. Determine whether the values of x are solutions of the inequality $5x - 12 > 0$.

(a) $x = 3$ (b) $x = -3$ (c) $x = \frac{5}{2}$ (d) $x = \frac{3}{2}$

Solution

(a) $5(3) - 12 = 3 > 0 \Rightarrow x = 3$ satisfies the inequality.

(b) $5(-3) - 12 = -27 \not> 0 \Rightarrow x = -3$ does not satisfy the inequality.

(c) $5\left(\frac{5}{2}\right) - 12 = \frac{25}{2} - \frac{24}{2} = \frac{1}{2} > 0 \Rightarrow x = \frac{5}{2}$ satisfies the inequality.

(d) $5\left(\frac{3}{2}\right) - 12 = \frac{15}{2} - \frac{24}{2} = -\frac{9}{2} \not> 0 \Rightarrow x = \frac{3}{2}$ does not satisfy the inequality.

5. Match the inequality $-2 < x \le 5$ with its graph. [The graphs are labeled (a), (b), (c), (d), (e), (f), (g), and (h) in the textbook.]

Solution

$-2 < x \le 5$ represents all real numbers greater than -2 and less than or equal to 5; graph (f).

9. Match the inequality $|x - 5| > 2$ with its graph. [The graphs are labeled (a), (b), (c), (d), (e), (f), (g), and (h) in the textbook.]

Solution

$|x - 5| > 2$ indicates all points more than 2 units from $x = 5$; graph (b).

15. Solve $-10x < 40$ and sketch its graph.

Solution

$$-10x < 40$$
$$-\tfrac{1}{10}(-10x) > -\tfrac{1}{10}(40)$$
$$x > -4$$

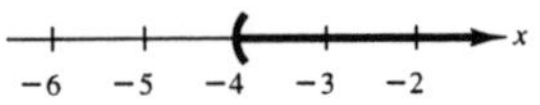

−6 −5 −4 −3 −2 x

19. Solve $4(x + 1) < 2x + 3$ and sketch its graph.

Solution

$$4(x + 1) < 2x + 3$$
$$4x + 4 < 2x + 3$$
$$2x < -1$$
$$x < -\tfrac{1}{2}$$

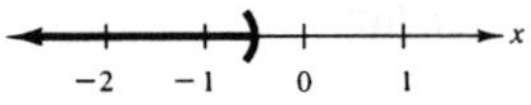

−2 −1 0 1 x

25. Solve $1 < 2x + 3 < 9$ and sketch its graph.

Solution

$$1 < 2x + 3 < 9$$
$$-2 < 2x < 6$$
$$-1 < x < 3$$

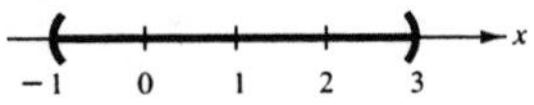

−1 0 1 2 3 x

29. Solve $\frac{3}{4} > x + 1 > \frac{1}{4}$ and sketch its graph.

Solution

$$\tfrac{3}{4} > x + 1 > \tfrac{1}{4}$$
$$-\tfrac{1}{4} > x > -\tfrac{3}{4} \Rightarrow -\tfrac{3}{4} < x < -\tfrac{1}{4}$$

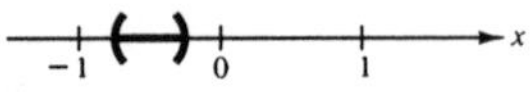

−1 0 1 x

33. Solve $\left|\frac{x}{2}\right| > 3$ and sketch its graph.

Solution

$$\frac{x}{2} < -3 \quad \text{or} \quad \frac{x}{2} > 3$$
$$x < -6 \qquad\qquad x > 6$$

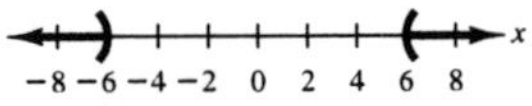

−8 −6 −4 −2 0 2 4 6 8 x

39. Solve $\left|\frac{x-3}{2}\right| \geq 5$ and sketch its graph.

Solution

$$\frac{x-3}{2} \leq -5 \quad \text{or} \quad \frac{x-3}{2} \geq 5$$
$$x - 3 \leq -10 \qquad x - 3 \geq 10$$
$$x \leq -7 \qquad\qquad x \geq 13$$

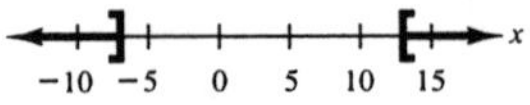

−10 −5 0 5 10 15 x

41. Solve $|9 - 2x| - 2 < -1$ and sketch its graph.

Solution

$$|9 - 2x| - 2 < -1$$
$$|9 - 2x| < 1$$
$$-1 < 9 - 2x < 1$$
$$-10 < -2x < -8$$
$$5 > x > 4 \Rightarrow 4 < x < 5$$

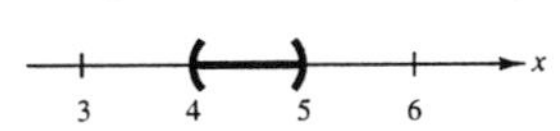

45. Solve $|x - 5| < 0$ and sketch its graph.

Solution

No solution. Absolute value is never negative.

49. Find the domain of $\sqrt{x + 3}$.

Solution

$$x + 3 \geq 0$$
$$x \geq -3$$
$$[-3, \infty)$$

53. Use absolute value notation to define each interval (or pair or intervals) on the real line. (See figure in the textbook.)

Solution

All real numbers no more than 2 units from 0 yields $|x - 0| \leq 2$ or $|x| \leq 2$.

57. Use absolute value notation to define "all real numbers within 10 units of 12."

Solution

$$|x - 12| \leq 10$$

61. *Comparative Shopping* You can rent a midsize car from Company A for \$250 per week with no extra charge for mileage. A similar car can be rented from Company B for \$150 per week, plus \$0.25 cents for each mile driven. How many miles must you drive in a week to make the rental fee for Company B *greater than* that for Company A?

Solution

$$B > A$$
$$150 + 0.25x > 250$$
$$0.25x > 100$$
$$x > 400 \text{ miles}$$

65. *Weight Loss Program* A person enrolls in a diet program that guarantees a loss of *at least* $1\frac{1}{2}$ pounds per week. The person's weight at the beginning of the program is 164 pounds. Find the maximum number of weeks that the person must be in the program before attaining a weight of 128 pounds.

Solution

$$164 - 1.5x \geq 128$$
$$36 \geq 1.5x$$
$$24 \geq x$$

The person must be in the program 24 or less weeks.

69. *Annual Operating Cost* A utility company has a fleet of vans. The annual operating cost per van is

$$C = 0.32m + 2,300$$

where m is the number of miles traveled by a van in a year. What number of miles will yield an annual operating cost that is *less than* \$10,000?

Solution

$$C = 0.32m + 2300 < 10,000$$

$$0.32m < 7700$$

$$m < 24,062.5$$

73. *Accuracy of Measurement* Suppose you buy six T-bone steaks at \$3.98 per pound. The weight that is listed on the package is 5.72 pounds. If the scale that weighed the package is accurate to within $\frac{1}{2}$ ounce, how much might you have been undercharged or overcharged?

Solution

$$|x - 5.72| \le \tfrac{1}{32}$$

$$-\tfrac{1}{32} \le x - 5.72 \le \tfrac{1}{32}$$

$$5.68875 \le x \le 5.75125$$

You could have been undercharged $5.75125(3.98) - 5.72(3.98) \approx \$0.12$ or you could have been overcharged $5.72(3.98) - 5.68875(3.98) \approx \$0.12$.

75. *Population Heights* The heights h of two-thirds of the members of a certain population satisfy the inequality

$$|h - 68.5| \le 2.7$$

where h is measured in inches. Determine the interval on the real line in which these heights lie.

Solution

$$|h - 68.5| \le 2.7$$

$$-2.7 \le h - 68.5 \le 2.7$$

$$65.8 \le h \le 71.2$$

In interval notation we have [65.8, 71.2].

1.7 Other Types of Inequalities

- You should be able to solve inequalities by factoring.
  - (a) Find the critical numbers—values that make the expression zero.
  - (b) Find values that make the expression undefined.
  - (c) Test one value in each interval on the real number line resulting from the critical numbers and values that make the expression undefined.
  - (d) Determine the solution intervals.

SOLUTIONS TO SELECTED EXERCISES

3. Solve the inequality $x^2 > 4$ and sketch its graph.

Solution

$$x^2 > 4$$

$$x^2 - 4 > 0$$

$$(x + 2)(x - 2) > 0$$

Critical numbers: $x = \pm 2$

Test intervals: $(-\infty, -2), (-2, 2), (2, \infty)$

Test: Is $(x + 2)(x - 2) > 0$?

Solution set: $(-\infty, -2) \cup (2, \infty)$

$x < -2, x > 2$

−3 −2 −1 0 1 2 3 x

5. Solve the ineqality $(x + 2)^2 < 25$ and sketch its graph.

Solution

$$(x + 2)^2 < 25$$

$$x^2 + 4x + 4 - 25 < 0$$

$$x^2 + 4x - 21 < 0$$

$$(x + 7)(x - 3) < 0$$

Critical numbers: $x = -7, x = 3$

Test intervals: $(-\infty, -7), (-7, 3), (3, \infty)$

Test: Is $(x + 7)(x - 3) < 0$?

Solution set: $(-7, 3)$

$-7 < x < 3$

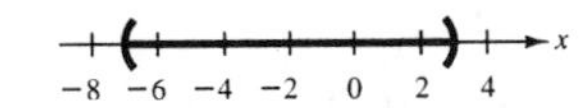

11. Solve $3(x-1)(x+1) > 0$ and sketch its graph.

Solution

$3(x-1)(x+1) > 0$

Critical numbers: $x = \pm 1$

Test intervals: $(-\infty, -1), (-1, 1), (1, \infty)$

Test: Is $3(x-1)(x+1) > 0$?

Solution set: $(-\infty, -1) \cup (1, \infty)$

$x < -1, x > 1$

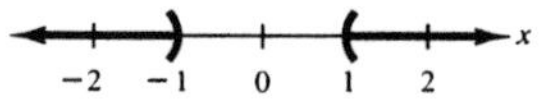

−2 −1 0 1 2 x

15. Solve $4x^3 - 6x^2 < 0$ and sketch its graph.

Solution

$4x^3 - 6x^2 < 0$

$2x^2(2x-3) < 0$

Critical numbers: $x = 0, x = \frac{3}{2}$

Test intervals: $(-\infty, 0), \left(0, \frac{3}{2}\right), \left(\frac{3}{2}, \infty\right)$

Test: Is $2x^2(2x-3) < 0$?

Solution set: $(-\infty, 0) \cup \left(0, \frac{3}{2}\right)$

$x < 0, 0 < x < \frac{3}{2}$

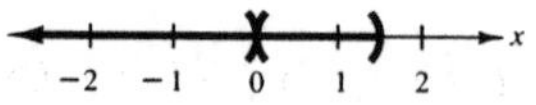

−2 −1 0 1 2 x

19. Solve the following inequality and sketch its graph.

$$\frac{1}{x} > x$$

Solution

$\frac{1}{x} > x$

$\frac{1}{x} - x > 0$

$\frac{1-x^2}{x} > 0$

Critical numbers: $x = 0, x = \pm 1$

Test intervals: $(-\infty, -1), (-1, 0), (0, 1), (1, \infty)$

Test: Is $\frac{1-x^2}{x} > 0$?

Solution set: $(-\infty, -1) \cup (0, 1)$

$x < -1, 0 < x < 1$

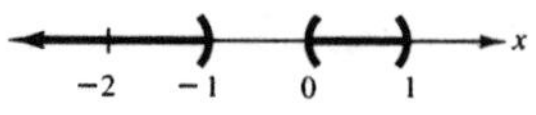

−2 −1 0 1 x

23. Solve the following inequality and sketch its graph.

$$\frac{3x-5}{x-5} > 4$$

Solution

$\frac{3x-5}{x-5} > 4$

$\frac{3x-5}{x-5} - 4 > 0$

$\frac{3x-5-4(x-5)}{x-5} > 0$

$\frac{15-x}{x-5} > 0$

Critical numbers: $x = 5, x = 15$

Test intervals: $(-\infty, 5), (5, 15), (15, \infty)$

Test: Is $\frac{15-x}{x-5} > 0$?

Solution set: $(5, 15)$

$5 < x < 15$

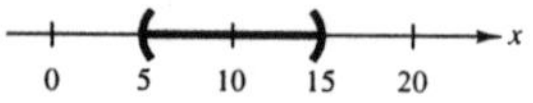

0 5 10 15 20 x

27. Solve the following inequality and sketch its graph.

$$\frac{1}{x-3} \le \frac{9}{4x+3}$$

Solution

$$\frac{1}{x-3} \le \frac{9}{4x+3}$$

$$\frac{1}{x-3} - \frac{9}{4x+3} \le 0$$

$$\frac{(4x+3) - 9(x-3)}{(x-3)(4x+3)} \le 0$$

$$\frac{30-5x}{(x-3)(4x+3)} \le 0$$

Critical numbers: $x = -\frac{3}{4}, x = 3, x = 6$

Test intervals: $\left(-\infty, -\frac{3}{4}\right), \left(-\frac{3}{4}, 3\right), (3, 6), (6, \infty)$

Test: Is $\frac{30-5x}{(x-3)(4x+3)} \le 0$?

Solution set: A test shows that the inequality is less than or equal to 0 on the intervals

$$\left(-\frac{3}{4}, 3\right) \cup [6, \infty)$$

$$-\frac{3}{4} < x < 3, x \ge 6$$

-2 0 2 4 6 8 x

31. Find the domain of $\sqrt{x^2 - 7x + 12}$.

Solution

$x^2 - 7x + 12 \ge 0$

$(x-3)(x-4) \ge 0$

$x \le 3$ or $x \ge 4$

$(-\infty, 3] \cup [4, \infty)$

35. Find the domain of $\sqrt{x^2 - 3x + 3}$.

Solution

$x^2 - 3x + 3 \ge 0$

Since $(-3)^2 - 4(1)(3) = -3 < 0$, there are no real roots and since $0^2 - 3(0) + 3 = 3 > 0$, the domain of x is all real numbers.

39. Use a calculator to solve $-0.5x^2 + 12.5x + 1.6 > 0$. (Round each number in your answer to two decimal places.)

Solution

The zeros are $x = \dfrac{-12.5 \pm \sqrt{(12.5)^2 - 4(-0.5)(1.6)}}{2(-0.5)}$

Critical numbers: $x \approx -0.13$

$x \approx 25.13$

$$-0.13 < x < 25.13$$

Test intervals: $(-\infty, -0.13)$, $(-0.13, 25.13)$, $(25.13, \infty)$

Solution set: $(-0.13, 25.13)$

45. *Geometry* A rectangular playing field with a perimeter of 100 meters is to have an area of at least 500 square meters. (See figure in the textbook.) Within what bounds must the length of the field lie?

Solution

$$2L + 2W = 100 \quad \Rightarrow \quad W = 50 - L$$

$$LW \geq 500$$

$$L(50 - L) \geq 500$$

$$-L^2 + 50L - 500 \geq 0$$

$$[L - (25 - 5\sqrt{5})][L + (25 + 5\sqrt{5})] \leq 0$$

13.8 meters $\approx 25 - 5\sqrt{5} \leq L \leq 25 + 5\sqrt{5} \approx 36.2$ meters

(Use the Quadratic Formula to find the critical numbers.)

47. *Company Profits* The revenue and cost equations for a product are given by

$$R = x(50 - 0.0002x)$$
$$C = 12x + 150,000$$

where R and C are measured in dollars and x represents the number of units sold. (See figure in the textbook.)

(a) How many units must be sold to obtain a profit of at least $1,650,000?

(b) The demand equation for the product is

$p = 50 - 0.0002x$

where p is the price per unit. What price per unit will produce a profit of at least $1,650,000?

Solution

(a)
$$\text{Profit} = \text{Revenue} - \text{Cost} \geq 1,650,000$$
$$x(50 - 0.0002x) - (12x + 150,000) \geq 1,650,000$$
$$-0.0002x^2 + 38x - 1,800,000 \geq 0$$
$$0.0002x^2 - 38x + 1,800,000 \leq 0$$

Critical numbers: $x = \dfrac{38 \pm \sqrt{38^2 - 4(0.0002)(1,800,000)}}{2(0.0002)}$

$x = 100,000$ or $x = 90,000$

90,000 units $\leq x \leq 100,000$ units

(b) $50 - 0.0002(100{,}000) \leq p \leq 50 - 0.0002(90{,}000)$

$$\$30 \leq p \leq \$32$$

51. *World Population* The world population (in millions) from 1980 to 1990 can be modeled by

$$\text{Population} = 4449 + 76.7t + 0.78t^2, \ 0 \leq t$$

where $t = 0$ represents 1980. (See figure in the textbook.) According to this model, when will the world population exceed 6,000,000,000? (*Source: Statistical Office of United Nations.*)

Solution

$4449 + 76.7t + 0.78t^2 \geq 6000$ (million)

$0.78t^2 + 76.7t - 1551 \geq 0$

Critical number: $t = \dfrac{-76.7 + \sqrt{(76.7)^2 - 4(0.78)(-1551)}}{2(0.78)}$

$$t \approx 17.21 \text{ years}$$

The population will exceed 6,000,000,000 in $1980 + 17.21 = 1997.21$, or in the year 1997.

Review Exercises for Chapter 1

1. Determine whether the following equation is an identity or a conditional equation.

$$5(x-3) = 2x + 9$$

Solution

$$5(x-3) = 2x + 9$$

$$5x - 15 = 2x + 9$$

$$3x = 24$$

$$x = 8$$

Conditional

5. Solve the following equation and check your answer.

$$4(x+3) - 3 = 2(4 - 3x) - 4$$

Solution

$$4(x+3) - 3 = 2(4 - 3x) - 4$$

$$4x + 12 - 3 = 8 - 6x - 4$$

$$4x + 9 = -6x + 4$$

$$10x = -5$$

$$x = -\frac{1}{2}$$

9. Solve the following equation and check your answer.

$$\frac{x}{x+3} - \frac{4}{x+3} + 2 = 0$$

Solution

$$\frac{x}{x+3} - \frac{4}{x+3} + 2 = 0$$

$$x - 4 + 2(x+3) = 0; \quad x \neq -3$$

$$3x + 2 = 0$$

$$x = -\frac{2}{3}$$

13. Three consecutive even integers have a sum of 42. Find the smallest of these integers.

Solution

Sum $=$ (first integer) $+$ (second integer) $+$ (third interger)

$$s = 2n + (2n+2) + (2n+4)$$

$$42 = 6n + 6$$

$$36 = 6n$$

$$n = 6$$

$$2n = 12$$

The smallest of these integers is 12.

17. *Geometry* A volleyball court is twice as long as it is wide, and its perimeter is 177 feet. Find the dimensions of the volleyball court.

Solution

$$2(\text{ Length}) + 2(\text{ Width}) = \text{ Perimeter}$$

$$2(2w) + 2w = 177$$

$$6w = 177$$

$$w = 29.5 \text{ feet}$$

$$l = 2w = 59 \text{ feet}$$

$l = 2w$

w

23. *Mixture* A car radiator contains 10 quarts of a 30% antifreeze solution. How many quarts will have to be replaced with pure antifreeze if the resulting 10 quart solution is to be 50% antifreeze?

Solution

	30% mixture	Pure antifreeze	50% mixture
Number of gallons	$10 - x$	x	10
Amount of antifreeze	$0.30(10 - x)$	$1.00x$	$0.50(10)$

$$0.30(10 - x) + 1.00x = 0.50(10)$$

$$3 - 0.30x + 1.00x = 5$$

$$0.70x = 2$$

$$x = \frac{2}{0.70} \approx 2.9 \text{ gallons}$$

27. Solve $x^2 - 11x + 24 = 0$ by factoring.

Solution

$$x^2 - 11x + 24 = 0$$

$$(x - 3)(x - 8) = 0$$

$$x - 3 = 0$$

$$x = 3 \quad \text{OR}$$

$$x - 8 = 0$$

$$x = 8$$

31. Solve $(x + 4)^2 = 18$ by extracting square roots. List both the exact and a decimal answer that has been rounded to two decimal places.

Solution

$$(x + 4)^2 = 18$$

$$x + 4 = \pm 3\sqrt{2}$$

$$x = -4 \pm 3\sqrt{2}$$

$$x \approx -8.24 \text{ OR } x \approx 0.24$$

35. *Total Revenue* The demand equation for a certain product is $p = 50 - 0.0001x$ where p is the price per unit and x is the number of units sold. The total revenue for selling x units is given by

$$\text{Revenue} = xp = x(50 - 0.0001x).$$

How many units must be sold to produce a revenue of \$6,000,000?

Solution

$$x(50 - 0.0001x) = 6,000,000$$

$$0 = 0.0001x^2 - 50x + 6,000,000$$

$$x = \frac{50 \pm \sqrt{50^2 - 4(0.0001)(6,000,000)}}{2(0.0001)}$$

$$x = 200,000 \text{ units or } x = 300,000 \text{ units}$$

39. Use the Quadratic Formula to solve $x^2 - 12x + 30 = 0$.

Solution

$$x = \frac{-(-12) \pm \sqrt{(-12)^2 - 4(1)(30)}}{2(1)} = \frac{12 \pm \sqrt{24}}{2}$$

$$= \frac{12 \pm 2\sqrt{6}}{2} = 6 \pm \sqrt{6}$$

43. Use the Quadratic Formula to solve the following equation.

$$x^2 + 6x - 3 = 0$$

Solution

$$x^2 + 6x - 3 = 0$$

$$x = \frac{-6 \pm \sqrt{6^2 - 4(1)(-3)}}{2(1)} = \frac{-6 \pm \sqrt{48}}{2}$$

$$= \frac{-6 \pm 4\sqrt{3}}{2} = -3 \pm 2\sqrt{3}$$

47. *On the Moon* An astronaut standing on the edge of a cliff on the moon drops a rock over the cliff. If the cliff is 100 feet high, how long will the rock remain in the air? If a rock were dropped off a cliff on Earth, how long would it remain in the air?

Solution

See example 6 in section 1.4.

On the moon: $-2.7t^2 + 100 = 0$

$$t^2 = \frac{100}{2.7}$$

$$t \approx 6.09 \text{ seconds}$$

On Earth: $-16t^2 + 100 = 0$

$$t^2 = \frac{100}{16}$$

$$t = 2.5 \text{ seconds}$$

51. Find all the solutions of $x^4 + 3x^3 - 5x - 15 = 0$. (Be sure to check your answer in the original equation.)

Solution

$$x^4 + 3x^3 - 5x - 15 = 0$$
$$x^3(x+3) - 5(x+3) = 0$$
$$(x+3)(x^3 - 5) = 0$$
$$x + 3 = 0$$
$$x = -3 \qquad \text{OR} \qquad x^3 - 5 = 0$$
$$x = \sqrt[3]{5}$$

55. Find all the solutions of $(x^2 - 5)^{2/3} = 9$. (Be sure to check your answer in the original equation.)

Solution

$$(x^2 - 5)^{2/3} = 9$$
$$x^2 - 5 = 9^{3/2}$$
$$x^2 - 5 = \pm 27$$
$$x^2 = 32 \qquad \text{OR} \qquad x^2 = -22$$
$$x = \pm\sqrt{32} = \pm 4\sqrt{2} \qquad \text{No real solution}$$

59. Find all the solutions of $\dfrac{5}{x+1} + \dfrac{3}{x+3} = 1$. (Be sure to check your answer in the original equation.)

Solution

$$\frac{5}{x+1} + \frac{3}{x+3} = 1$$
$$5(x+3) + 3(x+1) = (x+1)(x+3);\ x \neq -1,\ x \neq -3$$
$$5x + 15 + 3x + 3 = x^2 + 4x + 3$$
$$0 = x^2 - 4x - 15$$
$$x = \frac{-(-4) \pm \sqrt{(-4)^2 - 4(1)(-15)}}{2(1)}$$
$$= \frac{4 \pm \sqrt{76}}{2} = \frac{4 \pm 2\sqrt{19}}{2}$$
$$= 2 \pm \sqrt{19}$$

63. *Market Research* The demand equation for a product is given by

$$p = 42 - \sqrt{0.001x + 2}$$

where x is the number of units demanded per day and p is the price per unit. Find the demand if the price is set at \$29.95.

Solution

$$29.95 = 42 - \sqrt{0.001x + 2}$$

$$\sqrt{0.001x + 2} = 12.05$$

$$0.001x + 2 = 145.2025$$

$$x = \frac{143.2025}{0.001}$$

$$x \approx 143,203 \text{ units}$$

67. Solve the inequality $-1 \le -5 - 3x < 4$ and sketch its graph.

Solution

$$-1 \le -5 - 3x < 4$$

$$4 \le -3x < 9$$

$$-\frac{4}{3} \ge x > -3$$

$$-3 < x \le -\frac{4}{3}$$

73. Find the domain of $\sqrt{81 - 4x^2}$.

Solution

$(9 + 2x)(9 - 2x) \ge 0$

Critical numbers: $x = \pm\frac{9}{2}$

Test intervals:

$$\left(-\infty, -\tfrac{9}{2}\right), \left(-\tfrac{9}{2}, \tfrac{9}{2}\right), \left(\tfrac{9}{2}, \infty\right)$$

$-\frac{9}{2} \le x \le \frac{9}{2}$ or $\left[-\frac{9}{2}, \frac{9}{2}\right]$

75. *Accuracy of Measurement* Suppose you buy an 18-inch gold chain that costs \$8.95 per inch. If the chain is measured accurately to within $\frac{1}{16}$ of an inch, how much might you have been undercharged or overcharged?

Solution

$|x - 18| \le \frac{1}{16}$

$-\frac{1}{16} \le x - 18 \le \frac{1}{16}$

$17.9375 \le x \le 18.0625$

You could have been undercharged $(18.0625 - 18)(8.95) \approx \$0.56$ or you could have been overcharged $(18 - 17.9375)(8.95) \approx \$0.56$

79. Solve the following inequality and graph the solution on the real number line.

$$\frac{x+5}{x+8} \geq 2$$

Solution

$$\frac{x+5}{x+8} \geq 2$$

$$\frac{x+5}{x+8} - 2 \geq 0$$

$$\frac{x+5-2(x+8)}{x+8} \geq 0$$

$$\frac{-x-11}{x+8} \geq 0$$

Critical numbers: $x = -11, x = -8$

Test intervals: $(-\infty, -11), (-11, -8), (-8, \infty)$

Test: Is $\dfrac{-x-11}{x+8} \geq 0$?

Solution set: $[-11, \ -8)$

$-11 \leq x < 8$

x

−12 −11 −10 −9 −8 −7

Note: We do not include $x = -8$ since it would make the denominator zero.

85. Use a calculator to solve the following inequality. (Round each number in your answer to two decimal places.)

$$\frac{3}{5.4x - 2.7} < 8.9$$

Solution

$$\frac{3}{5.4x - 2.7} < 8.9$$

$$\frac{3}{5.4x - 2.7} - 8.9 < 0$$

$$\frac{3 - 8.9(5.4x - 2.7)}{5.4x - 2.7} < 0$$

$$\frac{-48.06x + 27.03}{5.4x - 2.7} < 0$$

Critical numbers: $x \approx 0.56, x \approx 0.50$

$x < 0.50$ or $x > 0.56$

91. *Price of a Product* The revenue and cost equations for a product are given by

$$R = x(80 - 0.0005x)$$

$$C = 20x + 300,000$$

where R and C are measured in dollars and x represents the number of units sold. The revenue equation is $R = x(80 - 0.0005x)$ which implies that the demand equation is $p = 80 - 0.0005x$ where p is the price per unit. What price per unit should be set to obtain a profit of at least $1,200,000?

Solution

$$R - C \geq 1,200,000$$

$$x(80 - 0.0005x) - (20x + 300,000) \geq 1,200,000$$

$$-0.0005x^2 + 60x - 1,500,000 \geq 0$$

$$0.0005x^2 - 60x + 1,500,000 \leq 0$$

Critical numbers: $x = \dfrac{-(-60) \pm \sqrt{(-60)^2 - 4(0.0005)(1,500,000)}}{2(0.0005)}$

$$x \approx 84,495, \quad x \approx 35,505$$

$35,505$ units $\leq x \leq 84,495$ units

When $x = 35,505$, $p = 80 - 0.0005(35,505) \approx \$62,25$

When $x = 84,495$, $p = 80 - 0.0005(84,495) \approx \$37.75$

Thus, $\$37.73 \leq p \leq \$62.25$.

Test for Chapter 1

1. Solve the following equation.

$$3(x + 2) - 8 = 4(2 - 5x) + 7$$

Solution

$$3(x + 2) - 8 = 4(2 - 5x) + 7$$

$$3x + 6 - 8 = 8 - 20x + 7$$

$$3x - 2 = -20x + 15$$

$$23x = 17$$

$$x = \frac{17}{23}$$

2. Solve the following equation.

$$0.875x + 0.375(300 - x) = 200$$

Solution

$$0.875x + 0.375(300 - x) = 200$$

$$0.875x + 112.5 - 0.375x = 200$$

$$0.5x = 87.5$$

$$x = 175$$

3. In May the total profit for a company was 20% less than it was in April. The total profit for the two months was $315,655.20. Find the profit for each month.

Solution

Model: April profit + May profit = Total profit

May profit = 80% of April profit

Let x = April profit.

$$x + 0.80x = 315,655.20$$

$$1.80x = 315,655.20$$

$$x = 175,364.00$$

$$0.80x = 140,291.20$$

April profit: $175,364.00

May profit: $140,291.20

4. Solve by factoring.

$$6x^2 + 7x = 5$$

Solution

$$6x^2 + 7x = 5$$

$$6x^2 + 7x - 5 = 0$$

$$(3x + 5)(2x - 1) = 0$$

$$3x + 5 = 0 \quad \text{or} \quad 2x - 1 = 0$$

$$3x = -5 \quad \text{or} \quad 2x = 1$$

$$x = -\tfrac{5}{3} \quad \text{or} \quad x = \tfrac{1}{2}$$

5. Solve by factoring.

$$12 + 5x - 2x^2 = 0$$

Solution

$$12 + 5x - 2x^2 = 0$$

$$(3 + 2x)(4 - x) = 0$$

$$3 + 2x = 0 \quad \text{or} \quad 4 - x = 0$$

$$2x = -3 \quad \text{or} \quad -x = -4$$

$$x = -\tfrac{3}{2} \quad \text{or} \quad x = 4$$

6. Extract square roots.

$$x^2 - 5 = 10$$

Solution

$$x^2 - 5 = 10$$

$$x^2 = 15$$

$$x = \pm\sqrt{15}$$

7. Use the Quadratic Formula to solve $(x + 5)^2 = -3x$.

Solution

$$(x + 5)^2 = -3x$$

$$x^2 + 10x + 25 = -3x$$

$$x^2 + 13x + 25 = 0$$

$$a = 1, b = 13, c = 25$$

$$x = \frac{-13 \pm \sqrt{13^2 - 4(1)(25)}}{2(1)}$$

$$= \frac{-13 \pm \sqrt{169 - 100}}{2}$$

$$= \frac{-13 \pm \sqrt{69}}{2}$$

8. Use the Quadratic Formula to solve $3x^2 - 11x = 2$.

Solution

$$3x^2 - 11x = 2$$

$$3x^2 - 11x - 2 = 0$$

$$a = 3,\ b = -11,\ c = -2$$

$$x = \frac{-(-11) \pm \sqrt{(-11)^2 - 4(3)(-2)}}{2(3)} = \frac{11 \pm \sqrt{121 + 24}}{6} = \frac{11 \pm \sqrt{145}}{6}$$

9. Use the Quadratic Formula to solve $5.4x^2 - 3.2x - 2.5 = 0$.

Solution

$$5.4x^2 - 3.2x - 2.5 = 0$$

$$a = 5.4,\ b = -3.2,\ c = -2.5$$

$$x = \frac{-(-3.2) \pm \sqrt{(-3.2)^2 - 4(5.4)(-2.5)}}{2(5.4)}$$

$$= \frac{3.2 \pm \sqrt{10.24 + 54}}{10.8}$$

$$= \frac{3.2 \pm \sqrt{64.24}}{10.8}$$

$$x = \frac{3.2 + \sqrt{64.24}}{10.8} \approx 1.038$$

$$x = \frac{3.2 - \sqrt{64.24}}{10.8} \approx -0.446$$

10. Solve $|3x + 2| = 8$.

$3x + 2 = -8$ $\qquad$ $3x + 2 = 8$

$3x = -10$ $\qquad$ $3x = 6$

$x = -\frac{10}{3}$ $\qquad$ $x = 2$

11. Find all solutions of $\sqrt{x - 3} + x = 5$.

Solution

$$\sqrt{x - 3} + x = 5$$

$$\sqrt{x - 3} = 5 - x$$

$$x - 3 = 25 - 10x + x^2$$

$$0 = x^2 - 11x + 28$$

$$0 = (x - 4)(x - 7)$$

$x - 4 = 0$ or $x - 7 = 0$

$x = 4$ $\qquad$ $x = 7$

$x = 4$ is the only solution.

(Note: $x = 7$ is extraneous.)

12. Solve $x^4 - 10x^2 + 9 = 0$.

Solution

$$x^4 - 10x^2 + 9 = 0$$
$$(x^2 - 1)(x^2 - 9) = 0$$
$$(x + 1)(x - 1)(x + 3)(x - 3) = 0$$
$$x = -1,\ 1,\ -3,\ 3$$

13. Solve $(x^2 - 9)^{2/3} = 9$.

Solution

$$(x^2 - 9)^{2/3} = 9$$
$$(x^2 - 9)^2 = 9^3$$
$$x^2 - 9 = \pm\sqrt{729}$$
$$x^2 - 9 = \pm 27$$
$$x^2 = 9 \pm 27$$
$$x^2 = 36 \quad \text{or} \quad x^2 = -18$$
$$x = \pm 6 \quad \text{No real solution}$$

14. The demand equation for a product is $p = 40 - 0.0001x$ where p is the price per unit and x is the number of units sold. The total revenue for selling x units is given by $R = xp$. How many units must be sold to produce a revenue of $2,000,000? Explain your reasoning.

Solution

$$\text{Revenue} = xp = x(40 - 0.0001x)$$
$$2,000,000 = 40x - 0.0001x^2$$
$$0.0001x^2 - 40x + 2,000,000 = 0$$
$$x = \frac{-(-40) \pm \sqrt{(-40)^2 - 4(0.0001)(2,000,000)}}{2(0.0001)} = \frac{40 \pm \sqrt{800}}{0.0002}$$
$$x = \frac{40 + 20\sqrt{2}}{0.0002} \approx 341,421 \text{ units or } x = \frac{40 - 20\sqrt{2}}{0.0002} \approx 58,579 \text{ units}$$

Since the equation is quadratic, it is possible to have two distinct solutions. In this case they would both have to be positive since x represents the number of units sold.

15. Solve the inequality and sketch its graph.

$$-2 \le \frac{3x + 1}{5} \le 2$$

Solution

$$-2 \le \frac{3x + 1}{5} \le 2$$
$$-10 \le 3x + 1 \le 10$$
$$-11 \le 3x \le 9$$
$$-\frac{11}{3} \le x \le 3$$

−4 −3 −2 −1 0 1 2 3 x

16. Solve $(x+3)^2 \le 5$ and sketch its graph.

Solution

$$(x+3)^2 \le 5$$

$$x^2 + 6x + 9 \le 5$$

$$x^2 + 6x + 4 \le 0$$

Critical numbers: $x = -3 \pm \sqrt{5}$

(Use the Quadratic Formula.)

Test intervals: $\left(-\infty, -3-\sqrt{5}\right)$, $\left(-3-\sqrt{5}, -3+\sqrt{5}\right)$, $\left(-3+\sqrt{5}, \infty\right)$

Solution set: $\left[-3-\sqrt{5}, -3+\sqrt{5}\right]$

$-3-\sqrt{5} \le x \le -3+\sqrt{5}$

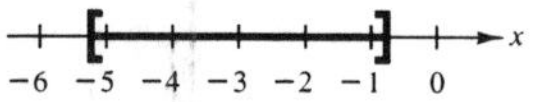

−6 −5 −4 −3 −2 −1 0 x

17. Solve the inequality and sketch its graph.

$$\frac{x+3}{x+7} \ge 2$$

Solution

$$\frac{x+3}{x+7} > 2$$

$$\frac{x+3}{x+7} - 2 > 0$$

$$\frac{x+3-2(x+7)}{x+7} > 0$$

$$\frac{-x-11}{x+7} > 0$$

Critical numbers: $x = -11, -7$

Test intervals: $(-\infty, -11)$, $(-11, -7)$, $(-7, \infty)$

Solution set: $(-11, -7)$

$-11 < x < -7$

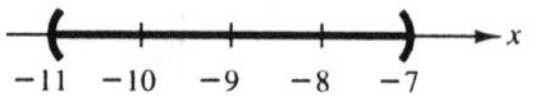

−11 −10 −9 −8 −7 x

18. Solve the inequality and sketch its graph.

$$3x^3 - 12x \le 0$$

Solution

$$3x^3 - 12x \le 0$$

$$3x(x^2 - 4) \le 0$$

$$3x(x+2)(x-2) \le 0$$

Critical numbers: $x = 0, \pm 2$

Test intervals: $(-\infty, -2)$, $(-2, 0)$, $(0, 2)$, $(2, \infty)$

Solution set: $(-\infty, -2] \cup [0, 2]$

$x \le -2,\ 0 \le x \le 2$

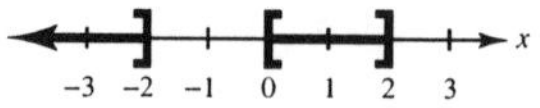

−3 −2 −1 0 1 2 3 x

19. The revenue and cost equations for a product are given by

$$R = x(100 - 0.0005x) \text{ and } C = 30x + 200,000$$

where R and C are measured in dollars and x represents the number of units sold. How many units must be sold to obtain a profit of at least \$500,000?

Solution

Profit = Revenue − Cost

$$R - C \geq 500,000$$

$$x(100 - 0.0005x) - (30x + 200,000) \geq 500,000$$

$$100x - 0.0005x^2 - 30x - 200,000 \geq 500,000$$

$$-0.0005x^2 + 70x - 700,000 \geq 0$$

$$0.0005x^2 - 70x + 700,000 \leq 0$$

Critical numbers: $x = \dfrac{-(-70) \pm \sqrt{(-70)^2 - 4(0.0005)(700,000)}}{2(0.0005)}$

$$= \frac{70 \pm \sqrt{3500}}{0.001}$$

$$x \approx 129,161 \text{ or}$$

$$x \approx 10,839$$

Test intervals: $(-\infty, 10,839)$, $(10,839,\ 129,161)$, $(129,161,\ \infty)$

Answer: $10,839 \text{ units} \leq x \leq 129,161$ units.

20. P dollars, invested at interest rate r compounded annually, increases to an amount $A = P(1 + r)^{10}$. If an investment of \$2,000 is to increase to an amount greater than \$5,000 in 10 years, then the interest rate must be greater than what percentage?

Solution

$$A = P(1 + r)^{10}$$

$$2000(1 + r)^{10} > 5000$$

$$(1 + r)^{10} > \tfrac{5}{2}$$

$$1 + r > \sqrt[10]{\tfrac{5}{2}}$$

$$r > \sqrt[10]{\tfrac{5}{2}} - 1$$

$$\approx 0.096$$

The rate must be greater than 9.6%.

Practice Test for Chapter 1

1. Solve $5x + 4 = 7x - 8$.

2. Solve $\frac{x}{3} - 5 = \frac{x}{5} + 1$.

3. Solve $\frac{3x + 1}{6x - 7} = \frac{2}{5}$.

4. Solve $(x - 3)^2 + 4 = (x + 1)^2$.

5. Solve $A = \frac{1}{2}(a + b)h$ for a.

6. Find three consecutive natural numbers whose sum is 132.

7. 301 is what percent of 4,300?

8. Cindy has \$6.05 in quarters and nickles. How many of each coin does she have if there are 53 coins in all?

9. Ed has \$15,000 invested in two funds paying $9\frac{1}{2}\%$ and 11% simple interest, respectively. How much is invested in each if the yearly interest is \$1,582.50?

10. Solve by factoring $28 + 5x - 3x^2 = 0$.

11. Solve $(x + 3)^2 = 49$ by extracting square roots.

12. Solve $(x - 2)^2 = 24$ by extracting square roots.

13. Solve $x^2 + 5x - 1 = 0$ by the quadratic formula.

14. Solve $3x^2 - 2x + 4 = 0$ by the quadratic formula.

15. The perimeter of a rectangle is 1,100 feet. Find the dimensions so that the enclosed area will be 60,000 square feet.

16. Find two consecutive even positive integers whose product is 624.

17. Solve by factoring $x^3 - 10x^2 + 24x = 0$.

18. Solve $\sqrt[3]{6 - x} = 4$.

19. Solve $(x^2 - 8)^{2/5} = 4$.

20. Solve $x^4 - x^2 - 12 = 0$.

21. Solve $4 - 3x > 16$.

22. Solve $4x^3 - 12x^2 > 0$.

23. Solve $\left|\frac{x - 3}{2}\right| < 5$.

24. Solve $\frac{x + 1}{x - 3} < 2$.

25. Solve $|3x - 4| \geq 9$.

CHAPTER TWO
The Cartesian Plane and Graphs

2.1 The Cartesian Plane

- You should be able to plot points.
- You should know that the distance between (x_1, y_1) and (x_2, y_2) in the plane is
  $d = \sqrt{(x_2 - x_1)^2 + (y_2 - y_1)^2}$.
- You should know that the midpoint of the line segment joining (x_1, y_1) and (x_2, y_2) is
  $$\left(\frac{x_1 + x_2}{2}, \frac{y_1 + y_2}{2}\right).$$

SOLUTIONS TO SELECTED EXERCISES

3. Sketch the square with vertices $(2, 4), (5, 1), (2, -2), (-1, 1)$.

Solution

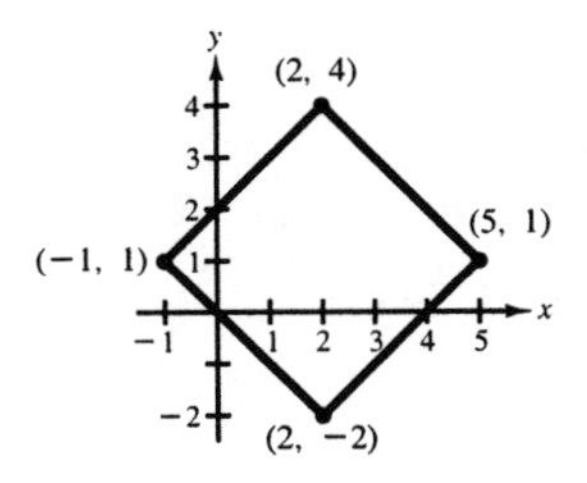

5. *Shifting a Graph* The figure in the textbook is shifted to a new location in the plane. Find the coordinates of the vertices of the figure in it *new* location.

Solution

$(1 + 2, -3 + 5) = (3, 2)$

$(-3 + 2, -4 + 5) = (-1, 1)$

$(-2 + 2, -1 + 5) = (0, 4)$

9. Plot the points and find the distance between them. $(-3, -1)$ and $(2, -1)$.

Solution

$|2-(-3)| = 5$

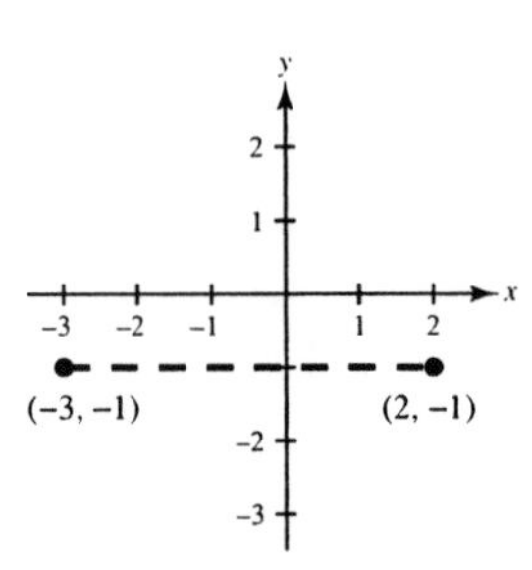

13. Find the length of the hypotenuse in two ways.

(a) Use the Pythagorean Theorem and

(b) Use the Distance Formula. (See figure in the textbook.)

Solution

(a) $a = 4, b = 3, c = \sqrt{4^2 + 3^2} = \sqrt{25} = 5$

(b) Use the points (0, 0) and (4, 3).

$$d = \sqrt{(4-0)^2 + (3-0)^2} = \sqrt{16+9}$$
$$= \sqrt{25} = 5$$

19. (a) Plot the points $(-4, 10)$ and $(4, -5)$,

(b) find the distance between the points, and

(c) find the midpoint of the line segment joining the points.

Solution

(a)

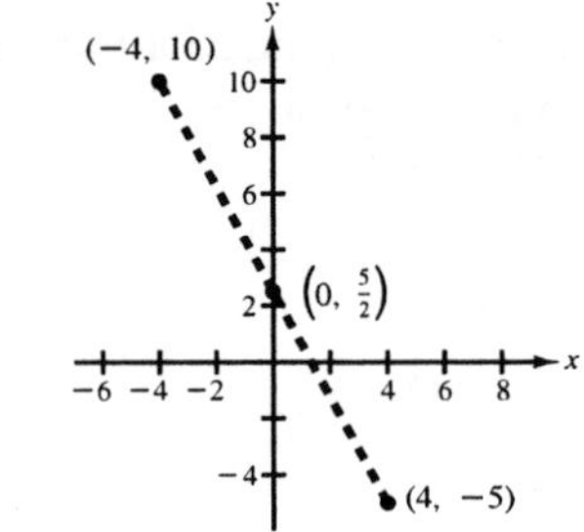

(b) $d = \sqrt{(4+4)^2 + (-5-10)^2} = \sqrt{64+225} = 17$

(c) $\left(\dfrac{4-4}{2}, \dfrac{-5+10}{2}\right) = \left(0, \dfrac{5}{2}\right)$

23. (a) Plot the points $\left(\frac{1}{2}, 1\right)$ and $\left(-\frac{5}{2}, \frac{4}{3}\right)$,

(b) find the distance between the points, and

(c) find the midpoint of the line segment joining the points.

Solution

(a)

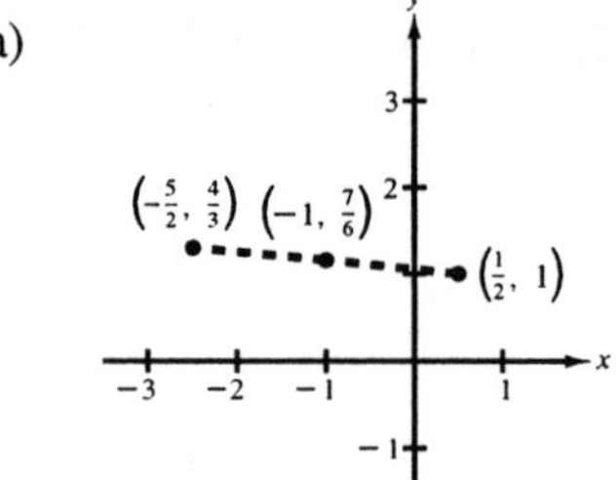

(b) $d = \sqrt{\left(\frac{1}{2}+\frac{5}{2}\right)^2 + \left(1-\frac{4}{3}\right)^2} = \sqrt{9+\frac{1}{9}} = \dfrac{\sqrt{82}}{3}$

(c) $\left(\dfrac{-\frac{5}{2}+\frac{1}{2}}{2}, \dfrac{\frac{4}{3}+1}{2}\right) = \left(-1, \dfrac{7}{6}\right)$

27. (a) Plot the points $(-36, -18)$ and $(48, -72)$, (b) find the distance between the points, and (c) find the midpoint of the line seqment joining the points.

Solution

(a)

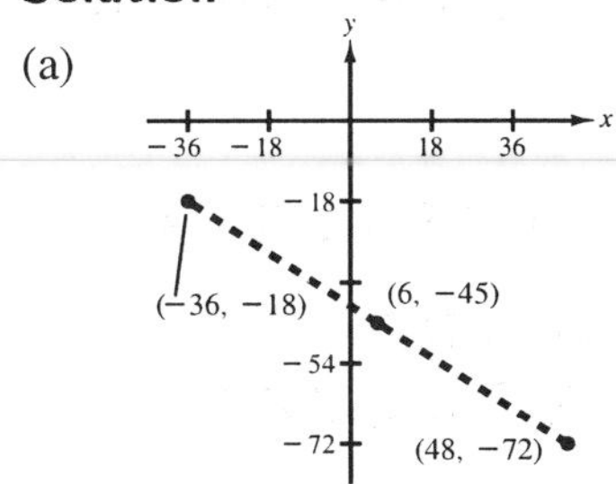

(b) $d = \sqrt{(48+36)^2 + (-72+18)^2} = \sqrt{7056 + 2916} = \sqrt{9972} = 6\sqrt{277}$

(c) $\left(\dfrac{-36+48}{2}, \dfrac{-18-72}{2}\right) = (6, -45)$

29. Use the Midpoint Formula to estimate the sales of a company for 1993, if the sales were \$520,000 in 1991 and \$740,000 in 1995. Assume the sales followed a linear pattern.

Solution

$$\left(\frac{1991+1995}{2}, \frac{\$520,000 + \$740,000}{2}\right) = (1993, \$630,000)$$

33. Show that the points $(0, 0)$, $(1, 2)$, $(2, 1)$ and $(3, 3)$ form the vertices of a rhombus. (A rhombus is a parallelogram whose sides are all the same length.)

Solution

$d_1 = \sqrt{(0-1)^2 + (0-2)^2} = \sqrt{5}$

$d_2 = \sqrt{(0-2)^2 + (0-1)^2} = \sqrt{5}$

$d_3 = \sqrt{(3-1)^2 + (3-2)^2} = \sqrt{5}$

$d_4 = \sqrt{(3-2)^2 + (3-1)^2} = \sqrt{5}$

$d_1 = d_2 = d_3 = d_4$

35. Find x so that the distance between $(1, 2)$ and $(x, -10)$ is 13.

Solution

$$\sqrt{(x-1)^2 + (-10-2)^2} = 13$$
$$(x-1)^2 + 144 = 169$$
$$(x-1)^2 = 25$$
$$x = 1 \pm 5 = -4, 6$$

37. Find y so that the distance between $(0, 0)$ and $(8, y)$ is 17.

Solution

$$\sqrt{(8-0)^2 + (y-0)^2} = 17$$
$$64 + y^2 = 289$$
$$y^2 = 225$$
$$y = \pm 15$$

41. Given $x > 0$ and $y > 0$, state the quadrant in which (x, y) lies.

Solution

$x > 0$ and $y > 0$ in Quadrant I

43. *Football Pass* In a football game, a quarterback throws a pass from the 15-yard line, 10 yards from the sideline. The pass is caught on the 40-yard line, 45 yards from the same sideline. How long was the pass? (Assume the 15-yard line and the 40-yard line are on the same side of mid-field.)

Solution

$(15, 10), (40, 45)$

$d = \sqrt{(40 - 15)^2 + (45 - 10)^2} = \sqrt{1850} = 5\sqrt{74} \approx 43$ yards

51. *Football Attendance* Use the figure in the textbook, which shows the average paid attendance at NFL football games for the years 1940, 1950, 1960, 1970, 1980, and 1990, to estimate the increase in attendance from 1940 to 1960. (*Source: National Football League.*)

Solution

Approximately $40,000 - 20,000 = 20,000$

2.2 Graphs of Equations

- You should be able to use the point-plotting method of graphing
  - (a) Try to write the equation so that one of the variables is isolated on one side of the equation. (This is not always possible.)
  - (b) Make up a chart of several solution points.
  - (c) Plot these points.
  - (d) Connect these points with a smooth curve.
- You should be able to find any intercepts.
  - (a) To find x-intercepts, let $y = 0$ and solve for x.
  - (b) To find y-intercepts, let $x = 0$ and solve for y.
- You should be able to test for symmetry.
  - (a) x-axis symmetry: replace y with $-y$
  - (b) y-axis symmetry: replace x with $-x$
  - (c) origin symmetry: replace x with $-x$ *and* replace y with $-y$.
- You should know the standard equation of a circle with center (h, k) and radius r:

  $$(x - h)^2 + (y - k)^2 = r^2$$

SOLUTIONS TO SELECTED EXERCISES

5. Determine whether $\left(1,\ \frac{1}{5}\right)$, and $\left(2,\ \frac{1}{2}\right)$ lie on the graph of $x^2y - x^2 + 4y = 0$.

Solution

(a) $1^2\left(\frac{1}{5}\right) - 1^2 + 4\left(\frac{1}{5}\right) = 0$, yes

(b) $2^2\left(\frac{1}{2}\right) - 2^2 + 4\left(\frac{1}{2}\right) = 0$, yes

9. Find the intercepts of the graph of $y = x^2 + x - 2$.

Solution

$y = x^2 + x - 2 = 0$

$(x + 2)(x - 1) = 0$

$x = -2,\ 1$

x-intercepts: $(-2,\ 0)$, $(1,\ 0)$

$y = 0^2 + 0 - 2 = -2$

y-intercept: $(0,\ -2)$

13. Find the intercepts of the graph of $xy - 2y - x + 1 = 0$.

Solution

$xy - 2y - x + 1 = 0$

$x(0) - 2(0) - x + 1 = 0$

$x = 1$

x-intercept: $(1, 0)$

$(0)y - 2y - 0 + 1 = 0$

$y = \frac{1}{2}$

y-intercept: $\left(0,\ \frac{1}{2}\right)$

17. Check $x - y^2 = 0$ for symmetry.

Solution

$x - (-y)^2 = 0 \quad \Rightarrow \quad x - y^2 = 0$

x-axis symmetry

21. Check for symmetry.

$$y = \frac{x}{x^2 + 1}.$$

Solution

$$-y = \frac{-x}{(-x)^2 + 1} \quad \Rightarrow \quad y = \frac{x}{x^2 + 1}$$

Origin symmetry

25. Use the origin symmetry to complete the graph of $y = -x^3 + x$.

Solution

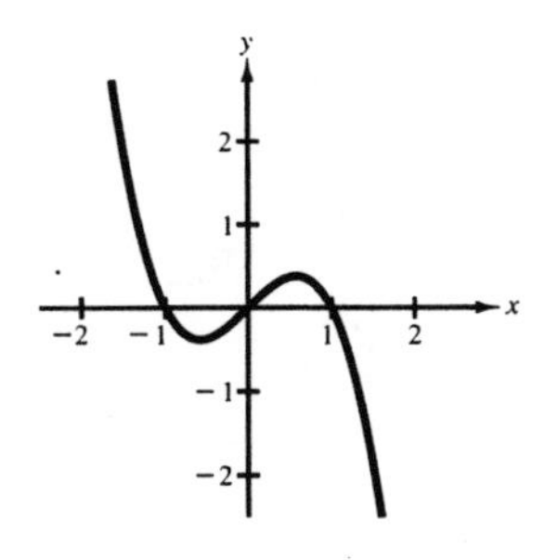

29. Match $y = \sqrt{4 - x^2}$ with its graph. [The graphs are labeled (a), (b), (c), (d), (e), and (f) in the textbook.]

Solution

$y = \sqrt{4 - x^2}$ has intercepts $(-2,\ 0)$, $(0,\ 2)$, and $(2, 0)$; graph (d)

35. Sketch the graph of $y = 1 - x^2$.
Identify any intercepts and test for symmetry.

Solution

Intercepts: $(-1, 0)$, $(0, 1)$, $(1, 0)$

y-axis symmetry

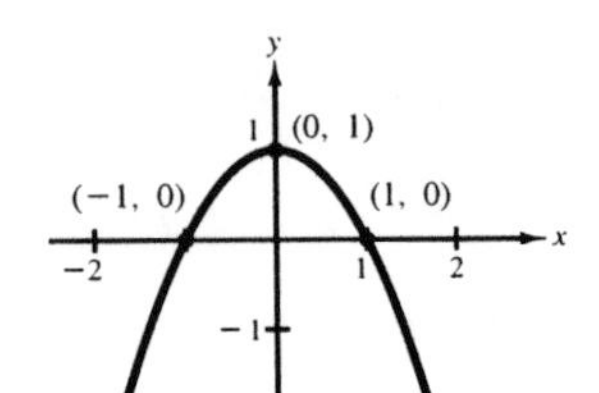

39. Sketch the graph of $y = x^3 + 2$.
Identify any intercepts and test for symmetry.

Solution

Intercepts: $(-\sqrt[3]{2}, 0)$, $(0, 2)$

No symmetry

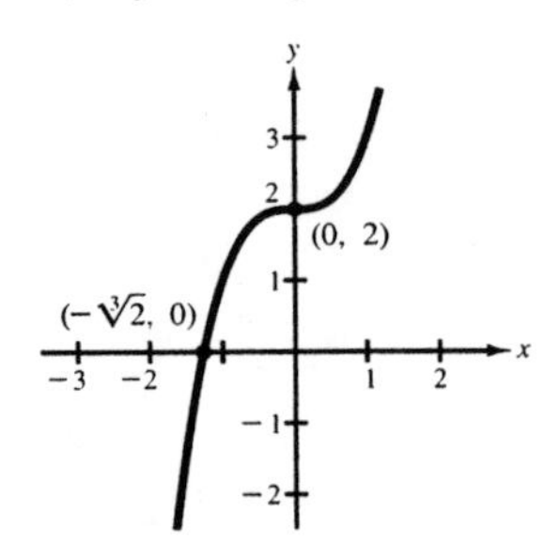

43. Sketch the graph of $y = \sqrt{x - 3}$.
Identify any intercepts and test for symmetry.

Solution

Domain: $[3, \infty)$

Intercept: $(3, 0)$

No symmetry

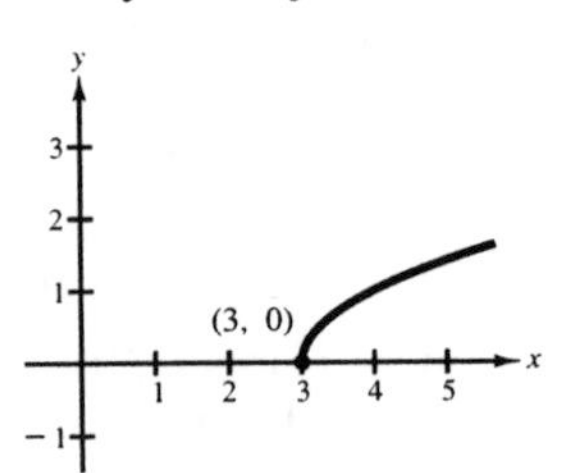

47. Sketch the graph of $y = |x - 2|$.
Identify any intercepts and test for symmetry.

Solution

Intercepts: $(0, 2)$, $(2, 0)$

No symmetry

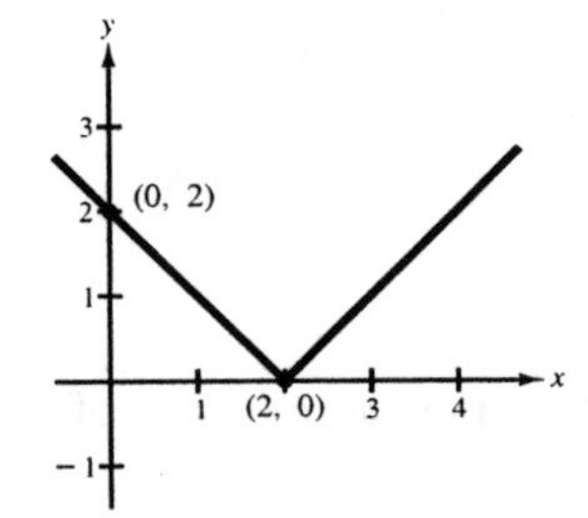

51. Sketch the graph of $x^2 + y^2 = 4$. Identify any intercepts and test for symmetry.

Solution

Intercepts: $(-2, 0)$, $(0, -2)$, $(0, 2)$, $(2, 0)$

x-axis, y-axis, and origin symmetry

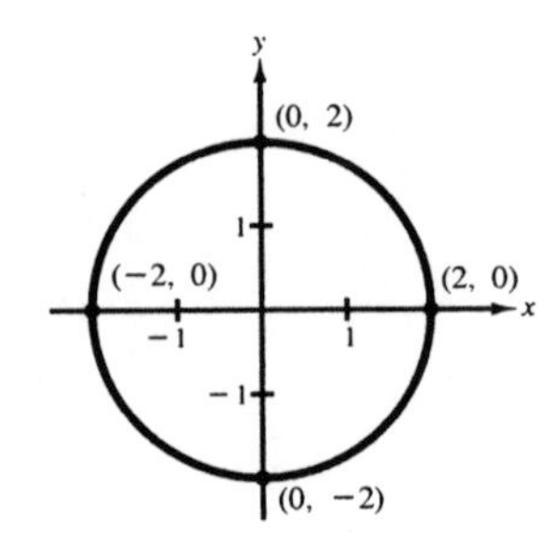

55. Find the standard form of the equation of a circle with center at $(2, -1)$, and radius of 4.

Solution

$$(x-2)^2 + [y-(-1)]^2 = 4^2$$

$$(x-2)^2 + (y+1)^2 = 16$$

59. Find the standard form of the equation a circle with the endpoints of a diameter $(0, 0)$ and $(6, 8)$.

Solution

$$r = \frac{1}{2}\sqrt{(6-0)^2 + (8-0)^2} = 5$$

$$\text{Center} = \left(\frac{0+6}{2}, \frac{0+8}{2}\right) = (3, 4)$$

$$(x-3)^2 + (y-4)^2 = 25$$

63. Write $x^2 + y^2 - 2x + 6y + 10 = 0$ in standard form. Then sketch the circle.

Solution

$$x^2 + y^2 - 2x + 6y + 10 = 0$$

$$(x^2 - 2x + 1) + (y^2 + 6y + 9) = -10 + 1 + 9$$

$$(x-1)^2 + (y+3)^2 = 0$$

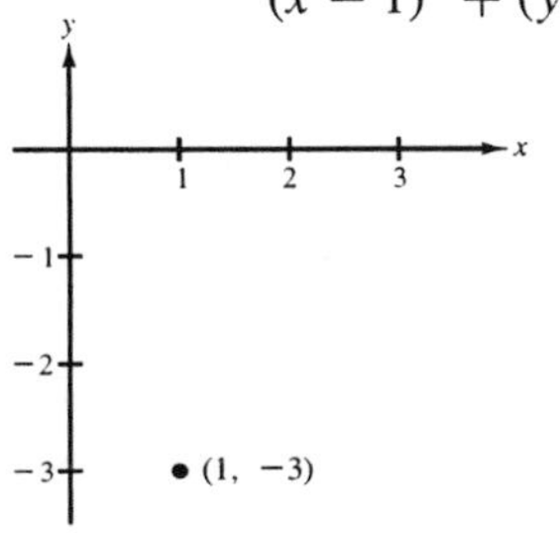

y
x
1 2 3
-1
-2
-3
(1, −3)

67. Write $16x^2 + 16y^2 + 16x + 40y - 7 = 0$ in standard form. Then sketch the circle.

Solution

$$16x^2 + 16y^2 + 16x + 40y - 7 = 0$$

$$\left(x^2 + x + \tfrac{1}{4}\right) + \left(y^2 + \tfrac{5}{2}y + \tfrac{25}{16}\right) = \tfrac{7}{16} + \tfrac{1}{4} + \tfrac{25}{16}$$

$$\left(x + \tfrac{1}{2}\right)^2 + \left(y + \tfrac{5}{4}\right)^2 = \tfrac{9}{4}$$

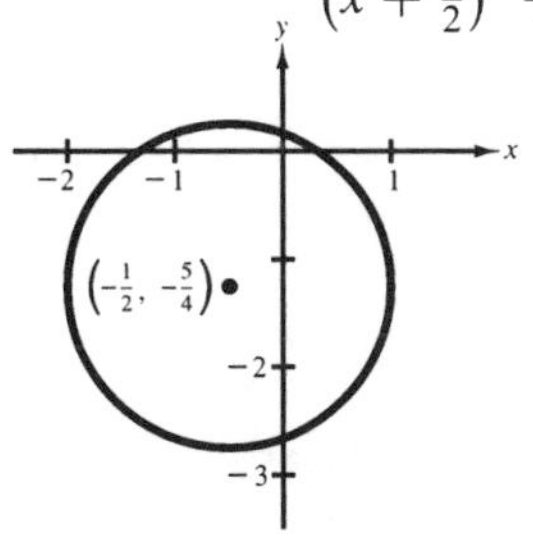

y
x
-2 -1 1
$\left(-\frac{1}{2}, -\frac{5}{4}\right)$
-2
-3

71. Find the constant C such that $(3, 8)$ is a solution point of $y = C\sqrt{x+1}$.

Solution

$$y = C\sqrt{x+1}$$

$$8 = C\sqrt{3+1}$$

$$C = 4$$

73. $(x-2)^2 + (y+3)^2 = 16$ is written in standard form. Indicate the coordinates of the center of the circle and determine the radius of the circle. Rewrite the equation of the circle in general form.

Solution

$$(x-2)^2 + (y-(-3))^2 = (4)^2$$

Center: $(2, -3)$

Radius: $r = 4$

$$x^2 - 4x + 4 + y^2 + 6y + 9 - 16 = 0$$

$$x^2 + y^2 - 4x + 6y - 3 = 0 \text{ General form}$$

77. *Earnings per Share* The earnings per share y (in dollars) for Bristol-Myers Squibb Corporation from 1984 to 1992 can be modeled by

$$y = 0.01t^2 + 0.128t + 1.03, \quad 4 \le t \le 12$$

where $t = 0$ represents 1980. Sketch the graph of this equation. (*Source: NYSE Stock Reports.*)

Solution

Year	1984	1985	1986	1987	1988	1989	1990	1991	1992
Earnings per share	1.70	1.92	2.16	2.42	2.69	2.99	3.31	3.65	4.01

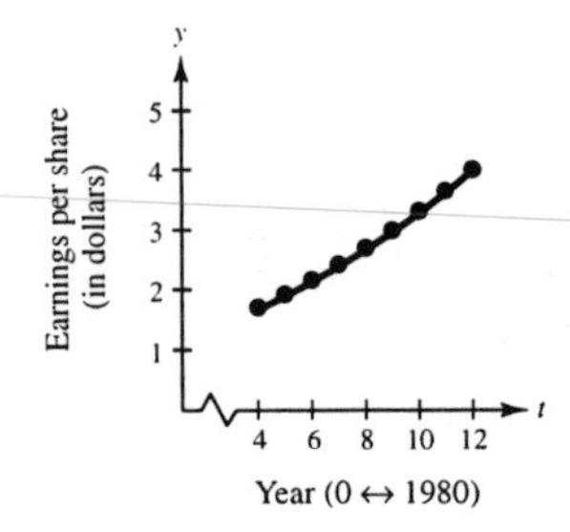

2.3 Graphing Calculators

- If you own a graphing calculator, study the owners manual to learn how to use the graphing features. (This section explains how to use the TI-82.)

SOLUTIONS TO SELECTED EXERCISES

3. Use a graphing utility to match $y = x^2$ with its graph. [The graphs are labeled (a), (b), (c), (d), (e), (f), (g), (h), (i), and (j) in the textbook.]

Solution

Matches graph (a)

9. Use a graphing utility to match $y = \sqrt{x}$ with its graph [The graphs are labeled (a), (b), (c), (d), (e), (f), (g), (h), (i), and (j) in the textbook.]

Solution

Matches graph (e)

15. Use a graphing utility to graph $2x - 3y = 4$. Use a standard setting for each graph.

Solution

$2x - 3y = 4$

$y = \frac{2}{3}x - \frac{4}{3}$

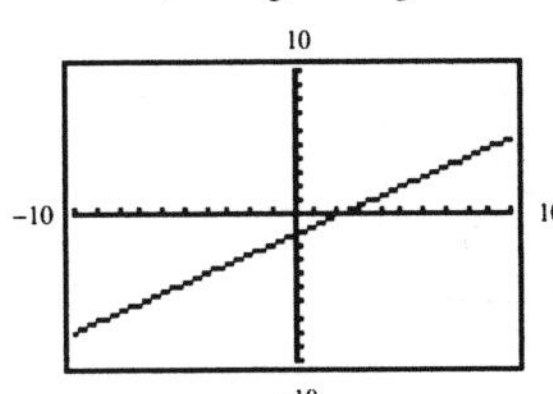

21. Use a graphing utility to graph $y = -x^2 + 2x + 1$. Use a standard setting for each graph.

Solution

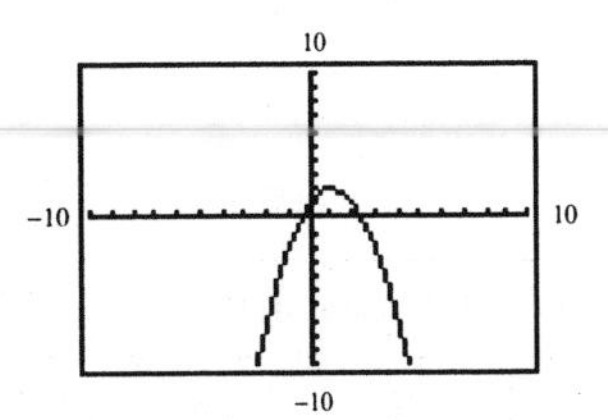

23. Use a graphing utility to graph $2y = x^2 + 2x - 3$. Use a standard setting for each graph.

Solution

$2y = x^2 + 2x - 3$

$y = \frac{1}{2}x^2 + x - \frac{3}{2}$

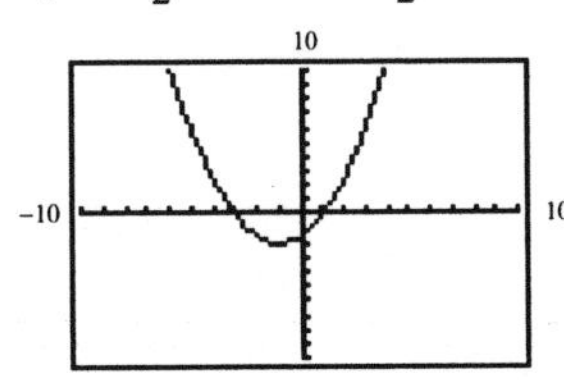

27. Use a graphing utility to graph $y = \sqrt{x^2 + 1}$. Use a standard setting for each graph.

Solution

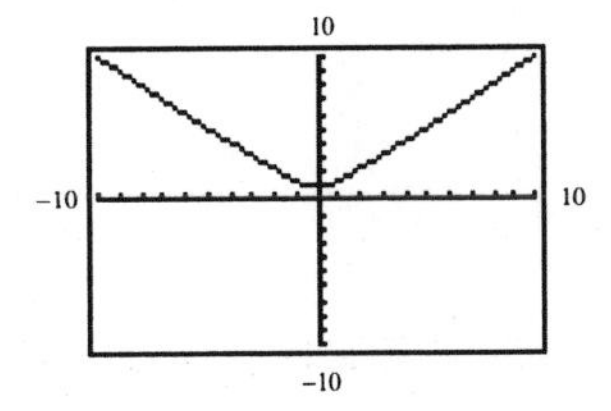

33. Use a graphing utility to graph $y = x^3 + 6x^2$. Use the setting indicated in the textbook. Does the setting give a good representation of the graph? Explain.

Solution

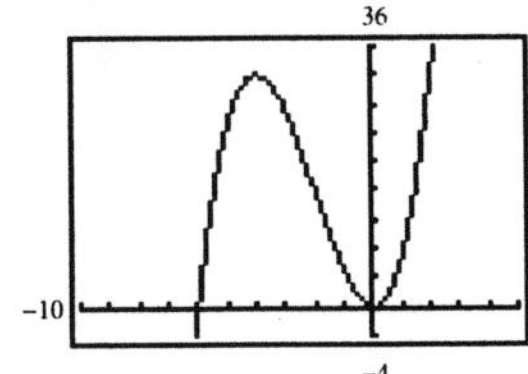

This setting gives a good representation of the graph since the key features (the maximum and the minimum) are clearly displayed.

37. Find a setting on a graphing utility so that the graph of $y = -x^3 + x^2 + 2x$ agrees with the graph shown in the textbook.

Solution

RANGE

Xmin=−10
Xmax=10
Xscl=1
Ymin=−10
Ymax=10
Yscl=1

41. Use a graphing utility to find the number of x-intercepts of

$y = 4x^3 - 20x^2 - 4x + 61$.

Solution

There are three x-intercepts.

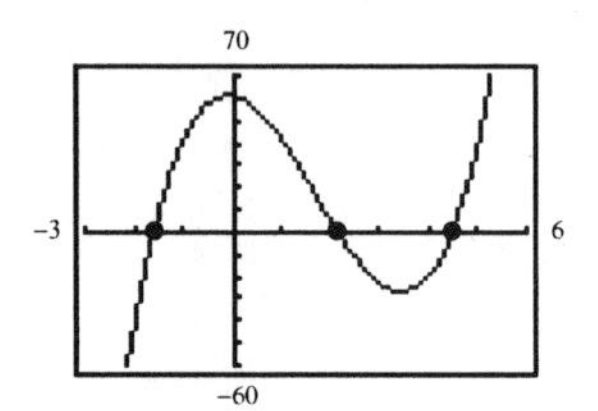

43. Use a graphing utility to sketch the graphs of $y = |x| - 4$ and $y = -|x| + 4$ on the same screen. Using a "square setting," what geometrical shape is bounded by the graphs?

Solution

A square is bounded by their graphs.

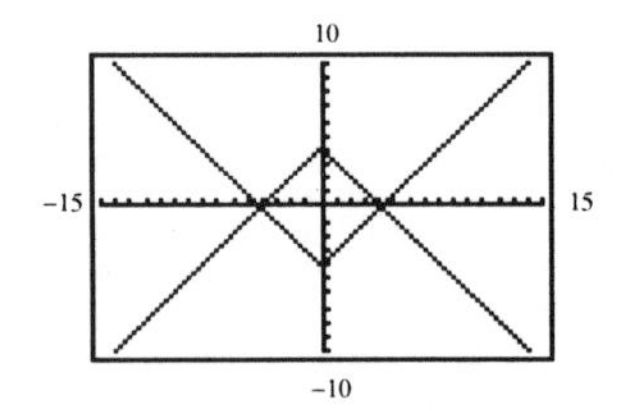

47. A mathematical model that approximates the purchasing power of the dollar is given by

$$y = 1.100 - 0.032t - \frac{0.016}{t}, \quad 5 \le t \le 13$$

where y represents the purchasing power of the dollar and $t = 5$ represents 1985.

(a) Use this model to find the purchasing power of the dollar in 1993.

(b) The purchasing power of the dollar in 1993 was 0.692. What does this tell you about the model?

Solution

(a) For 1993, use $t = 13$.

$$y = 1.100 - 0.032(13) - \frac{0.016}{13} \approx 0.683$$

(b) Since the model was off by less than one cent, it is a pretty good model.

51. Use the model

$$y = \frac{47.07 - 0.7487x}{1 - 0.0738x + 0.00205x^2}, \quad 20 \le x \le 50$$

which relates ages to the percents, y (in decimal form), of American males who have never been married. Suppose that an American male is chosen at random from the population. If the person is 25 years old, what is the probability that he has never been married?

Solution

$$y = \frac{47.07 - 0.7487(25)}{1 - 0.0738(25) + 0.00205(25)^2} \approx 65 \text{ percent}$$

53. *Earnings and Dividends* Use the model

$$y = -0.166 + 0.502x - 0.0953x^2, \quad 0.25 \le x \le 2,$$

which approximates the relationship between dividends per share and earnings per share for the Pall Corporation between 1982 and 1989. In this model y is the dividends per share (in dollars) and x is the earnings per share (in dollars). (*Source: Standard ASE Stock Reports.*) Use a graphing utility to graph this model.

Solution

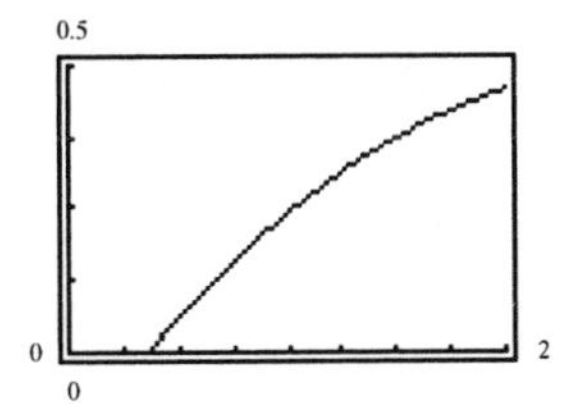

Mid-Chapter Quiz for Chapter 2

1. Given $(-3, 2)$, $(4, -5)$.

(a) Plot the points,

(b) find the distance between the points, and

(c) find the midpoint of the line segment joining the points.

Solution

(a) (−3, 2) (4, −5)

(b) $d = \sqrt{(4-(-3))^2 + (-5-2)^2}$

$= \sqrt{(7)^2 + (-7)^2}$

$= \sqrt{49 + 49}$

$= \sqrt{98}$

$= 7\sqrt{2}$

(c) $\left(\frac{-3+4}{2}, \frac{2+(-5)}{2}\right) = \left(\frac{1}{2}, -\frac{3}{2}\right)$

2. Given $(1.3, -4.5)$, $(-3.7, 0.7)$.

(a) Plot the points,

(b) find the distance between the points, and

(c) find the midpoint of the line segment joining the points.

Solution

(a)

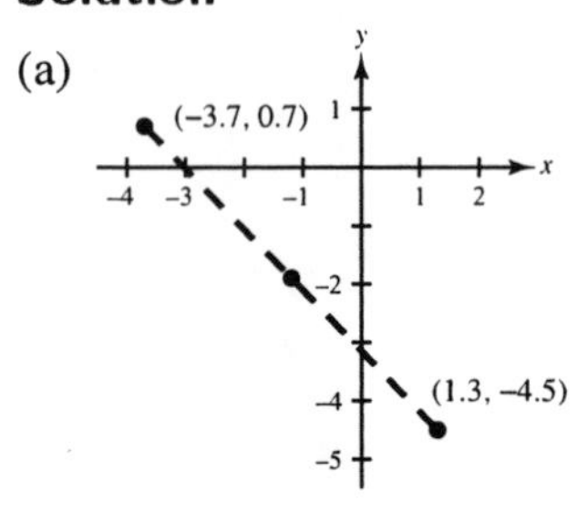

(b) $$d = \sqrt{(-3.7 - 1.3)^2 + (0.7 + 4.5)^2}$$
$$= \sqrt{(5)^2 + (5.2)^2}$$
$$= \sqrt{52.04}$$
$$\approx 7.214$$

(c) $$\left(\frac{1.3 + (-3.7)}{2}, \frac{-4.5 + 0.7}{2}\right) = (-1.2, -1.9)$$

3. A business had sales of \$460,000 in 1991 and \$580,000 in 1995. Estimate the sales in 1993. Explain your reasoning.

Solution

If we assume that the sales growth is linear, we can approximate the sales for 1993 by finding the midpoint between the sales amounts.

$$\frac{460{,}000 + 580{,}000}{2} = \$520{,}000$$

4. One plane is 200 miles due north of a transmitter and another plane is 150 miles due east of the transmitter. What is the distance between the planes?

Solution

By the Pythagorean Theorem

$$d = \sqrt{200^2 + 150^2} = 250 \text{ miles}$$

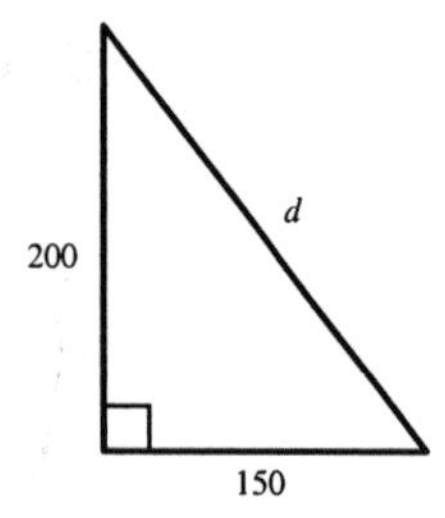

5. Describe the figure with the vertices $(-1, -1)$, $(10, 7)$, and $(2, 18)$.

Solution

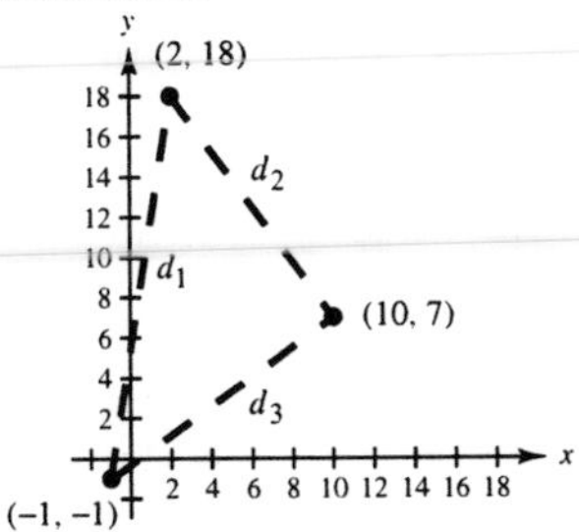

$$d_1 = \sqrt{(2-(-1))^2 + (18-(-1))^2} = \sqrt{370}$$

$$d_2 = \sqrt{(10-2)^2 + (7-18)^2} = \sqrt{185}$$

$$d_3 = \sqrt{(10-(-1))^2 + (7-(-1))^2} = \sqrt{185}$$

Since $d_2 = d_3$ and ${d_1}^2 = {d_2}^2 + {d_3}^2$, the figure is a right isosceles triangle.

6. Describe the figure with the vertices $(-3, -1)$, $(0, 2)$, $(4, 2)$, and $(1, -1)$.

Solution

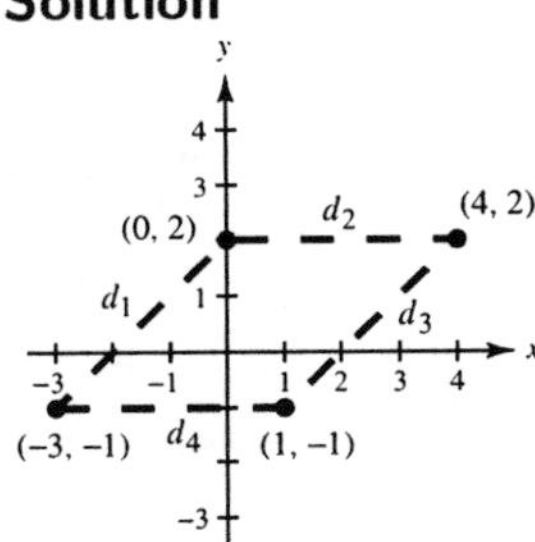

$$d_1 = \sqrt{(-3-0)^2 + (-1-2))^2} = \sqrt{18} = 3\sqrt{2}$$

$$d_2 = |4-0| = 4$$

$$d_3 = \sqrt{(1-4)^2 + (-1-2)^2} = \sqrt{18} = 3\sqrt{2}$$

$$d_4 = |1-(-3)| = 4$$

$$d_1 = d_3 \text{ and } d_2 = d_4$$

The figure is a parallelogram.

7. Sketch the graph of $y = 9 - x^2$. Identify any intercepts and symmetry.

Solution

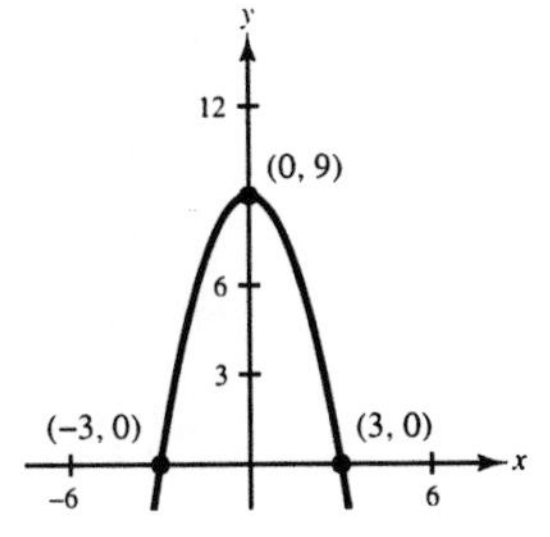

Intercepts: $(\pm 3, 0)$, $(0, 9)$

Symmetry: y-axis

8. Sketch the graph of $y = x\sqrt{x+4}$. Identify any intercepts and symmetry.

Solution

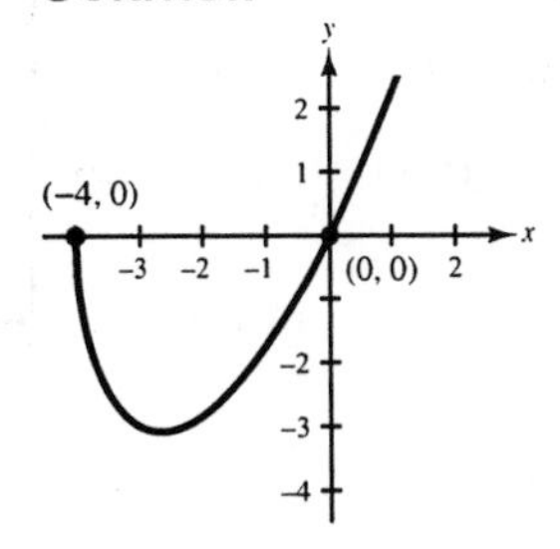

Domain: $[-4, \infty)$

Intercepts: $(-4, 0)$, $(0, 0)$

Symmetry: None

9. Sketch the graph of $xy = 9$. Identify any intercepts and symmetry.

Solution

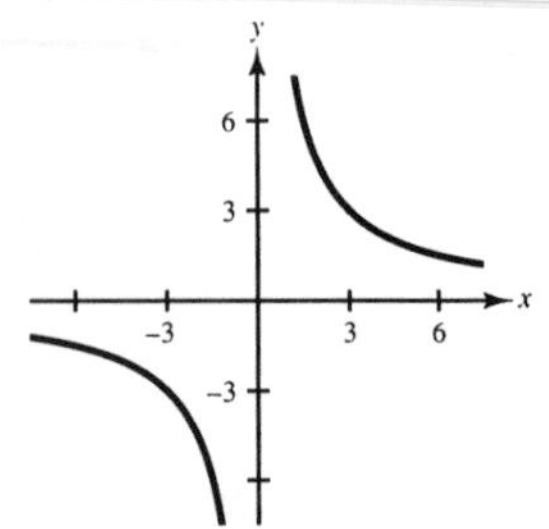

Domain: $x \neq 0$

Intercepts: None

Symmetry: Origin

10. Sketch the graph of $y = \sqrt{16 - x^2}$. Identify any intercepts and symmetry.

Solution

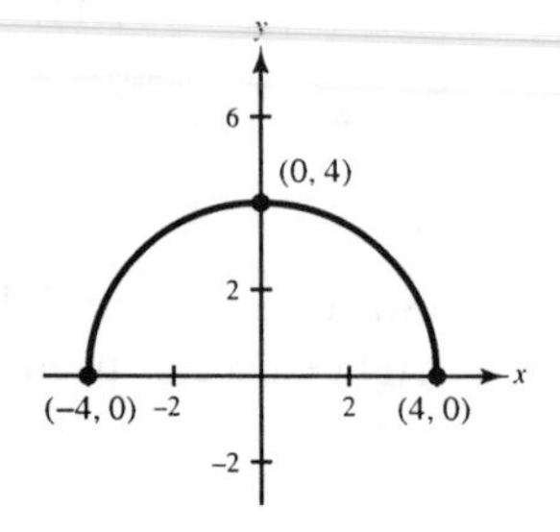

Domain: $[-4, 4]$

Intercepts: $(\pm 4, 0)$, $(0, 4)$

Symmetry: y-axis

11. Sketch the graph of $x^2 + y^2 = 9$. Identify any intercepts and symmetry.

Solution

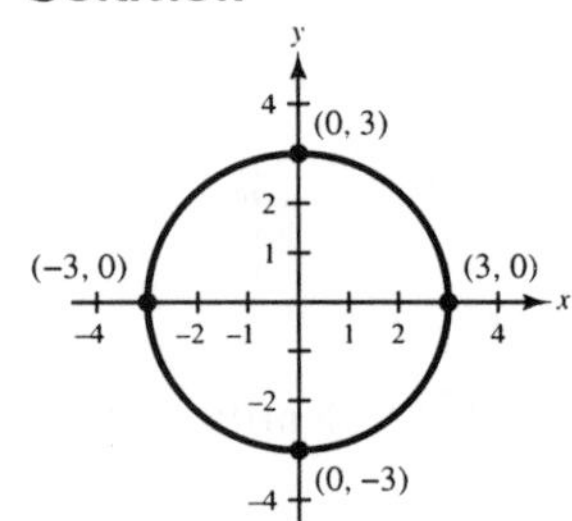

Intercepts: $(\pm 3, 0)$, $(0, \pm 3)$

Symmetry: x-axis, y-axis, and origin

12. Sketch the graph of $y = |x - 3|$. Identify any intercepts and symmetry.

Solution

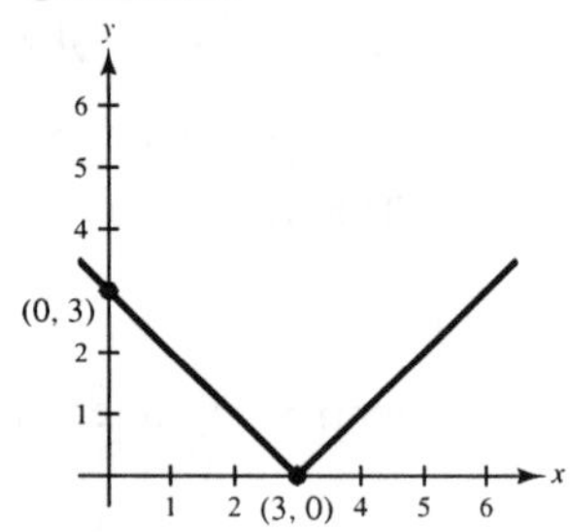

Intercepts: $(3, 0)$, $(0, 3)$

Symmetry: None

13. Find the standard form of the equation of the circle with center $(2, -3)$ and radius 4.

Solution

$$(x - 2)^2 + (y - (-3))^2 = 4^2$$

$$(x - 2)^2 + (y + 3)^2 = 16$$

14. Find the standard form of the equation of the circle with center $\left(0, -\frac{1}{2}\right)$ and radius 2.

Solution

$$(x - 0)^2 + \left(y - \left(-\tfrac{1}{2}\right)\right)^2 = 2^2$$

$$x^2 + \left(y + \tfrac{1}{2}\right)^2 = 4$$

15. Write the equation $x^2 + y^2 - 2x + 4y - 4 = 0$ in standard form. Then sketch the circle.

Solution

$$x^2 + y^2 - 2x + 4y - 4 = 0$$
$$(x^2 - 2x + 1) + (y^2 + 4y + 4) = 4 + 1 + 4$$
$$(x - 1)^2 + (y + 2)^2 = 9$$

Center: $(1, -2)$

Radius: 3

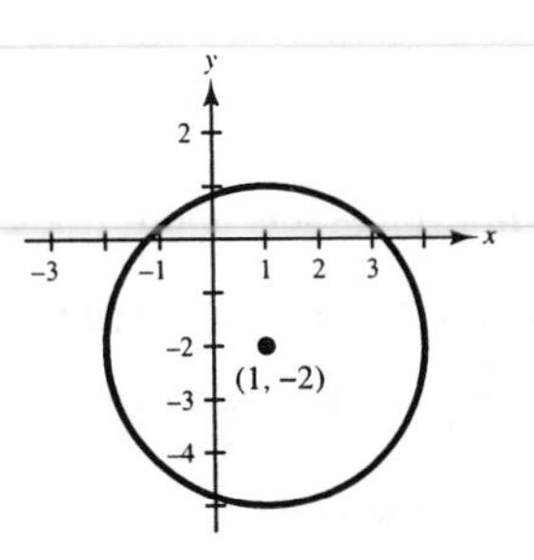

16. Use a graphing utility to find the number of x-intercepts of the equation.

$$y = \frac{1}{4}(-4x^2 + 16x - 14)$$

Two x-intercepts

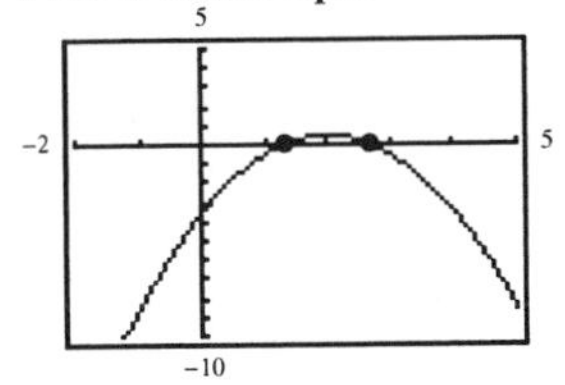

17. Use a graphing utility to find the number of x-intercepts of the equation.

$$y = \frac{1}{8}(4x^2 - 32x + 63)$$

Two x-intercepts

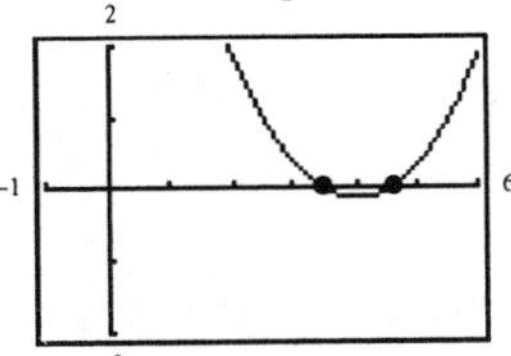

In Exercises 18–20, use the following model. The earnings per share, y (in dollars), for Warner-Lambert Company from 1986 to 1993 can be modeled by

$$y = 0.478t - 1.195, \quad 6 \le t \le 13$$

where $t = 0$ represents 1980.

18. Determine a convenient graphing utility viewing window.

Solution

RANGE

Xmin=6
Xmax=13
Xscl=1
Ymin=0
Ymax=7
Yscl=1

19. Use a graphing utility to graph this model.

Solution

$y_1 = (0.478x - 1.195)(x \ge 6)(x \le 13)$

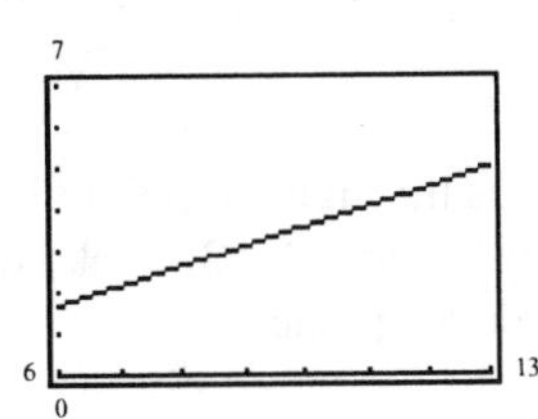

20. Use the trace feature to predict the earnings per share in 1995.

Solution

Remove the inequality constraints. When $x = 15$, $y_1 \approx \$5.98$.

2.4 Lines in the Plane

You should know the following important facts about lines.

- The slope of the line throught (x_1, y_1) and (x_2, y_2) is
$$m = \frac{y_2 - y_1}{x_2 - x_1} = \frac{\text{Change in } y}{\text{Change in } x}, \text{ where } x_1 \neq x_2$$
- (a) If $m > 0$, the line rises from left to right.
  (b) If $m = 0$, the line is horizontal.
  (c) If $m < 0$, the line falls from left to right.
  (d) If m is undefined, the line is vertical.
- Equations of lines
  (a) Point-Slope Form: $y - y_1 = m(x - x_1)$
  (b) Two-Point Form: $y - y_1 = \dfrac{y_2 - y_1}{x_2 - x_1}(x - x_1)$
  (c) Slope-Intercept Form: $y = mx + b$
  (d) Vertical: $x = a$
  (e) Horizontal: $y = b$
  (f) General Form: $Ax + By + C = 0$
- Given two distinct nonvertical lines
$$L_1 : y = m_1x + b_1 \quad \text{and} \quad L_2 : y = m_2x + b_2$$
  (a) L_1 is parallel to L_2 if and only if $m_1 = m_2$.
  (b) L_1 is perpendicular to L_2 if and only if $m_1 = -1/m_2$.

SOLUTIONS TO SELECTED EXERCISES

3. Estimate the slope of the line from its graph in the textbook.

Solution

$$\text{Slope} = \frac{\text{rise}}{\text{run}} = \frac{0}{1} = 0$$

7. Sketch the graph of the lines through the point (2, 3) with slopes (a) 0, (b) 1, (c) 2, and (d) -3. Make the sketches on the same coordinate plane.

Solution

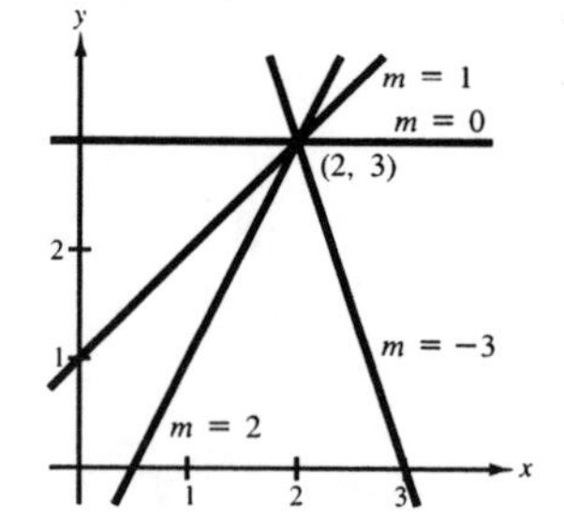

11. Plot the points $(-6, -1)$ and $(-6, 4)$ and find the slope of the line passing through the points.

Solution

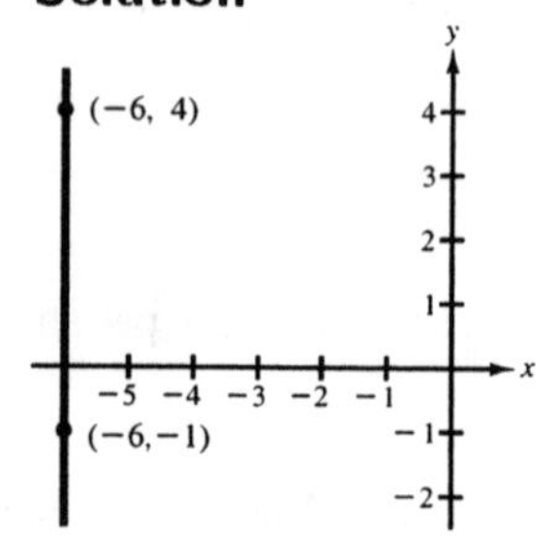

Slope is undefined

15. Determine if the lines L_1 and L_2 passing through the following pairs of points are parallel, perpendicular, or neither.

$L_1 : (0, -1), (5, 9)$
$L_2 : (0, 3), (4, 1)$

Solution

$$m_{L_1} = \frac{9+1}{5-0} = 2$$

$$m_{L_2} = \frac{1-3}{4-0} = -\frac{1}{2} = -\frac{1}{m_{L_1}}$$

L_1 and L_2 are perpendicular.

19. Use the point (2, 1) and the slope of the line $m = 0$ to find three additional points that the line passes through. (The problem has more than one correct answer.)

Solution

Since $m = 0$, y does not change. Three points are (0, 1), (−1, 1), and (3, 1).

23. Use the point (−8, 1) and the undefined slope m to find three additional points that the line passes through. (The problem has more than one correct answer.)

Solution

Since m is undefined, x does not change. Three points are (−8, 0), (−8, 2), and (−8, 3).

29. Find the slope and y-intercept of the line specified by $7x + 6y - 30 = 0$.

Solution

$$7x + 6y - 30 = 0$$

$$y = -\tfrac{7}{6}x + 5$$

Slope: $m = -\frac{7}{6}$

y-intercept: (0, 5)

33. Find an equation of the line passing through the points $\left(2, \frac{1}{2}\right)$ and $\left(\frac{1}{2}, \frac{5}{4}\right)$ and use a graphing utility to verify your answer.

Solution

$$y - \frac{1}{2} = \frac{\frac{5}{4} - \frac{1}{2}}{\frac{1}{2} - 2}(x - 2)$$

$$y = -\frac{1}{2}(x - 2) + \frac{1}{2}$$

$$y = -\frac{1}{2}x + \frac{3}{2} \quad \Rightarrow \quad x + 2y - 3 = 0$$

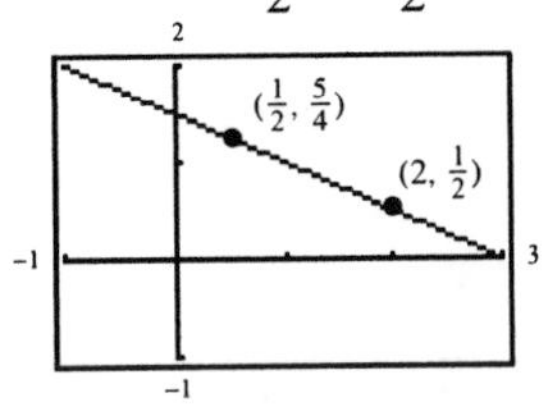

37. Find an equation of the line passing through the points (1, 0.6) and (−2, −0.6) and use a graphing utility to verify your answer.

Solution

$$y - 0.6 = \frac{-0.6 - 0.6}{-2 - 1}(x - 1)$$

$$y = 0.4(x - 1) + 0.6$$

$$y = 0.4x + 0.2 \quad \Rightarrow \quad 2x - 5y + 1 = 0$$

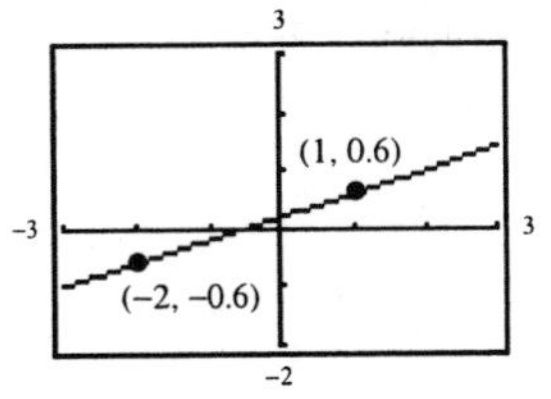

41. Find an equation of the line that passes through the point $(-3, 6)$ and has the slope $m = -2$. Sketch the line.

Solution

$y - 6 = -2(x + 3)$

$y = -2x \quad \Rightarrow \quad 2x + y = 0$

45. Find the equation of the line that passes through the point $(6, -1)$ with an undefined slope m. Sketch the line.

Solution

$x = 6$

$x - 6 = 0$

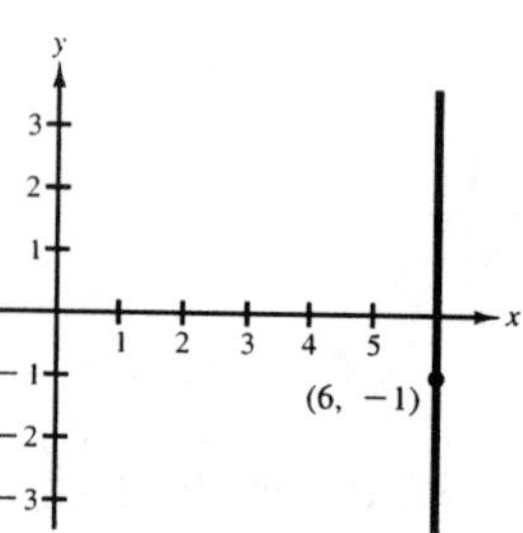

49. Use the following **intercept form** to find the equation of the line with x-intercept $(2, 0)$ and y-intercept $(0, 3)$. The intercept form of the equation of a line with intercepts at $(a, 0)$ and $(0, b)$ is

$$\frac{x}{a} + \frac{y}{b} = 1, \ a \neq 0, \ b \neq 0.$$

Solution

$$\frac{x}{2} + \frac{y}{3} = 1$$

$3x + 2y - 6 = 0$

53. Write an equation of the line through the point $(2, 1)$ (a) parallel and (b) perpendicular to $4x - 2y = 3$. Use a graphing utility to verify your answer.

Solution

$y = 2x - \frac{3}{2}$

slope: $m = 2$

(a) $y - 1 = 2(x - 2)$

$y = 2x - 3$

$2x - y - 3 = 0$

(b) $y - 1 = -\frac{1}{2}(x - 2)$

$y = -\frac{1}{2}x + 2$

$x + 2y - 4 = 0$

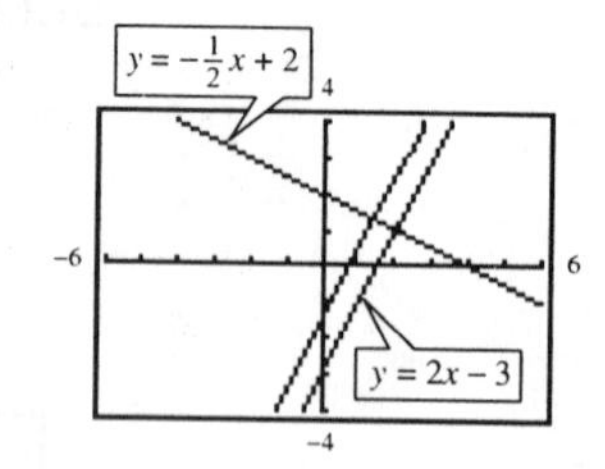

57. Write an equation of the line through the point $(-1,\ 0)$ (a) parallel and (b) perpendicular to $y = -3$.

Solution

slope: $m = 0$

(a) $y = 0$

(b) $x = -1$

$x + 1 = 0$

61. *Simple Interest* A person deposits P dollars into an account that pays simple interest. After two months the balance in the account is \$759, and after three months the balance in the account is \$763.50. Find an equation that gives the relationship between the balance A and the time t in months.

Solution

(2,759)(3,763.50)

$$A - 759 = \frac{763.50 - 759}{3 - 2}(t - 2)$$

$$A = 4.5t + 750$$

65. *Annual Salary* Suppose your salary was \$23,500 in 1992 and \$26,800 in 1995. If your salary follows a linear growth pattern, what salary will you be making in 1999?

Solution

Let $t = 0$ represent 1992.

$(0, 23, 500), (3, 26, 800)$

$$S - 23,500 = \frac{26,800 - 23,500}{3 - 0}(t - 0)$$

$$S = 1100t + 23,500$$

When $t = 7$, we have $S = 1100(7) + 23,500 = \$31,200$.

67. *Turner Broadcasting Revenue* In 1988 Turner Broadcasting Systems had a total revenue of \$810 million. By 1991 the total revenue had increased \$1480 million. If the total revenue continued to increase in a linear pattern, how much would the revenue have been in 1993? The actual revenue in 1993 was \$1920 million. Do you think the increase in revenue was approximately linear? (*Source: Turner Broadcasting System Inc.*)

Solution

Let $t = 0$ represent 1988.

$(0, 810), (3, 1480)$

$$R - 810 = \frac{1480 - 810}{3 - 0}(t - 0)$$

$$R = \frac{670}{3}t + 810$$

When $t = 5$, we have $R = \frac{670}{3}(5) + 810 \approx \$1927$ million. Since the actual revenue in 1993 was \$1920 million, the increase in revenue *was* approximately linear.

2.5 Linear Modeling

- Know the two basic types of linear models.
  (a) General model: $y = mx + b$
  (b) Simpler model: $y = mx$
    This model is called direct variation.
    (1.) y varies directly as x.
    (2.) y is directly proportional to x.
  Note: m is called the constant of variation or the constant of proportionality.

SOLUTIONS TO SELECTED EXERCISES

1. *Employment* The total number of employed (in thousands) in the United States from 1987 to 1993 is given by the ordered pairs listed in the textbook. A linear model that approximates this data is

$$y = 113{,}336.2 + 1265.0t, \quad 7 \le t \le 13$$

where y represents the number of employed (in thousands) and $t = 0$ represents 1980. Plot the actual data and the model on the same graph. How closely does the model represent the data? (*Source: U.S. Bureau of Labor Statistics.*)

Solution

Year	Actual Number	Model Number
1987	121,602	122,191
1988	123,378	123,456
1989	125,557	124,721
1990	126,424	125,986
1991	126,867	127,251
1992	128,548	128,516
1993	129,525	129,781

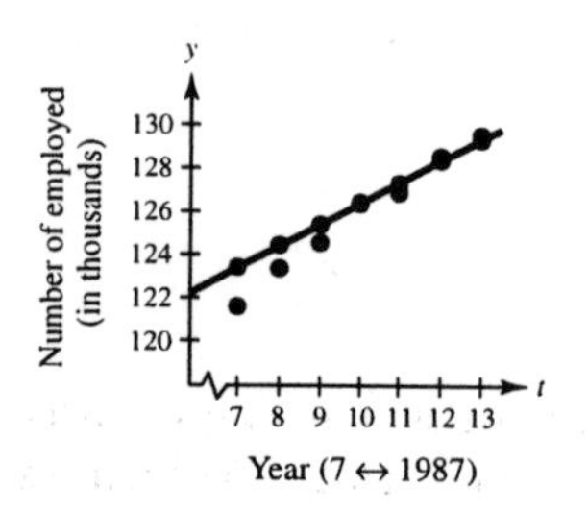

The model is a "good fit" for the actual data.

5. *Direct Variation* Assume that y is proportional to x. Use $x = 10$ and $y = 2050$ to find a linear model that relates x and y.

Solution

$$y = mx$$
$$2050 = m(10)$$
$$205 = m$$
$$y = 205x$$

9. *Simple Interest* The simple interest that a person receives for an investment is directly proportional to the amount of the investment. Suppose that by investing $2,500 in a bond issue, you obtained an interest payment of $187.50 at the end of one year. Find a mathematical model that gives the interest I at the end of one year in terms of the amount invested P.

Solution

$$I = kP$$

$$187.59 = k(2500)$$

$$0.075 = k$$

$$I = 0.075P$$

13. *Centimeters and Inches* On a yardstick you notice that 13 inches is the same length as 33 centimeters. (a) Use this information to find a mathematical model that relates centimeters to inches. (b) Use the model to complete the table in the textbook.

Solution

(a) $y = kx$

$33 = k(13)$

$\frac{33}{13} = k$

$y = \frac{33}{13}x$

(b)

Inches	5.00	10.00	20.00	25.00	30.00
Centimeters	12.69	25.38	50.77	63.46	76.15

17. The dollar value of a product in 1994 is $20,400. The value of the item is expected to decrease $2,000 per year during the next five years. Use this information to write a linear equation that gives the dollar value of the product in terms of the year. (Let $t = 0$ represent 1994.)

Solution

$y = -2,000t + 20,400, \quad 0 \le t \le 5$

21. *Straight-Line Depreciation* A small business purchases a piece of equipment for $875. After five years the equipment will have no value. Write a linear equation giving the value V of the equipment during the five years.

Solution

$V = 875 - \left(\frac{875}{5}\right)t = 875 - 175t, \quad 0 \le t \le 5$

25. *Hourly Wages* A manufacturer pays its assembly line workers $11.50 per hour. In addition, workers receive a piecework rate of $0.75 per unit produced. Write a linear equation for the hourly wages W in terms of the number of units x produced per hour.

Solution

$W = 0.75x + 11.50$

27. *Height of a Parachutist* After opening the parachute, the descent of a parachutist follows a linear model. At 2:08 P.M. the height of the parachutist is 7,000 feet. At 2:10 P.M. the height is 4,600 feet. (a) Write a linear equation that gives the height of the parachutist in terms of the time. (Let $t = 0$ represent 2:08 P.M. and let t be measured in seconds.) (b) Use the equation to find the time when the parachutist will reach the ground.

Solution

(a) 2 minutes = 120 seconds

$(0, 7{,}000), (120, 4{,}600)$

$$h - 7,000 = \frac{4,600 - 7,000}{120 - 0}(t - 0)$$

$$h = 7{,}000 - 20t$$

(b) When the parachutist reaches the ground, we have

$$h = 0$$

$$0 = -20t + 7,000$$

$$t = \frac{7,000}{20} = 350 \text{ seconds} \approx 5.833 \text{ minutes.}$$

The time will be

(2 hours + 8 minutes) + (5 minutes + 50 seconds) = 2:13:50 P.M.

31. Match the description, "a sales representative receives \$20 per day for food, plus \$0.25 for each mile traveled," with one of the graphs in the textbook. Also determine the slope and how it is interpreted in the given situation. [The graphs are labeled (a), (b), (c), and (d).]

Solution

$y = 20 + 0.25x,\ x \geq 0$ where y is the amount paid to the sales representative for traveling x miles. Matches graph (a)

35. State whether the data represented by the scatter plot could be approximated by a linear model. (See figure in the textbook.)

Solution

Yes, can be approximated by a linear model. The slope would be negative.

39. Sketch the line that you think best approximates the date represented by the scatter plot. Then find an equation of the line. (See figure in the textbook.)

Solution

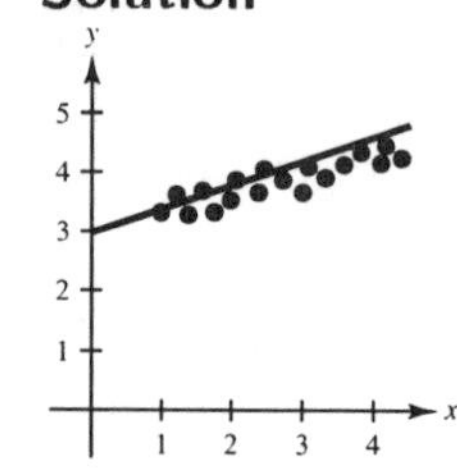

Two approximate points on the line are (1, 3.4) and (3, 4.25). Thus,

$$y - 3.4 = \frac{4.25 - 3.4}{3 - 1}(x - 1)$$

$$y - 3.4 = 0.425x - 0.425$$

$$y = 0.425x + 2.975 \approx 0.4x + 3.$$

47. *Contracting Purchase* A contractor purchases a piece of equipment for $36,500. The equipment costs $5.25 per hour for fuel and maintenance, and the operator is paid $11.50 per hour.

(a) Write an equation giving the cost C of operating the equipment for t hours. (Include the purchase cost.)

(b) If customers are charged $27 per hour, write an equation for the revenue R derived from t hours of use.

(c) Write an equation for the profit derived from t hours of use.

(d) *Break-Even Point* Use the result of part (c) to find the number of hours this equipment must be used to yield a profit of 0 dollars.

Solution

(a) $C = 36,500 + 5.25t + 11.50t = 16.75t + 36,500$

(b) $R = 27t$

(c) $P = R - C = 27t - (16.75t + 36,500) = 10.25t - 36,500$

(d)
$$0 = 10.25t - 36,500$$
$$-10.25t = -36,500$$
$$t \approx 3561 \text{ hours}$$

Review Exercises for Chapter 2

SOLUTIONS TO SELECTED EXERCISES

3. Find (a) the length of the two legs of the right triangle and use the Pythagorean Theorem to find the length of the hypotenuse and (b) use the Distance Formula to find the length of the hypotenuse of the triangle. (See figure in the textbook.)

Solution

Let $A = (1, 0)$, $B = (5, 0)$, and $C = (5, 3)$

(a)
$$d(AB) = |5 - 1| = 4$$
$$d(BC) = |3 - 0| = 3$$
$$d^2(AC) = d^2(AB) + d^2(BC)$$
$$= 16 + 9$$
$$= 25$$
$$d(AC) = 5$$

(b)
$$d = \sqrt{(5-1)^2 + (3-0)^2}$$
$$= \sqrt{4^2 + 3^2}$$
$$= \sqrt{25}$$
$$= 5$$

7. (a) Plot the points $(-6.7,\ -3.9)$ and $(5.1,\ 8.2)$, (b) find the distance between the points, and (c) find the midpoint of the line seqment joining the points.

Solution

(a)

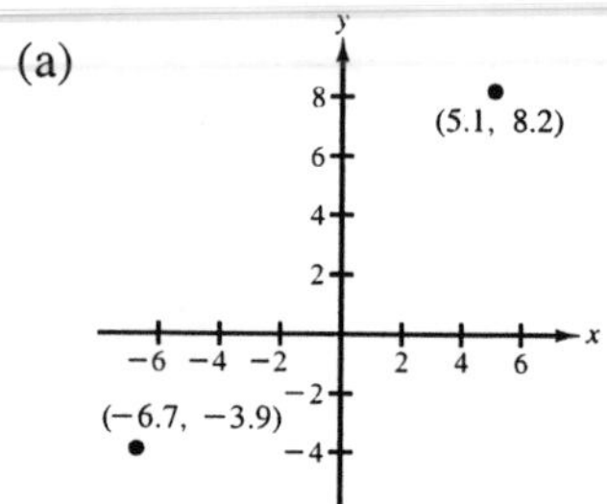

(b) $d = \sqrt{(5.1+6.7)^2 + (8.2+3.9)^2} \approx 16.90$

(c) Midpoint: $\left(\dfrac{-6.7+5.1}{2}, \dfrac{-3.9+8.2}{2}\right) = (-0.8, 2.15)$

11. Show that the points $(1, 1)$, $(8, 2)$, $(9, 5)$, and $(2, 4)$ form the vertices of a parallelogram.

Solution

By Distance Formula

$d_1 = \sqrt{(8-1)^2 + (2-1)^2} = \sqrt{50}$

$d_2 = \sqrt{(2-1)^2 + (4-1)^2} = \sqrt{10}$

$d_3 = \sqrt{(9-2)^2 + (5-4)^2} = \sqrt{50}$

$d_4 = \sqrt{(9-8)^2 + (5-2)^2} = \sqrt{10}$

$d_1 = d_3$ and $d_2 = d_4$

OR

by slopes

$(1, 1)$ and $(8, 2) \Rightarrow m_1 = \frac{1}{7}$

$(8, 2)$ and $(9, 5) \Rightarrow m_2 = 3$

$(9, 5)$ and $(2, 4) \Rightarrow m_3 = \frac{1}{7}$

$(2, 4)$ and $(1, 1) \Rightarrow m_4 = 3$

$m_1 = m_3$ and $m_2 = m_4$

15. Find y so that the distance between the points $(6,\ y)$ and $(-6,\ -2)$ is 13.

Solution

$\sqrt{(-6-6)^2 + (-2-y)^2} = 13$

$144 + 4 + 4y + y^2 = 169$

$y^2 + 4y - 21 = 0$

$(y+7)(y-3) = 0$

$y = -7$ or $y = 3$

19. Find the x- and y-intercepts of the graph of $y = (x+3)(x-2)$.

Solution

x-intercepts: $0 = (x+3)(x-2)$

$x = -3$ or $x = 2$

$(-3, 0)$ or $(2, 0)$

y-intercept: $y = (0+3)(0-2)$

$y = -6$

$(0, -6)$

23. Check the following for symmetry with respect to both axes and the origin.

$$y = \frac{x}{x^2 - 2}$$

Solution

Origin Symmetry

$$-y = \frac{-x}{(-x)^2 - 2}$$

$$-y = -\frac{x}{x^2 - 2}$$

$$y = \frac{x}{x^2 - 2}$$

27. Sketch the graph of $y = x^2 + 3$.
Identify any intercepts and test for symmetry.

Solution

y-intercept (0,3)

y-axis symmetry

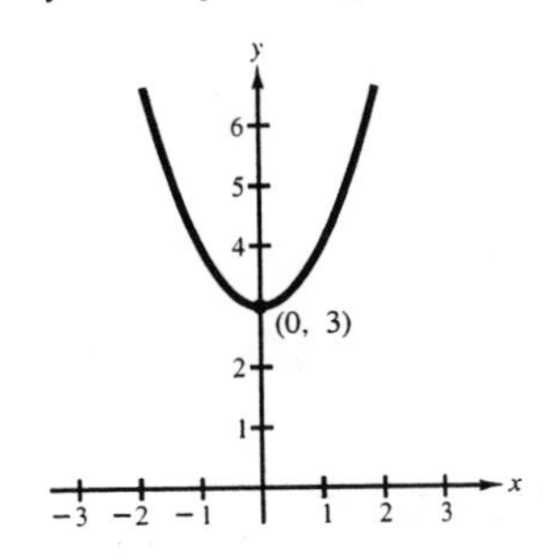

31. Sketch the graph of $y = x^3 + 1$.
Identify any intercepts and test for symmetry.

Solution

x-intercept: $(-1, 0)$

y-intercept: $(0, 1)$

Symmetry: None

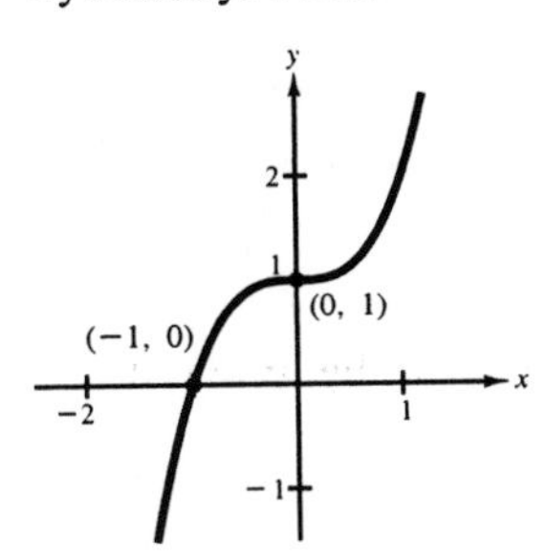

35. Find the standard form of the circle with endpoints of the diameter, $(-1, \ -2)$ and $(3, 1)$.

Solution

Center: $\left(\frac{-1+3}{2}, \frac{-2+1}{2}\right) = \left(1, -\frac{1}{2}\right)$

Radius: $r = \frac{1}{2}\sqrt{(3+1)^2 + (1+2)^2} = \frac{1}{2}\sqrt{16+9} = \frac{5}{2}$

$$(x-1)^2 + \left(y + \frac{1}{2}\right)^2 = \left(\frac{5}{2}\right)^2 = \frac{25}{4}$$

37. Write the equation of the circle $x^2 + y^2 - 6x + 4y - 3 = 0$ in standard form and sketch its graph.

Solution

$$x^2 + y^2 - 6x + 4y - 3 = 0$$

$$(x^2 - 6x + 9) + (y^2 + 4y + 4) = 3 + 9 + 4$$

$$(x-3)^2 + (y+2)^2 = 16$$

Center: $(3, -2)$

Radius: 4

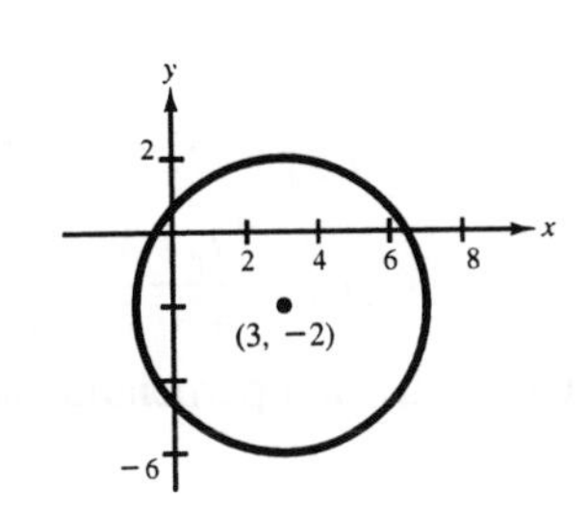

41. Use a graphing utility to match $y = \sqrt{x}$ with its graph. [The graphs are labeled (a), (b), (c), (d), (e), and (f) in the textbook.]

Solution

$y = \sqrt{x}$ Matches (e)

47. Use a graphing utility to sketch the graph of $y = |x + 4|$. Use a standard setting on each graph.

Solution

$y = |x + 4|$

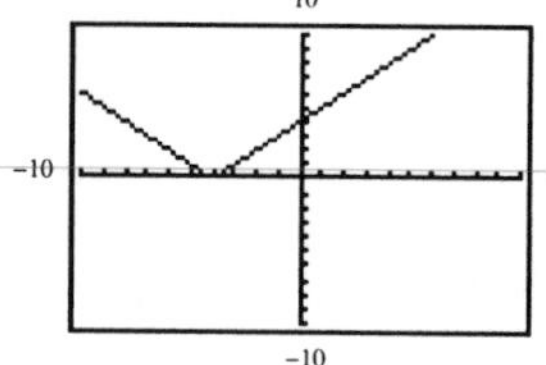

49. Find the constant C such that (2, 12) is a solution point of $y = Cx^2$.

Solution

$$y = Cx^2$$

$$12 = C(2)^2$$

$$3 = C$$

51. Use a graphing utility to find the number of x-intercepts of
$y = \frac{1}{3}(3x^2 - 27x + 55)$.

Solution

$y = \frac{1}{3}(3x^2 - 27x + 55)$

Two x-intercepts

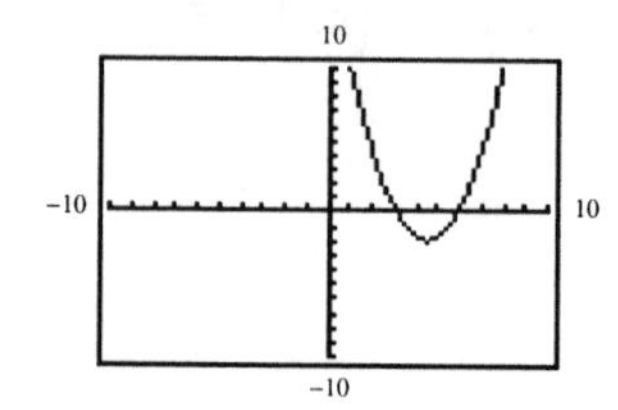

55. Plot the points (1, 5) and (−3, −6) and find the slope of the line passing through the points.

Solution

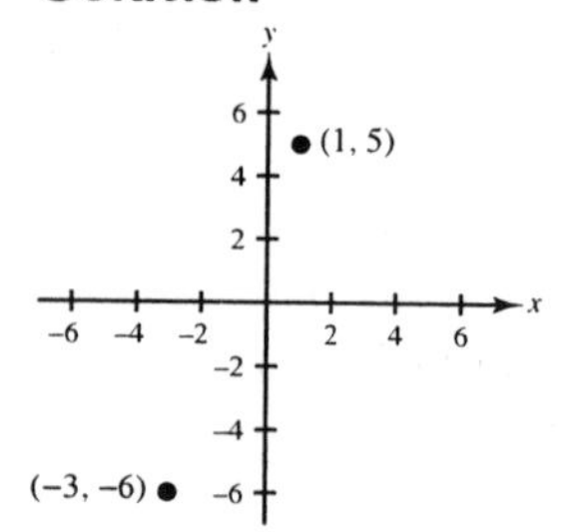

$$m = \frac{-6 - 5}{-3 - 1} = \frac{-11}{-4} = \frac{11}{4}$$

59. Determine whether the lines L_1 and L_2 passing through the following pairs of points are parallel, perpendicular, or neither.

$L_1 : (3,\ 6), (-1,\ -5)$

$L_2 : (-2,\ 3), (4,\ 7)$

Solution

$L_1 : (3, 6), (-1, -5)$ $\qquad$ $L_2 : (-2, 3), (4, 7)$

$$m_1 = \frac{-5 - 6}{-1 - 3} = \frac{11}{4} \qquad m_2 = \frac{7 - 3}{4 + 2} = \frac{4}{6} = \frac{2}{3}$$

The lines are neither parallel nor perpendicular.

63. Find an equation of the line passing through the points (3, 7) and (2, −1) and sketch the line.

Solution

$$y - 7 = \frac{-1 - 7}{2 - 3}(x - 3)$$

$$y - 7 = 8(x - 3)$$

$$y = 8x - 17$$

$$8x - y - 17 = 0$$

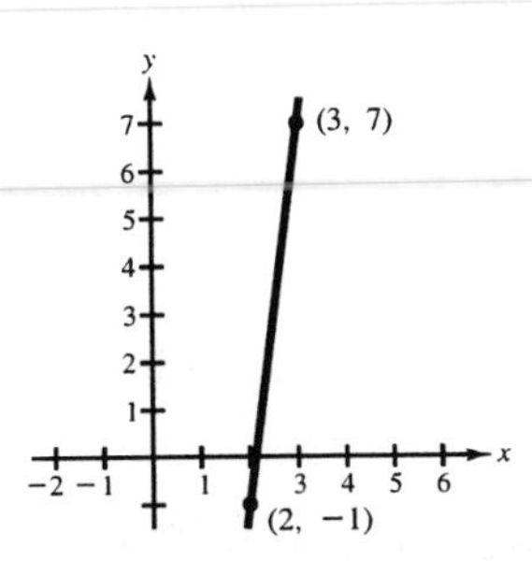

67. Find an equation of the line that passes through the point (0, −5) and has the slope $m = \frac{3}{2}$. Sketch the graph of the line.

Solution

$$y + 5 = \tfrac{3}{2}(x - 0)$$

$$y = \tfrac{3}{2}x - 5$$

$$3x - 2y - 10 = 0$$

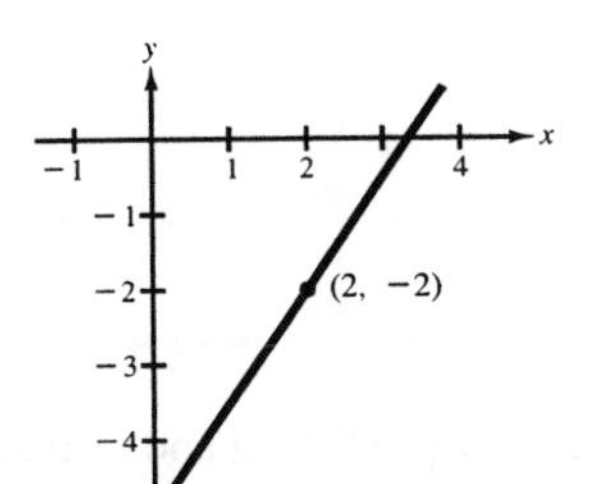

71. Write an equation of the line through the point (3, −2) (a) parallel and (b) perpendicular to $5x - 4y = 8$.

Solution

$$5x - 4y = 8$$

$$y = \tfrac{5}{4}x - 2$$

$$m_1 = \tfrac{5}{4}$$

(a) $m_2 = \frac{5}{4}$, (3, −2)

$$y + 2 = \tfrac{5}{4}(x - 3)$$

$$4y + 8 = 5x - 15$$

$$0 = 5x - 4y - 23$$

(b) $m_2 = -\frac{4}{5}$, (3, −2)

$$y + 2 = -\tfrac{4}{5}(x - 3)$$

$$5y + 10 = -4x + 12$$

$$4x + 5y - 2 = 0$$

73. *Fourth Quarter Sales* During the second and third quarters of the year, a business had sales of $160,000 and $185,000, respectively. If the growth of sales follows a linear pattern, what will sales be during the fourth quarter?

Solution

(2, 160,000), (3, 185,000)

$$y - 160,000 = \frac{185,000 - 160,000}{3 - 2}(x - 2)$$

$$y - 160,000 = 25,000(x - 2)$$

$$y = 25,000x + 110,000$$

When $x = 4$, we have $y = 25,000(4) + 110,000 = \$210,000$

77. *Direct Variation* Assume that y is proportional to x. Use $x = 10$ and $y = 3,480$ to find a linear model that relates x and y.

Solution

$$y = kx$$

$$3,480 = k(10)$$

$$348 = k$$

$$y = 348x$$

81. *Grams and Ounces* Suppose you are buying cherry pie filling and notice that 21 ounces of filling is the same as 595 grams. Use this information to find a linear model that relates ounces to grams. Then, use the model to complete the table in the textbook.

Solution

$$y = kx$$

$$595 = k(21)$$

$$\tfrac{85}{3} = k$$

$$y = \tfrac{85}{3}x$$

Ounces	6	8	12	16	24
Grams	170	226.7	340	453.3	680

87. *Vehicle Sales* During November 1–10, 1989, the 10 major United States auto makers sold cars and trucks at an average daily rate of 26,194. During the same period in 1990, they sold cars and trucks at an average daily rate of 26,817. (*Source: Dallas Times Herald,* November 15, 1990.) Write a linear equation for the average daily sales rate of vehicles during November 1–10. (Let $t = 0$ represent 1989.) Use the result to estimate the average daily sales for November 1–10,1991.

Solution

Let $t = 0$ represent 1989.

(0, 26,194), (1, 26,817)

$$y - 26,194 = \frac{26,817 - 26,194}{1 - 0}(t - 0)$$

$$y = 623t + 26,194$$

For 1991 we let $t = 2$, and then we have $y = 623(2) + 26,194 = 27,440$ cars and trucks per day.

Test for Chapter 2

1. For the points $(-3,\ 2)$ and $(5,\ -2)$, find the distance between the two points, and the midpoint of the line segment joining the points.

Solution

$$d = \sqrt{(5-(-3))^2 + (-2-2)^2} = \sqrt{(8)^2 + (-4)^2} = \sqrt{80} = 4\sqrt{5} \approx 8.94$$

$$\text{Midpoint: } \left(\frac{-3+5}{2}, \frac{2+(-2)}{2}\right) = \left(\frac{2}{2}, \frac{0}{2}\right) = (1, 0)$$

2. For the points $(3.25,\ 7.05)$ and $(-2.37,\ 1.62)$, find the distance between the two points, and the midpoint of the line segment joining the points.

Solution

$$d = \sqrt{(-2.37-3.25)^2 + (1.62-7.05)^2} = \sqrt{(-5.62)^2 + (-5.43)^2}$$

$$= \sqrt{61.0693} \approx 7.81$$

$$\text{Midpoint: } \left(\frac{3.25+(-2.37)}{2}, \frac{7.05+1.62}{2}\right) = \left(\frac{0.88}{2}, \frac{8.67}{2}\right) \approx (0.44, 4.34)$$

3. Find the intercepts of the graph of $y = (x+5)(x-3)$.

Solution

x-intercepts:

$0 = (x+5)(x-3) \Rightarrow x = -5, 3$

$(-5, 0), (3, 0)$

y-intercept: $y = (0+5)(0-3) = -15$

$(0, -15)$

4. Find the intercepts of the graph of $y = x\sqrt{x-2}$.

Solution

Domain: $x \geq 2$

x-intercept: $0 = x\sqrt{x-2} \Rightarrow x = 2$

$(2, 0)$

y-intercept: $y = 0\sqrt{0-2}$ Not defined.

5. Check $x^2 + y^2 = 16$ for symmetry.

Solution

$x^2 + (-y)^2 = 16 \Rightarrow x^2 + y^2 = 16$ x-axis symmetry

$(-x)^2 + y^2 = 16 \Rightarrow x^2 + y^2 = 16$ y-axis symmetry

$(-x)^2 + (-y)^2 = 16 \Rightarrow x^2 + y^2 = 16$ Origin symmetry

6. Check $y = x/(x^2-4)$ for symmetry.

Solution

$(-y) = \dfrac{x}{x^2-4} \Rightarrow y = -\dfrac{x}{x^2-4}$ No x-axis symmetry

$y = \dfrac{(-x)}{(-x)^2-4} \Rightarrow y = -\dfrac{x}{x^2-4}$ No y-axis symmetry

$(-y) = \dfrac{(-x)}{(-x)^2-4} \Rightarrow y = \dfrac{x}{x^2-4}$ Origin symmetry

7. Find an equation of the line passing through (3, 4) and (−2, 4).

Solution

$$m = \frac{4-4}{-2-3} = \frac{0}{-5} = 0 \Rightarrow \text{line is horizontal.}$$

$y = 4$ or $y - 4 = 0$

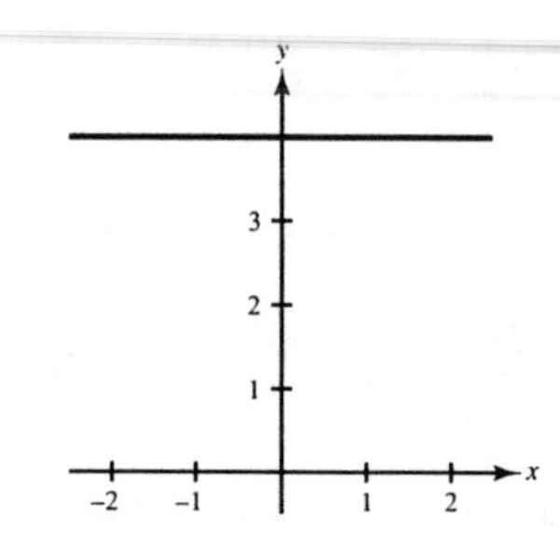

8. Find an equation of the line through (4, −3) with slope $\frac{3}{4}$.

Solution

$m = \frac{3}{4}$ and $(4, -3)$

$$y - (-3) = \tfrac{3}{4}(x - 4)$$

$$y + 3 = \tfrac{3}{4}x - 3$$

$$4y + 12 = 3x - 12$$

$$0 = 3x - 4y - 24$$

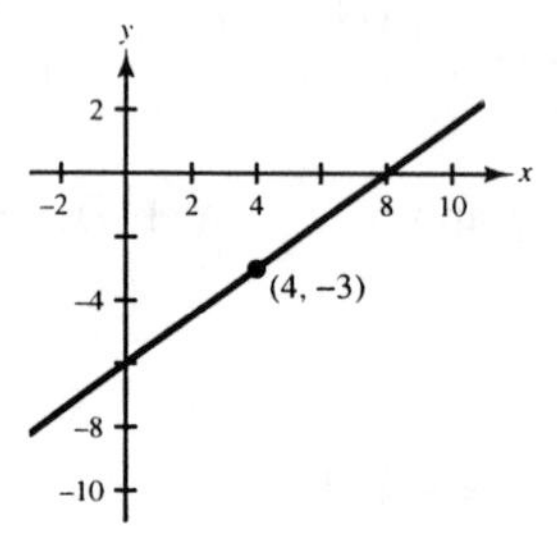

9. Find an equation of the line passing through $(-1,\ 5)$ and $(2,\ 1)$.

Solution

$$m = \frac{1-5}{2-(-1)} = \frac{-4}{3}$$

$$y - 1 = -\frac{4}{3}(x-2)$$

$$y - 1 = -\frac{4}{3}x + \frac{8}{3}$$

$$3y - 3 = -4x + 8$$

$$4x + 3y - 11 = 0$$

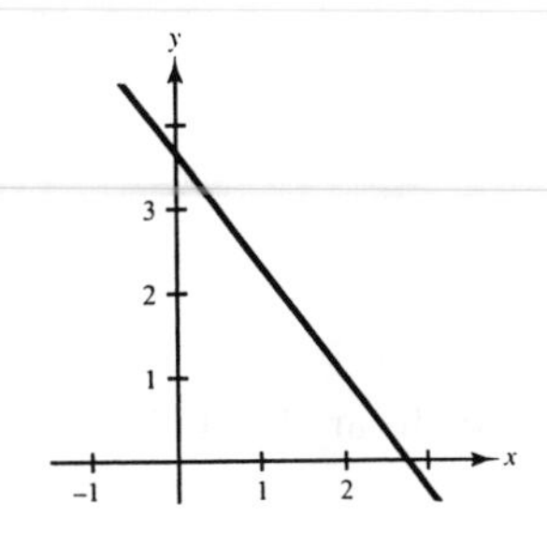

10. Find an equation of the vertical line through $(2,\ -7)$.

Solution

Since the line is vertical, $x = 2$ or $x - 2 = 0$.

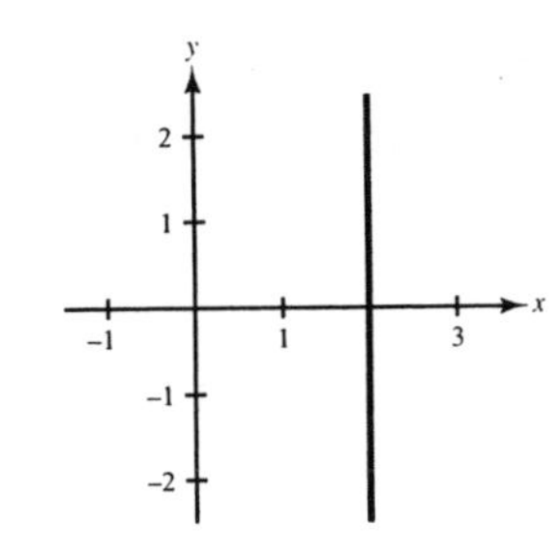

11. Find an equation of the line with intercepts $(3, 0)$ and $(0,\ -4)$.

Solution

x-intercept: $(3, 0)$

y-intercept: $(0, -4)$ $\Rightarrow$

$$y = mx - 4$$

$$0 = m(3) - 4 \Rightarrow m = \tfrac{4}{3}$$

$$y = \tfrac{4}{3}x - 4$$

$$3y = 4x - 12$$

$$4x - 3y - 12 = 0$$

12. Match $y = \sqrt{x}$ to its graph. [The graphs are labeled (a), (b), (c), and (d) in the textbook.]

Solution

Domain: $[0,\ \infty)$

Range: $[0\ \infty)$

Matches (a)

13. Match $y = -\sqrt{x-2}$ to its graph. [The graphs are labeled (a), (b), (c), and (d) in the textbook.]

Solution

Domain: $[2, \infty)$

Range: $(-\infty, 0]$

Matches (b)

14. Match $y = |x-2|$ to its graph. [The graphs are labeled (a), (b), (c), and (d) in the textbook.]

Solution

Domain: $(-\infty, \infty)$

Range: $[0, \infty)$

Matches (c)

15. Match $y = -x^3$ to its graph. [The graphs are labeled (a), (b), (c), and (d) in the textbook.]

Solution

Domain: $(-\infty, \infty)$

Range: $(-\infty, \infty)$

Matches (d)

16. Write $x^2 + y^2 - 4x - 2y - 4 = 0$ in standard form and sketch its graph.

Solution

$$x^2 + y^2 - 4x - 2y - 4 = 0$$

$$x^2 - 4x + \underline{4} + y^2 - 2y + \underline{1} = 4 + 4 + 1$$

$$(x-2)^2 + (y-1)^2 = 9$$

Center: $(2, 1)$

Radius: $r = 3$

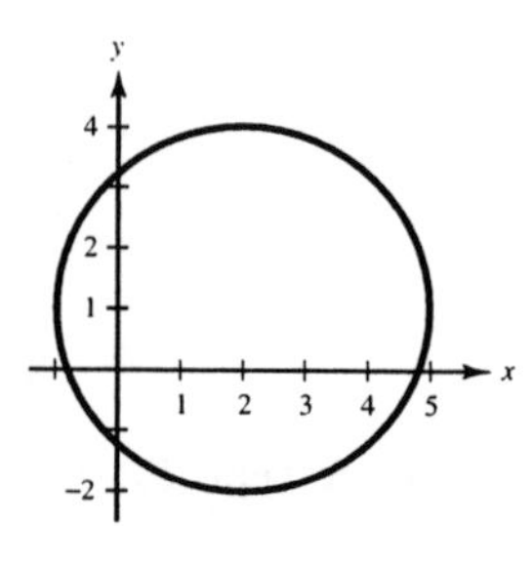

17. During the first and second quarters of the year, a business had sales of \$150,000 and \$185,000, respectively. If the growth of sales follows a linear pattern, what will sales be during the third quarter?

Solution

$(1, 150,000), (2, 185,000)$

$$m = \frac{185,000 - 150,000}{2-1} = 35,000$$

$$y - 150,000 = 35,000(x-1)$$

$$y - 150,000 = 35,000x - 35,000$$

$$y = 35,000x + 115,000$$

When $x = 3$, $y = \$220,000$

18. An assembly line worker is paid \$8.75 per hour plus a piecework rate of \$0.55 per unit. Write a linear equation for the hourly wages W in terms of the number of units x produced per hour.

Solution

$W = 0.55x + 8.75$

19. A small business purchases a piece of equipment for \$25,000. After five years the equipment will be worth only\$5,000. Write a linear equation that gives the value V of the equipment during the five years.

Solution

$(0, 25{,}000), (5, 5000)$

$$m = \frac{5000 - 25,000}{5 - 0} = -4000$$

$V = 25{,}000 - 4000t$

20. A store is offering a 30% discount on all summer apparel. Write a linear equation giving the sale price S of an item in terms of its list price L.

Solution

$S = L - 0.30L$

$S = 0.70L$

Practice Test for Chapter 2

1. Find the distance between $(4, -1)$ and $(0, 3)$.

2. Find the midpoint of the line segment joining $(4, -1)$ and $(0, 3)$.

3. Find x so that the distance from the origin to $(x, -2)$ is 6.

4. Given $y = \dfrac{x-2}{x+3}$, find the intercepts.

5. Given $y = x\sqrt{9-x^2}$, find the intercepts.

6. Given $xy^2 = 6$, list all symmetries.

7. Given $y = \dfrac{x}{x^2+3}$, list all symmetries.

8. Graph $y = x^3 - 4x$.

9. Write the equation of the circle with center $(3, -2)$ and radius 4.

10. Find the center and radius of the circle $x^2 + y^2 - 6x + 2y + 6 = 0$.

11. Sketch the graph of $y = x^2 - 5$.

12. Sketch the graph of $y = |x + 3|$.

13. Sketch the graph of $y = \sqrt{4 - x}$.

14. Find the equation of the line through $(2, 4)$ and $(3, -1)$.

15. Find the equation of the line with slope $m = \frac{4}{3}$ and y-intercept $b = -3$.

16. Find the equation of the line through $(4, 1)$ and perpendicular to the line $2x + 3y = 0$.

17. Find the equation through $(6, -5)$ parallel to the line $x = -1$.

18. If it costs a company \$32 to produce 5 units of a product and \$44 to produce 9 units, how much does it cost to produce 20 units? (Assume that the cost function is linear.)

19. Find the equation: y varies directly as x and $y = 30$ when $x = 5$.

20. Lois receives a monthly salary of \$2,100 plus a commission of 8% of her sales. Write a linear equation for her monthly wages W in terms of her monthly sales S.

Cumulative Test for Chapters P–2

1. Simplify $5\sqrt{x} - 11\sqrt{x}$.

Solution

$5\sqrt{x} - 11\sqrt{x} = (5-11)\sqrt{x} = -6\sqrt{x}$

2. Simplify $7^{1/3} \cdot 7^{5/3}$.

Solution

$$7^{1/3} \cdot 7^{5/3} = 7^{1/3+5/3}$$
$$= 7^{6/3} = 7^2 = 49$$

3. Simplify $\sqrt{24x^5}$.

Solution

$\sqrt{24x^5} = \sqrt{4x^4 6x} = 2x^2\sqrt{6x}$

4. Simplify $\dfrac{1}{3-\sqrt{5}}$.

Solution

$$\frac{1}{3-\sqrt{5}} = \frac{1}{3-\sqrt{5}} \cdot \frac{3+\sqrt{5}}{3+\sqrt{5}} = \frac{3+\sqrt{5}}{9-5} = \frac{3+\sqrt{5}}{4}$$

5. Solve $4y^2 - 64 = 0$.

Solution

$$4y^2 - 64 = 0$$
$$4(y^2 - 16) = 0$$
$$4(y+4)(y-4) = 0$$
$$y = \pm 4$$

6. Solve $2(x+4) - 5 = 11 + 5(2-x)$.

Solution

$$2(x+4) - 5 = 11 + 5(2-x)$$
$$2x + 8 - 5 = 11 + 10 - 5x$$
$$2x + 3 = -5x + 21$$
$$7x = 18$$
$$x = \tfrac{18}{7}$$

7. Solve $x^4 - 17x^2 + 16 = 0$.

Solution

$$x^4 - 17x^2 + 16 = 0$$
$$(x^2 - 1)(x^2 - 16) = 0$$
$$(x+1)(x-1)(x+4)(x-4) = 0$$
$$x = \pm 1,\ \pm 4$$

8. Solve $|2x - 3| = 5$.

Solution

$$|2x - 3| = 5$$
$$2x - 3 = -5 \quad \text{or} \quad 2x - 3 = 5$$
$$2x = -2 \quad \text{or} \quad 2x = 8$$
$$x = -1 \quad \text{or} \quad x = 4$$

9. Solve $\sqrt{y+3} + y = 3$.

Solution

$$\sqrt{y+3} + y = 3$$
$$\sqrt{y+3} = 3 - y$$
$$y + 3 = 9 - 6y + y^2$$
$$0 = y^2 - 7y + 6$$
$$0 = (y-1)(y-6)$$
$$y = 1$$

(Note: $y = 6$ is extraneous.)

10. Solve $2x^2 + x = 5$.

Solution

$$2x^2 + x = 5$$
$$2x^2 + x - 5 = 0$$
$$x = \frac{-1 \pm \sqrt{(1)^2 - 4(2)(-5)}}{2(2)}$$
$$x = \frac{-1 \pm \sqrt{41}}{4}$$

11. You deposit \$5,000 in an account that pays 8.5%, compounded monthly. Find the balance after 10 years.

Solution

$$B = 5000\left(1 + \frac{0.085}{12}\right)^{(12)(10)} \approx \$11,663.24$$

12. You just bought a new camcorder. The manufacturer made 300,000 camcorders of your model and of these, 20,000 have a defective zoom lens. What is the probability that your new camcorder has a defective zoom lens? Explain what this tells you about the probability that your new camcorder has a properly working zoom lens.

Solution

$p = \frac{20,000}{300,000} = \frac{1}{15} \approx 0.0667$

The probability of the zoom lens working properly is $\frac{14}{15} \approx 0.9333$.

13. Solve the following inequality.

$$-3 \le \frac{2x-5}{4} \le 3$$

Solution

$$-3 \le \frac{2x-5}{4} \le 3$$

$$-12 \le 2x - 5 \le 12$$

$$-7 \le 2x \le 17$$

$$-\frac{7}{2} \le x \le \frac{17}{2}$$

14. Solve $(x+4)^2 \le 4$.

Solution

$$(x+4)^2 \le 4$$

$$x^2 + 8x + 16 - 4 \le 0$$

$$x^2 + 8x + 12 \le 0$$

$$(x+2)(x+6) \le 0$$

Critical Numbers: $-2, -6$

Test Intervals:

$(-\infty, -6), (-6, -2), (-2, \infty)$

Answer: $-6 \le x \le -2$

15. Find an equation of the line passing through $(3, -2)$ and $(-1, 5)$.

Solution

$$m = \frac{5-(-2)}{-1-3} = \frac{7}{-4}$$

$$y - (-2) = -\frac{7}{4}x + \frac{21}{4}$$

$$y + 2 = -\frac{7}{4}x + \frac{21}{4}$$

$$4y + 8 = -7x + 21$$

$$7x + 4y - 13 = 0$$

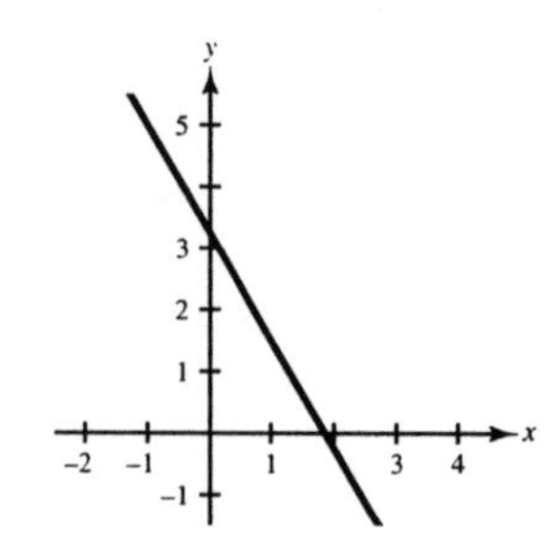

16. Find an equation of the line with slope $\frac{2}{3}$ and passing through $(2, -1)$.

Solution

$(2, -1), m = \frac{2}{3}$

$$y - (-1) = \tfrac{2}{3}(x - 2)$$

$$y + 1 = \tfrac{2}{3}x - \tfrac{4}{3}$$

$$3y + 3 = 2x - 4$$

$$2x - 3y - 7 = 0$$

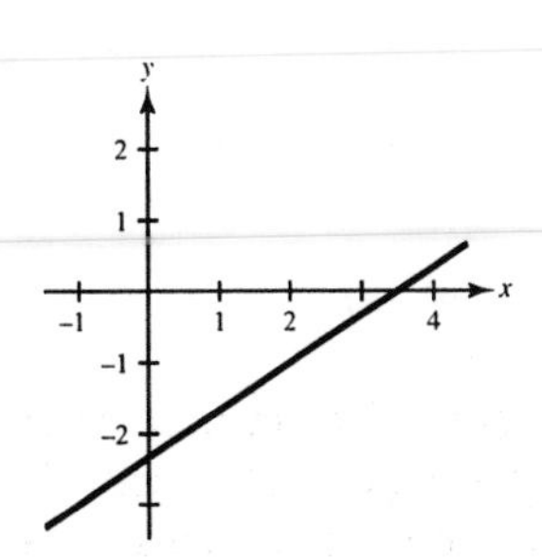

17. Find an equation of the line with zero slope and passing through $(-1, -3)$.

Solution

$(-1, -3), m = 0 \Rightarrow$ Line is horizontal.

$y = -3$ or $y + 3 = 0$

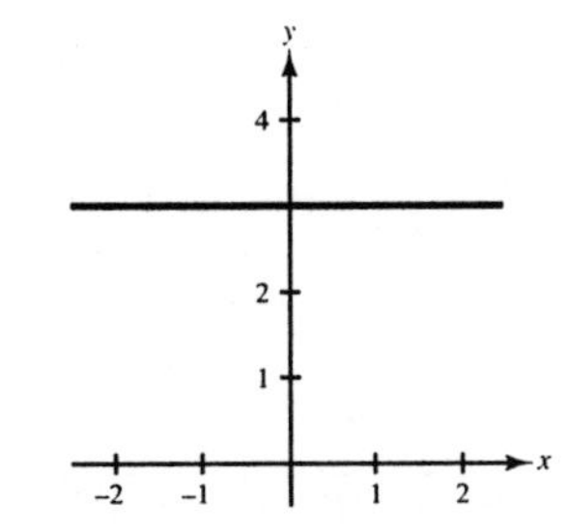

18. Write $x^2 + y^2 - 6x + 4y - 3 = 0$ in standard form and sketch its graph.

Solution

$$x^2 + y^2 - 6 + 4y - 3 = 0$$

$$x^2 - 6x + \underline{9} + y^2 + 4y + \underline{4} = 3 + 9 + 4$$

$$(x - 3)^2 + (y + 2)^2 = 16$$

Center: $(3, -2)$

Radius: $r = 4$

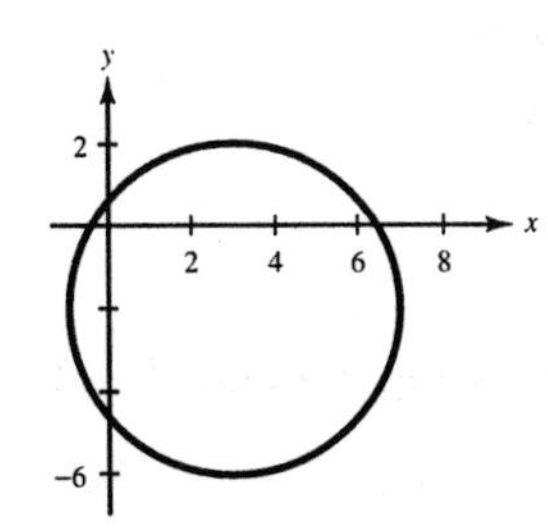

19. During the second and third quarters of the year, a business had sales of $210,000 and $230,000, respectively. If the growth of sales follows a linear pattern, what will sales be during the fourth quarter?

Solution

(2, 210,000), (3, 230,000)

$$m = \frac{230,000 - 210,000}{3 - 2} = 20,000$$

$$y - 210,000 = 20,000(x - 2)$$

$$y - 210,000 = 20,000x - 40,000$$

$$y = 20,000x + 170,000$$

When $x = 4, \; y = 250,000$

20. The revenue and cost equations for a product are $R = x(100 - 0.001x)$ and $C = 20x + 30,000$ where R and C are measured in dollars and x represents the number of units sold. How many units must be sold for revenue to equal cost?

Solution

$$R = C$$

$$x(100 - 0.001x) = 20x + 30,000$$

$$100x - 0.001x^2 = 20x + 30,000$$

$$0 = 0.001x^2 - 80x + 30,000$$

$$x = \frac{-(-80) \pm \sqrt{(-80)^2 - 4(0.001)(30,000)}}{2(0.001)} = \frac{80 \pm \sqrt{6,280}}{0.002}$$

$$x = \frac{80 - \sqrt{6,280}}{0.002} \approx 377 \text{ units}$$

or

$$x = \frac{80 + \sqrt{6,280}}{0.002} \approx 79,623 \text{ units}$$

CHAPTER THREE
Functions and Graphs

3.1 Functions

- Know the definition of a function.
  (a) Discrete mathematics: A function f from a set A to a set B is a rule of correspondence that assigns to each element in set A exactly one element in set B.
  (b) Algebra: A function, $y = f(x)$, is a relationship between two variables such that to each value of the independent variable there corresponds exactly one value of the dependent variable.
- Given a rule of correspondence or an equation, you should be able to determine if it represents a function.
- Given a function, you should be able to do the following:
  (a) Evaluate it at specific values.
  (b) Find the domain.

SOLUTIONS TO SELECTED EXERCISES

3. Evaluate the function $g(x) = \dfrac{1}{x^2 - 2x}$.

(a) $g(1) = \dfrac{1}{(\quad)^2 - 2(\quad)}$ (b) $g(-3) = \dfrac{1}{(\quad)^2 - 2(\quad)}$

(c) $g(t) = \dfrac{1}{(\quad)^2 - 2(\quad)}$ (d) $g(t+1) = \dfrac{1}{(\quad)^2 - 2(\quad)}$

Solution

(a) $g(1) = \dfrac{1}{(1)^2 - 2(1)} = -1$ (b) $g(-3) = \dfrac{1}{(-3)^2 - 2(-3)} = \dfrac{1}{15}$

(c) $g(t) = \dfrac{1}{t^2 - 2t}$ (d) $g(t+1) = \dfrac{1}{(t+1)^2 - 2(t+1)} = \dfrac{1}{t^2 + 2t + 1 - 2t - 2} = \dfrac{1}{t^2 - 1}$

7. Evaluate the function $h(t) = t^2 - 2t$ and simplify the results.

(a) $h(2)$ (b) $h(-1)$ (c) $h(x+2)$ (d) $h(1.5)$

Solution

(a) $h(2) = 2^2 - 2(2) = 0$ (b) $h(-1) = (-1)^2 - 2(-1) = 3$

(c) $h(x+2) = (x+2)^2 - 2(x+2) = x^2 + 2x$ (d) $h(1.5) = (1.5)^2 - 2(1.5) = -0.75$

13. Evaluate the following function and simplify the results.

$$f(x) = \frac{|x|}{x}$$

(a) $f(2)$ (b) $f(-2)$ (c) $f(x^2)$ (d) $f(x-1)$

Solution

(a) $f(2) = \dfrac{|2|}{2} = 1$

(b) $f(-2) = \dfrac{|-2|}{-2} = -1$

(c) $f(x^2) = \dfrac{|x^2|}{x^2} = 1$

(d) $f(x-1) = \dfrac{|x-1|}{x-1}$

15. Evaluate the following function and simplify the results.

$$f(x) = \begin{cases} 2x+1, & x < 0 \\ 2x+2, & x \geq 0 \end{cases}$$

(a) $f(-1)$ (b) $f(0)$ (c) $f(1)$ (d) $f(2)$

Solution

(a) $f(-1) = 2(-1) + 1 = -1$

(b) $f(0) = 2(0) + 2 = 2$

(c) $f(1) = 2(1) + 2 = 4$

(d) $f(2) = 2(2) + 2 = 6$

19. Find all real values of x such that $f(x) = 0$ for the equation $f(x) = x^2 - 9$.

Solution

$$x^2 - 9 = 0$$
$$x^2 = 9$$
$$x = \pm 3$$

21. Find all real values of x such that $f(x) = 0$ for the following equation.

$$f(x) = \frac{3}{x-1} + \frac{4}{x-2}$$

Solution

$$\frac{3}{x-1} + \frac{4}{x-2} = 0$$
$$3(x-2) + 4(x-1) = 0$$
$$7x - 10 = 0$$
$$x = \frac{10}{7}$$

25. Find the domain of $h(t) = 4/t$.

Solution

All real numbers except $t = 0$

29. Find the domain of $f(x) = \sqrt[4]{1 - x^2}$.

Solution

$$1 - x^2 \geq 0$$
$$x^2 \leq 1$$
$$-1 \leq x \leq 1$$

33. Determine if y is a function of x in the equation $x^2 + y^2 = 4$.

Solution

$x^2 + y^2 = 4 \quad \Rightarrow \quad y = \pm\sqrt{4 - x^2}$

Thus, y *is not* a function of x.

37. Determine if y is a function of x in the equation $2x + 3y = 4$.

Solution

$2x + 3y = 4 \quad \Rightarrow \quad y = \frac{1}{3}(4 - 2x)$

Thus, y *is* a function of x.

41. Determine if y is a function of x in the equation $x^2y - x^2 + 4y = 0$.

Solution

$$x^2y - x^2 + 4y = 0$$
$$y(x^2 + 4) = x^2$$
$$y = \frac{x^2}{x^2 + 4}$$

Thus, y *is* a function of x.

45. Determine if the set of ordered pairs $\{(0, 0), (1, 0), (2, 0), (3, 0)\}$ represents a function from A to B when $A = \{0, 1, 2, 3\}$ and $B = \{-2, -1, 0, 1, 2\}$. Give a reason for your answer.

Solution

Since each element of A is matched with exactly one element of B, $\{(0, 0), (1, 0), (2, 0), (3, 0)\}$ represents a function from A to B.

49. Determine if the set of orders pairs $\{(1, a), (0, a), (2, c), (3, b)\}$ represents a function from A to B when $A = \{a, b, c\}$ and $B = \{0, 1, 2, 3\}$. Give a reason for your answer.

Solution

Since $0, 1, 2, 3, \notin A$ and $a, b, c, \notin B$, $\{(1, a), (0, a), (2, c), (3, b)\}$ does not represent a function from A to B. (Note: It does define a function from B to A.)

53. Determine if the set of ordered pairs shown in the textbook represents a function from A to B when $A = \{a, b, c\}$ and $B = \{1, 2, 3, 4\}$. Give a reason for your answer.

Solution

Since $b \in A$ is not matched with an element of B, $\{(a, 1), (c, 3)\}$ does not represent a function from A to B.

57. Assume that the domain of f is the set $A = \{-2, -1, 0, 1, 2\}$. Determine the set of order pairs representing the function $f(x) = \sqrt{x + 2}$.

Solution

$(-2, 0), (-1, 1), (0, \sqrt{2}), (1, \sqrt{3}), (2, 2)$

61. *Height of a Balloon* A balloon carrying a transmitter ascends vertically from a point 2,000 feet from the receiving station (see figure in the textbook). Let d be the distance between the balloon and the receiving station. Express the height of the balloon as a function of d. What is the domain of this function?

Solution

$$h^2 + 2000^2 = d^2$$
$$h^2 = d^2 - 2000^2$$
$$h = \sqrt{d^2 - 2000^2}$$

Domain: $d^2 - 2000^2 \geq 0$

$$d^2 \geq 2000^2$$
$$d \geq 2000$$

65. *Revenue, Cost and Profit* A company produces a product for which the variable cost is \$12.30 per unit and the fixed costs are \$98,000. The product sells for \$17.98. Let x be the number of units produced.

(a) Write the total cost C as a function of the number of units produced.

(b) Write the revenue R as a function of the number of units produced.

(c) Write the profit P as a funtion of the number of units produced.
(*Note*: $P = R - C$.)

Solution

(a) $C = 12.30x + 98{,}000$

(b) $R = 17.98x$

(c) $P = R - C = 17.98x - (98{,}000 + 12.30x) = 5.68x - 98{,}000$

69. *Charter Bus Fares* For groups of 80 or more people, a charter bus company determines the rate per person according to the formula

$$\text{Rate} = 8 - 0.05(n - 80), \quad n \geq 80$$

where the rate is given in dollars and n is the number of people.

(a) Express the total revenue R for the bus company as a function of n.

(b) Use the function from part (a) to complete the table in the textbook.

(c) Write a paragraph analyzing the data in the table.

Solution

(a) $R = n(\text{Rate}) = n[8.00 - 0.05(n - 80)], \quad n \geq 80$

$R = 12.00n - 0.05n^2, \quad n \geq 80$

(b)

n	90	100	110	120	130	140	150
R	\$675	\$700	\$715	\$720	\$715	\$700	\$675

(c) The maximum profit (\$720) occurs when 120 people charter the bus.

3.2 Graphs of Functions

- You should be able to determine the domain and range of a function from its graph.
- You should be able to use the vertical line test for functions.
- You should be able to determine when a function is constant, increasing, or decreasing.
- You should know that f is
  (a) Odd if $f(-x) = -f(x)$. The graph of an odd function has origin symmetry.
  (b) Even if $f(-x) = f(x)$. The graph of an even function has y-axis symmetry.
- Know what the graphs of the six common functions listed on page 249 look like.

SOLUTIONS TO SELECTED EXERCISES

3. Find the range of $f(x) = \sqrt{x^2 - 4}$. (See figure in the textbook.)

Solution

Range: $y \geq 0$

$[0, \infty)$

11. Use the vertical line test to determine whether y is a function of x in the equation $x - y^2 = 0$. (See figure in the textbook.)

Solution

Some vertical lines intersect the graph more than once, so y is **not** a function of x.

13. Use the vertical line test to determine whether y is a function of x in the equation $x^2 = xy - 1$. (See figure in the textbook.)

Solution

A vertical line intersects the graph just once, so y **is** a function of x.

17. Describe the increasing and decreasing behavior of the function $f(x) = x^3 - 3x^2$. (See figure in the textbook.)

Solution

Increasing on $(-\infty, 0)$ and $(2, \infty)$

Decreasing on $(0, 2)$

21. Describe the increasing and decreasing behavior of $f(x) = x\sqrt{x+3}$. (See figure in the textbook.)

Solution

Decreasing on $(-3, -2)$

Increasing on $(-2, \infty)$

23. Determine whether the function $f(x) = x^6 - 2x^2 + 3$ is even, odd, or neither.

Solution

$f(-x) = (-x)^6 - 2(-x)^2 + 3 = x^6 - 2x^2 + 3 = f(x)$

f is even.

27. Determine whether the function $f(t) = t^2 + 2t - 3$ is even, odd, or neither.

Solution

$f(-t) = (-t)^2 + 2(-t) - 3 = t^2 - 2t - 3 \neq f(t) \neq -f(t)$

f is neither even nor odd.

29. Sketch the graph of $f(x) = 3$ and determine whether the function is even, odd, or neither.

Solution

$f(x) = 3$, even

The graph has y-axis symmetry.

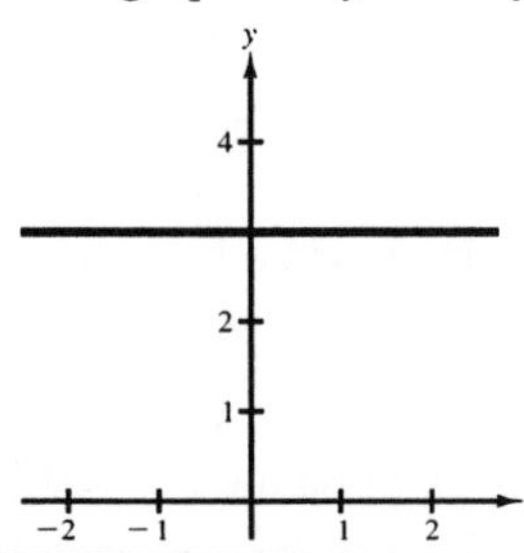

33. Sketch the graph of the following function and determine whether the function is even, odd, or neither.

$$g(s) = \frac{s^3}{4}$$

Solution

$g(s) = \frac{s^3}{4}$, odd

The graph has origin symmetry.

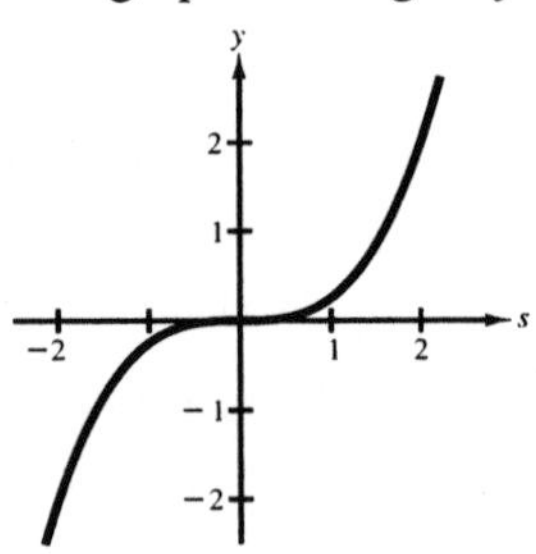

37. Sketch the graph of $g(t) = \sqrt[3]{t-1}$ and determine whether the function is even, odd, or neither.

Solution

$g(t) = \sqrt[3]{t-1}$, neither even nor odd

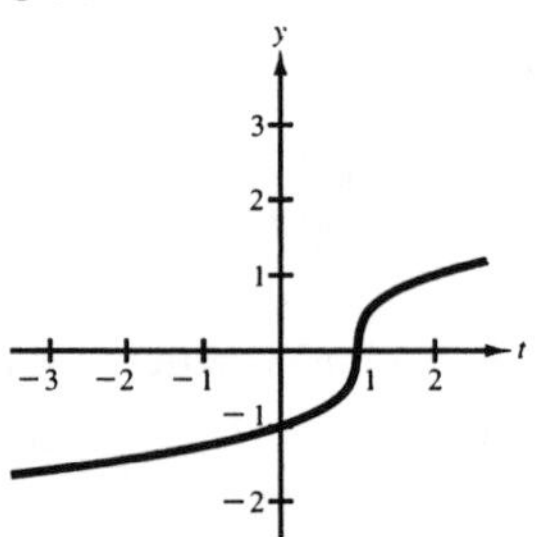

39. Sketch the graph of the following function and determine whether the function is even, odd, or neither.

$$f(x) = \begin{cases} x+3, & x \le 0 \\ 3, & 0 < x \le 2 \\ 2x-1, & x > 2 \end{cases}$$

Solution

Neither even nor odd

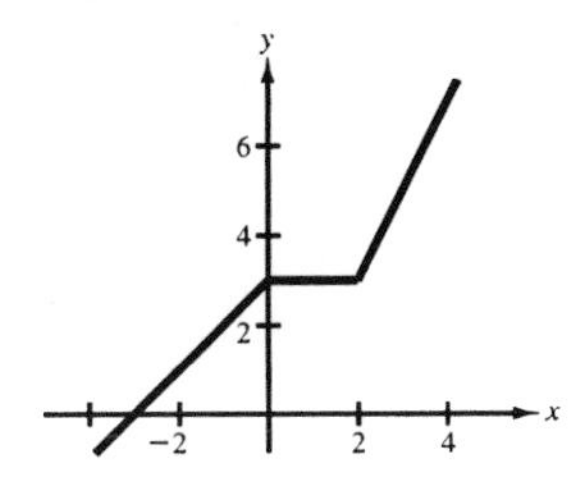

45. Sketch the graph of $f(x) = 1 - x^4$.

Solution

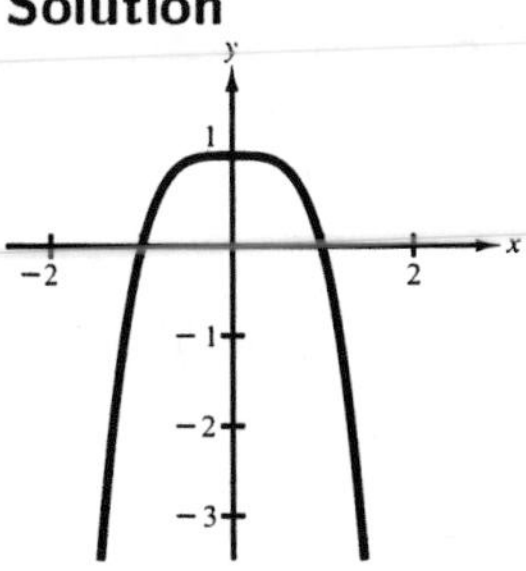

51. Sketch the graph of $f(x) = -[[x]]$.

Solution

Use the greatest integer function.

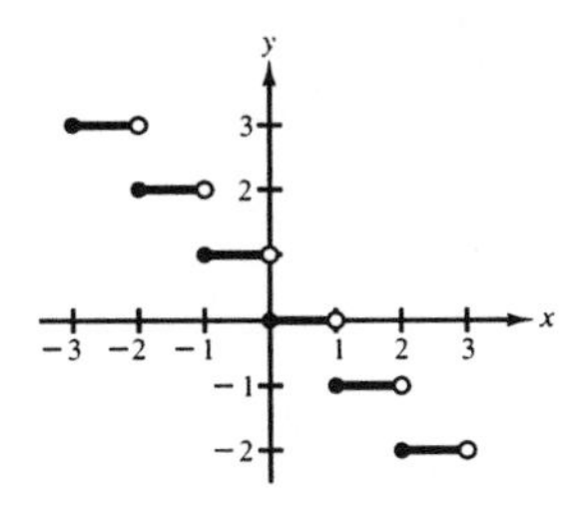

55. Use a graphing utility to sketch the graph of $f(x) = x^2 - 4x + 1$ and then estimate the open intervals on which the function is increasing or decreasing.

Solution

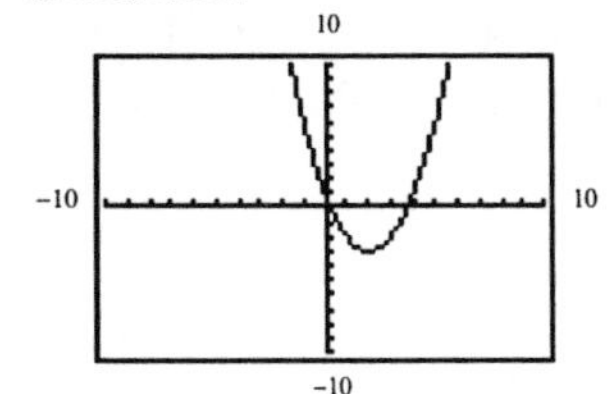

Decreasing on $(-\infty, 2)$; $x < 2$

Increasing on $(2, \infty)$; $x > 2$

59. *Population of Oklahoma* The population of the state of Oklahoma fell during the late part of the 1980's and then began to increase. A model that approximates the population of Oklahoma is

$$P = 7.8t^2 - 151t + 3878, \quad 6 \le t \le 13$$

where P represents the population in thousands and $t = 6$ represents 1986. Use the figure to estimate the years that the population was decreasing and the years that the population was increasing. (See figure in the textbook.) (*Source: U.S. Census Bureau.*)

Solution

From the graph in the textbook we see that the population was decreasing during 1986, 1987, 1988, and 1989 and increasing during 1990, 1991, 1992, and 1993.

63. *Maximum Profit* The marketing department of a company estimates that the demand for a product is given by $p = 120 - 0.0001x$, where p is the price per unit and x is the number of units. The cost of producing x units is given by $C = 450{,}000 + 50x$, and the profit for producing x units is given by $P = R - C = xp - C$. Sketch the graph of the profit function and estimate the number of units that would produce a maximum profit.

Solution

$P = R - C = xp - C = x(120 - 0.0001x) - (450{,}000 + 50x) = -0.0001x^2 + 70x - 450{,}000$

The profit is maximum when $x = 350{,}000$ units.

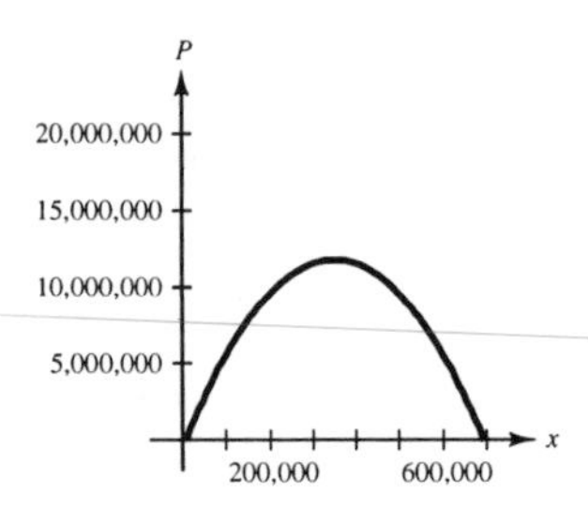

69. *Grade Level Salaries* The 1988 average salary for federal employees can be approximated by the model

$$S = \begin{cases} 6874.7 + 2162.1x, & x = 1, 2, \ldots, 12 \\ -66{,}957 + 8706.9x, & x = 13, 14, 15, 16 \end{cases}$$

where S is the salary in dollars and x represents the "GS" level. Sketch a bar graph that represents this function. (*Source: U.S. Office of Personnel Management.*)

Solution

x	1	2	3	4	5	6	7	8
S	\$9,036.80	\$11,198.90	\$13,361.00	\$15,523.10	\$17,685.20	\$19,847.30	\$22,009.40	\$24,171.50
x	9	10	11	12	13	14	15	16
S	\$26,333.60	\$28,495.70	\$30,657.80	\$32,819.90	\$46,232.70	\$54,939.60	\$63,646.50	\$72,353.40

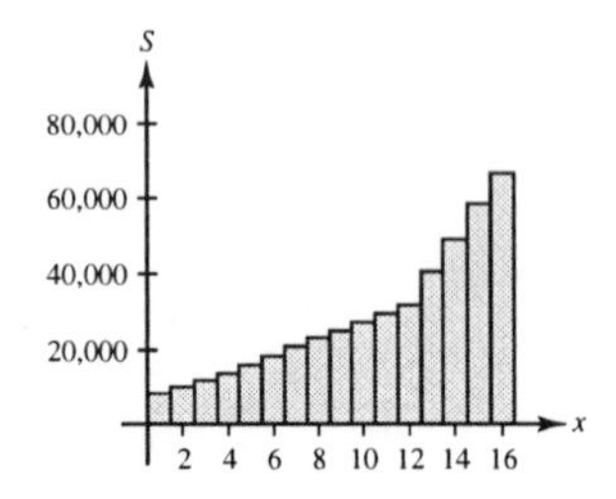

3.3 Translations and Combinations of Functions

- You should know the basic types of translations. Let $c > 0$.
  (a) Vertical shift c units upward: $f(x) + c$
  (b) Vertical shift c units downward: $f(x) - c$
  (c) Horizontal shift c units to the right: $f(x - c)$
  (d) Horizontal shift c units to the left: $f(x + c)$
  (e) Reflection in the x-axis: $-f(x)$
  (f) Reflection in the y-axis: $f(-x)$
  (g) Vertical stretch: $cf(x)$ where $c > 1$
  (h) Vertical shrink: $cf(x)$ where $0 < c < 1$
- Given two functions, f and g, you should be able to form the following functions (if defined):
  1. Sum: $(f + g)(x) = f(x) + g(x)$
  2. Difference: $(f - g)(x) = f(x) - g(x)$
  3. Product: $(fg)(x) = f(x)g(x)$
  4. Quotient: $(f/g)(x) = f(x)/g(x),\ g(x) \neq 0$
  5. Composition of f with g: $(f \circ g)(x) = f(g(x))$
  6. Composition of g with f: $(g \circ f)(x) = g(f(x))$

SOLUTIONS TO SELECTED EXERCISES

3. Use the graph of $f(x) = x^2$ to sketch the graph of the function $g(x) = (x + 3)^2$.

Solution

$g(x) = (x + 3)^2 = f(x + 3)$

Horizontal shift 3 units to the left.

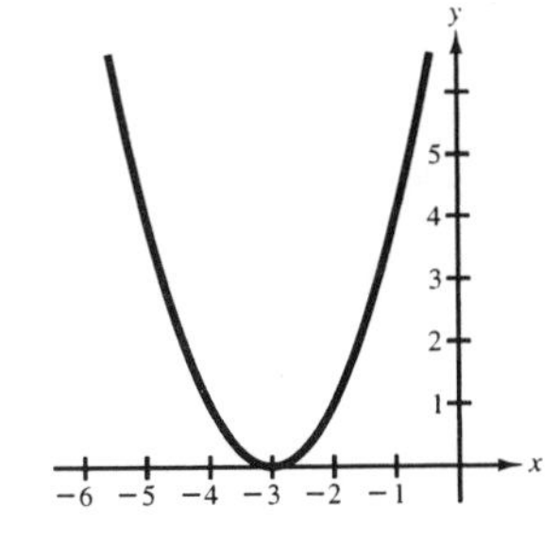

7. Use the graph of $f(x) = x^2$ to sketch the graph of the function $g(x) = -x^2 + 1$.

Solution

$g(x) = -x^2 + 1 = -f(x) + 1$

A reflection in the x-axis and a vertical shift 1 unit upward.

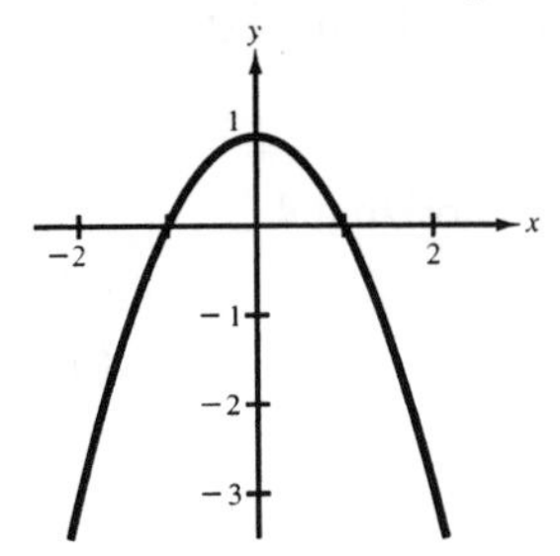

11. Use the graph of $f(x) = \sqrt{x}$ to sketch the graph of the function $y = \sqrt{x-2}$.

Solution

$y = \sqrt{x-2} = f(x-2)$

Horizontal shift two units to the right

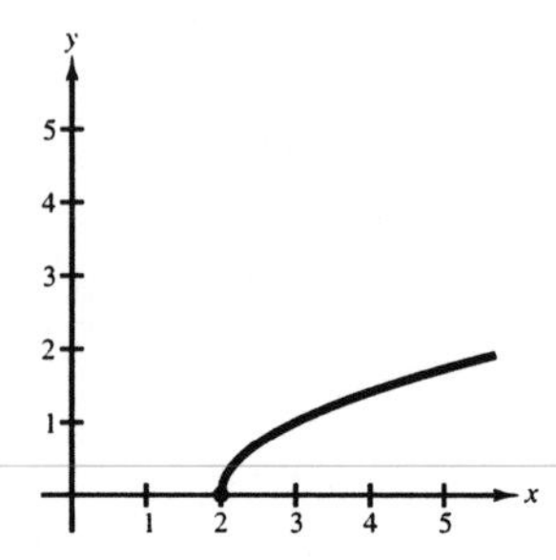

15. Use the graph of $f(x) = \sqrt[3]{x}$ to sketch the graph of the function $y = \sqrt[3]{x} - 1$.

Solution

$y = \sqrt[3]{x} - 1 = f(x) - 1$

Vertical shift one unit downward

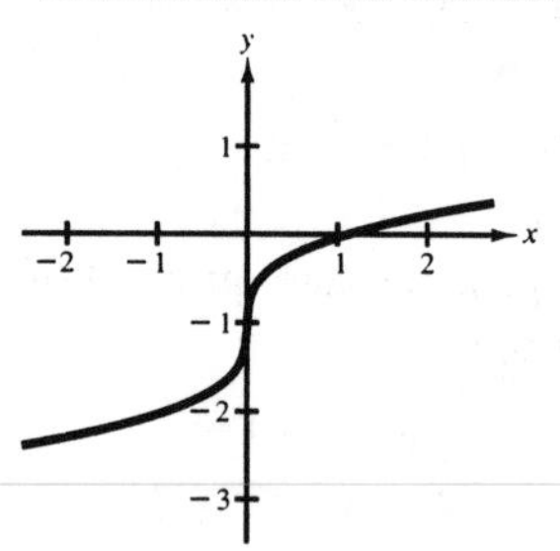

19. Use the graph of $f(x) = \sqrt[3]{x}$ to sketch the graph of the function $y = \sqrt[3]{x+1} - 1$.

Solution

$y = \sqrt[3]{x+1} - 1 = f(x+1) - 1$

Horizontal shift one unit to the left and vertical shift one unit downward.

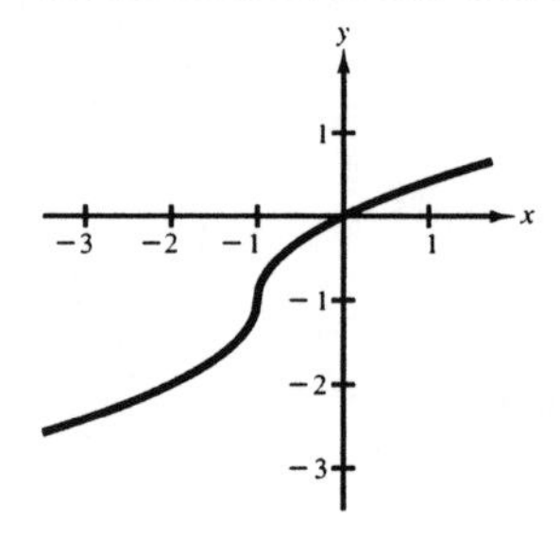

23. Use the graph of $f(x) = x^2$ to write formulas for the functions whose graphs are shown in parts (a) and (b). (See figures in the textbook.)

Solution

Vertical shift 1 unit upward; horizontal shift 1 unit to the right.

(a) $g(x) = f(x-1) + 1$

$g(x) = (x-1)^2 + 1$

Reflection in the x-axis; horizontal shift one unit to the left.

(b) $g(x) = -f(x+1)$

$g(x) = -(x+1)^2$

27. Find the sum, difference, product, and quotient of f and g. What is the domain of f/g given the functions $f(x) = x^2$ and $g(x) = 1 - x$?

Solution

$$(f+g)(x) = f(x) + g(x) = x^2 + (1 - x) = x^2 - x + 1$$
$$(f-g)(x) = f(x) - g(x) = x^2 - (1 - x) = x^2 + x - 1$$
$$(fg)(x) = f(x) \cdot g(x) = x^2(1 - x) = x^2 - x^3$$
$$\left(\frac{f}{g}\right)(x) = \frac{f(x)}{g(x)} = \frac{x^2}{1-x}, \quad x \neq 1$$

31. Find the sum, difference, product, and quotient of f and g. What is the domain of f/g given the following functions?

$$f(x) = \frac{1}{x}, \qquad g(x) = \frac{1}{x^2}$$

Solution

$$(f+g)(x) = f(x) + g(x) = \frac{1}{x} + \frac{1}{x^2} = \frac{x+1}{x^2}$$
$$(f-g)(x) = f(x) - g(x) = \frac{1}{x} - \frac{1}{x^2} = \frac{x-1}{x^2}$$
$$(fg)(x) = f(x) \cdot g(x) = \frac{1}{x}\left(\frac{1}{x^2}\right) = \frac{1}{x^3}$$
$$\left(\frac{f}{g}\right)(x) = \frac{f(x)}{g(x)} = \frac{1/x}{1/x^2} = \frac{x^2}{x} = x, \quad x \neq 0$$

35. Evaluate the function $(f - g)(2t)$ for $f(x) = x^2 + 1$ and $g(x) = x - 4$.

Solution

$$(f-g)(2t) = f(2t) - g(2t) = [(2t)^2 + 1] - (2t - 4) = 4t^2 - 2t + 5$$

39. Evaluate the function $(f/g)(5)$ for $f(x) = x^2 + 1$ and $g(x) = x - 4$.

Solution

$$\left(\frac{f}{g}\right)(5) = \frac{f(5)}{g(5)} = \frac{5^2+1}{5-4} = 26$$

43. Evaluate the function $(f/g)(-1) - g(3)$ for $f(x) = x^2 + 1$ and $g(x) = x - 4$.

Solution

$$\left(\frac{f}{g}\right)(-1) - g(3) = \frac{f(-1)}{g(-1)} - g(3) = \frac{(-1)^2+1}{-1-4} - (3-4) = -\frac{2}{5} + 1 = \frac{3}{5}$$

47. Find (a) $f \circ g$ and (b) $f \circ f$ given the functions $f(x) = 3x + 5$ and $g(x) = 5 - x$.

Solution

(a) $(f \circ g)(x) = f(g(x)) = f(5 - x) = 3(5 - x) + 5 = 20 - 3x$

(b) $(f \circ f)(x) = f(f(x)) = f(3x + 5) = 3(3x + 5) + 5 = 9x + 20$

51. Find (a) $f \circ g$, and (b) $g \circ f$ given the functions $f(x) = \frac{1}{3}x - 3$ and $g(x) = 3x + 1$.

Solution

(a) $(f \circ g)(x) = f(g(x)) = f(3x+1) = \frac{1}{3}(3x+1) - 3 = x - \frac{8}{3}$

(b) $(g \circ f)(x) = g(f(x)) = g\left(\frac{1}{3}x - 3\right) = 3\left(\frac{1}{3}x - 3\right) + 1 = x - 8$

55. Find (a) $f \circ g$, and (b) $g \circ f$ given the functions $f(x) = |x|$ and $g(x) = x + 6$.

Solution

(a) $(f \circ g)(x) = f(g(x)) = f(x+6) = |x+6|$

(b) $(g \circ f)(x) = g(f(x)) = g(|x|) = |x| + 6$

57. Evaluate the functions (a) $(f+g)(3)$ and (b) $(f/g)(2)$. Use the graphs of f and g shown in the textbook.

Solution

(a) $(f+g)(3) = f(3) + g(3) = 2 + 1 = 3$

(b) $\left(\dfrac{f}{g}\right)(2) = \dfrac{f(2)}{g(2)} = \dfrac{0}{2} = 0$

59. Evaluate the functions (a) $(f \circ g)(2)$ and (b) $(g \circ f)(2)$. Use the graphs of f and g shown in the textbook.

Solution

(a) $(f \circ g)(2) = f(g(2)) = f(2) = 0$

(b) $(g \circ f)(2) = g(f(2)) = g(0) = 4$

63. Determine the domain of (a) f, (b) g, and (c) $f \circ g$ given the following functions.

$$f(x) = \frac{3}{x^2 - 1}, \qquad g(x) = x + 1$$

Solution

(a) $f(x) = \dfrac{3}{x^2 - 1}$

Domain: all real numbers except $x = \pm 1$

(b) $g(x) = x + 1$

Domain: all real numbers

(c) $(f \circ g)(x) = \dfrac{3}{(x+1)^2 - 1} = \dfrac{3}{x^2 + 2x} = \dfrac{3}{x(x+2)}$

Domain: all real numbers except $x = -2,\ 0$

65. *Comparing Profits* A company has two manufacturing plants, one in New Jersey and the other in California. From 1990 to 1995, the profits for the manufacturing plant in New Jersey have been decreasing according to the function

$$P_1 = 14.82 - 0.43t, \; t = 0, \; 1, \; 2, \; 3, \; 4, \; 5$$

where P_1 represents the profit in millions of dollars and $t = 0$ represents 1990. On the other hand, the profits for the manufacturing plant in California have been increasing according to the function

$$P_2 = 16.16 + 0.56t, \; t = 0, \; 1, \; 2, \; 3, \; 4, \; 5$$

Write a function that represents the overall company profits during the 6-year period. Use the figure in the textbook, which represents the total profit for the company during this 6-year period, to determine whether the overall company profits have been increasing or decreasing.

Solution

$$P_1 + P_2 = (14.82 - 0.43t) + (16.16 + 0.56t) = 30.98 + 0.13t, \quad t = 0, 1, 2, 3, 4, 5$$

Overall profits have been increasing.

71. *Troubled Waters* A pebble is dropped into a calm pond, causing ripples in the form of concentric circles. The radius (in feet) of the outer ripple is given by $r(t) = 0.6t$, where t is the time in seconds after the pebble strikes the water. The area of the circle is given by the function $A(r) = \pi r^2$. Find and interpret $(A \circ r)(t)$. (See figure in the textbook.)

Solution

$$(A \circ r)(t) = A(r(t)) = \pi(0.6t)^2 = 0.36\pi t^2$$

$A \circ r$ represents the area of the circle at time t.

3.4 Inverse Functions

- Two functions f and g are inverses of each other if $f(g(x)) = x$ for every x in the domain of g and $g(f(x)) = x$ for every x in the domain of f.
- Be able to determine if a function has an inverse.
- Be able to find the inverse of a function, if it exists.
- Be able to use the horizontal line test for inverse functions.

SOLUTIONS TO SELECTED EXERCISES

3. Show that f and g are inverse algebraically and by using a graphing utility.

$$f(x) = 5x + 1, \qquad g(x) = \frac{x-1}{5}$$

Solution

$$f(g(x)) = f\left(\frac{x-1}{5}\right) = 5\left(\frac{x-1}{5}\right) + 1 = x$$

$$g(f(x)) = g(5x+1) = \frac{(5x+1)-1}{5} = x$$

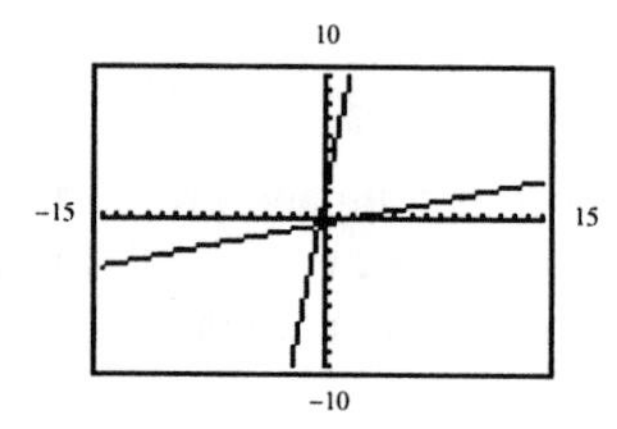

7. Show that f and g are inverse algebraically and by using a graphing utility.

$f(x) = \sqrt{x-4}$ and $g(x) = x^2 + 4,\ x \geq 0.$

Solution

$$f(g(x)) = f(x^2+4), \quad x \geq 0$$
$$= \sqrt{(x^2+4)-4} = x$$

$$g(f(x)) = g(\sqrt{x-4})$$
$$= (\sqrt{x-4})^2 + 4 = x$$

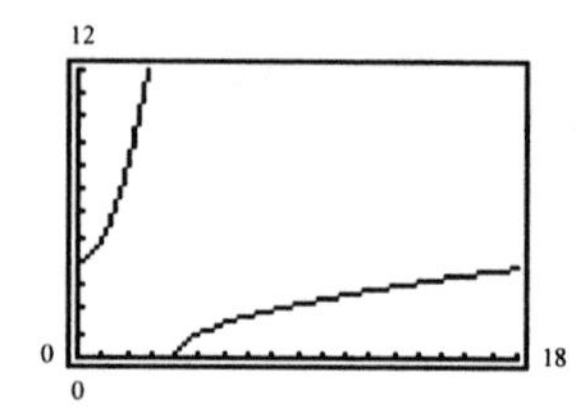

11. Use the graph of $y = f(x)$ to determine whether the function has an inverse. (See figure in the textbook.)

Solution

Since no horizontal line crosses the graph of f at more than one point, f **has** an inverse.

13. Use the graph of $y = f(x)$ to determine whether the function has an inverse. (See figure in the textbook.)

Solution

Since some horizontal lines cross the graph of f twice, f does **not** have an inverse.

17. Use the graph of $y = h(x)$ to determine whether $h(x) = |x + 4|$ has an inverse.

Solution

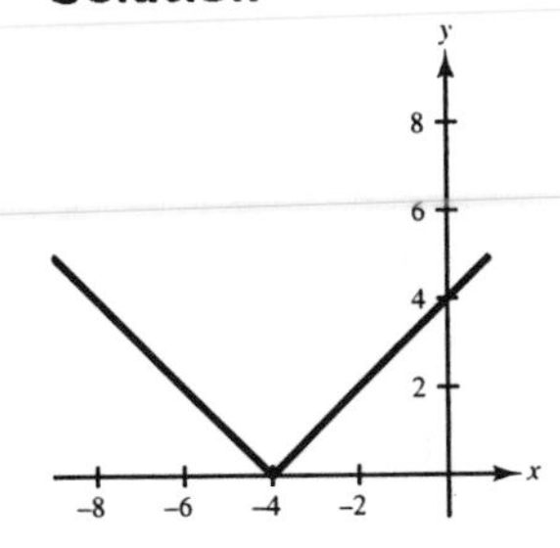

h fails the horizontal line test, so h does **not** have an inverse.

19. Use the graph of $y = f(x)$ to determine whether $f(x) = -\sqrt{16 - x^2}$ has an inverse.

Solution

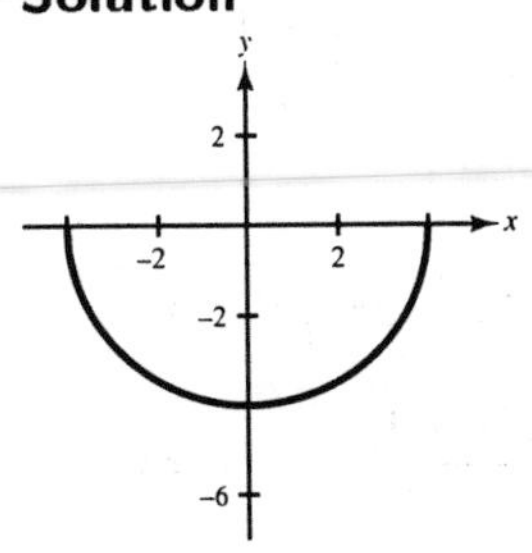

f fails the horizontal line test, so f does **not** have an inverse.

23. Find the inverse of $f(x) = x^5$. Then use a graphing utility to graph both f and f^{-1} on the same screen.

Solution

$$f(x) = x^5$$
$$y = x^5$$
$$x = y^5$$
$$y = \sqrt[5]{x}$$
$$f^{-1}(x) = \sqrt[5]{x}$$

27. Find the inverse of the function $f(x) = \sqrt{4 - x^2}$, $0 \le x \le 2$. Then use a graphing utility to graph both f and f^{-1} on the same screen.

Solution

$$f(x) = \sqrt{4 - x^2}, \quad 0 \le x \le 2$$
$$y = \sqrt{4 - x^2}$$
$$x = \sqrt{4 - y^2}$$
$$f^{-1}(x) = \sqrt{4 - x^2}, \quad 0 \le x \le 2$$

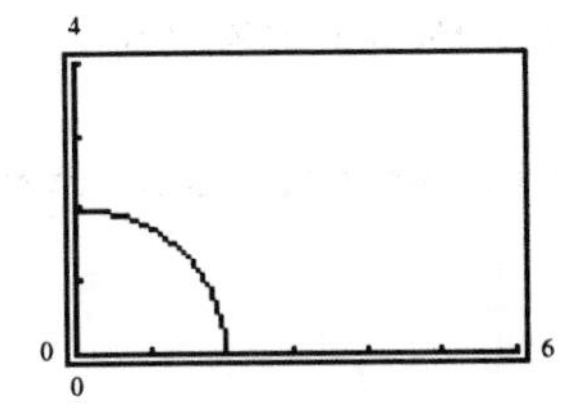

31. Determine whether $f(x) = x^4$ has an inverse.

Solution

$f(x) = x^4$

$y = x^4$

$x = y^4$

$y = \pm\sqrt[4]{x}$

This does not represent y as a function of x. f does not have an inverse. Also, a graph of $f(x) = x^4$ will show that it fails the horizontal line test.

37. Determine whether $f(x) = (x+3)^2,\ x \geq -3$ has an inverse. If it does, find its inverse.

Solution

$f(x) = (x+3)^2,\ x \geq -3 \quad \Rightarrow \quad y \geq 0$

$y = (x+3)^2,\ x \geq -3,\ y \geq 0$

$x = (y+3)^2,\ y \geq -3,\ x \geq 0$

$\sqrt{x} = y + 3,\ y \geq -3,\ x \geq 0$

$y = \sqrt{x} - 3,\ x \geq 0,\ y \geq -3$

This is a function of x, so f has an inverse.

$f^{-1}(x) = \sqrt{x} - 3,\ x \geq 0$

41. Determine whether $f(x) = \sqrt{2x+3}$ has an inverse. If it does, find its inverse.

Solution

$f(x) = \sqrt{2x+3} \quad \Rightarrow \quad x \geq -\frac{3}{2},\ y \geq 0$

$y = \sqrt{2x+3},\ x \geq -\frac{3}{2},\ y \geq 0$

$x = \sqrt{2y+3},\ y \geq -\frac{3}{2},\ x \geq 0$

$x^2 = 2y + 3,\ x \geq 0,\ y \geq -\frac{3}{2}$

$y = \frac{x^2-3}{2},\ x \geq 0,\ y \geq -\frac{3}{2}$

This is a function of x, so f has an inverse.

$f^{-1}(x) = \frac{x^2-3}{2},\ x \geq 0$

43. Determine whether $g(x) = x^2 - x^4$ has an inverse.

Solution

The graph fails the horizontal line test, so g does not have an inverse.

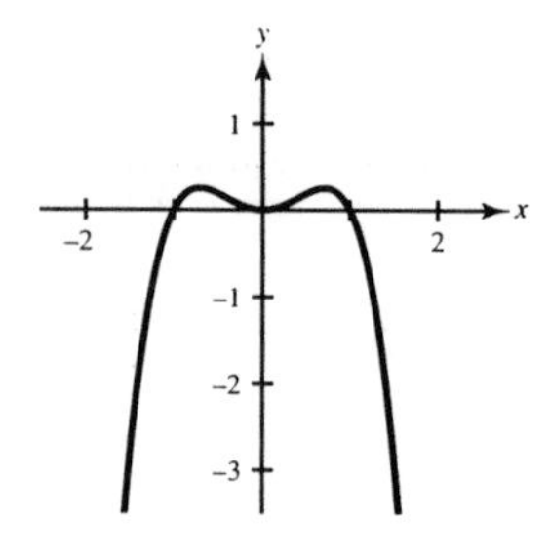

47. Use the graph of the function f to complete the table and to sketch the graph of f^{-1}. (See figure in the textbook.)

Solution

Locate the points on the graph of f and switch x and y.

x	0	1	2	3	4
$f^{-1}(x)$	−2	0	1	2	4

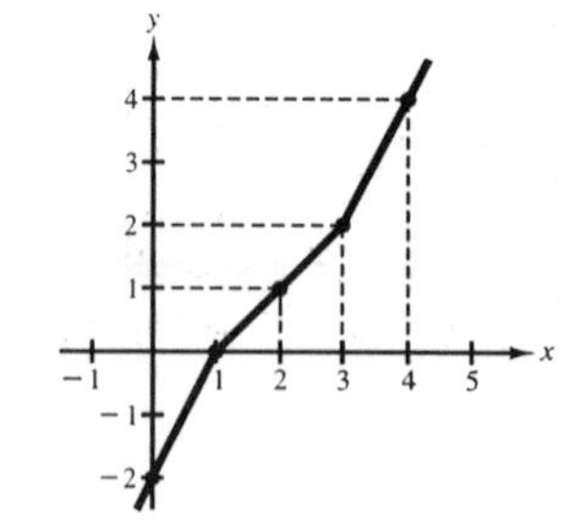

51. Use the functions $f(x) = \frac{1}{8}x - 3$ and $g(x) = x^3$ to find the value $(f^{-1} \circ f^{-1})(6)$.

Solution

$f^{-1}(x) = 8(x+3)$

$(f^{-1} \circ f^{-1})(6) = f^{-1}(f^{-1}(6)) = f^{-1}(8[6+3]) = 8[8(6+3)+3] = 600$

55. Use the functions $f(x) = x + 4$ and $g(x) = 2x - 5$ to find the function $(f \circ g)^{-1}$.

Solution

$$(f \circ g)(x) = f(g(x)) = f(2x - 5) = (2x - 5) + 4 = 2x - 1$$

$$(f \circ g)^{-1}(x) = \frac{x + 1}{2}$$

61. *Earnings-Dividend Ratio* From 1983 to 1993, the earnings per share for Coca-Cola Corporation were approximately related to the dividends per share by the function

$$f(x) = \sqrt{0.214 + 0.169x}, \quad 0.22 \le x \le 1.75$$

where $f(x)$ represents the dividends per share (in dollars) and x represents the earnings per share (in dollars). In 1992, Coca-Cola paid dividends of \$0.56 per share. What were the earnings per share in 1992? (*Source: Coca-Cola.*)

Solution

$$0.56 = \sqrt{0.214 + 0.169x}$$

$$0.3136 = 0.214 + 0.169x$$

$$0.0996 = 0.169x$$

$$x \approx \$0.59 \text{ per share (earnings)}$$

Mid-Chapter Quiz for Chapter 3

In Exercises 1–4, find the value of $f(x) = 2x^2 - x + 1$.

Solutions

1. $f(3) = 2(3)^2 - 3 + 1 = 16$

2. $f(-1) = 2(-1)^2 - (-1) + 1 = 4$

3. $f(0) = 2(0)^2 - 0 + 1 = 1$

4. $f(a) = 2a^2 - a + 1$

5. Find all real values of x such that $f(x) = 0$ for $f(x) = x^3 - 4x$.

Solution

$$x^3 - 4x = 0$$

$$x(x + 2)(x - 2) = 0$$

$$x = 0, -2, 2$$

6. Find all real values of x such that $f(x) = 0$ for $f(x) = 3 - \frac{4}{x}$.

Solution

$$3 - \frac{4}{x} = 0$$

$$3 = \frac{4}{x}$$

$$3x = 4$$

$$x = \frac{4}{3}$$

7. A company produces a product for which the variable cost is \$11.40 per unit and fixed costs are \$85,000. The product sells for \$16.89. Let x be the number of units produced. Write the total profit P as a function of x.

Solution

Cost: $C = 11.40x + 85{,}000$

Revenue: $R = 16.89x$

Profit: $P = R - C = (16.89x) - (11.40x + 85{,}000)$

$= 5.49x - 85{,}000$

8. Find the domain of the function $f(x) = \dfrac{3}{x^2 - x}$.

Solution

Since the denominator, $x^2 - x = x(x - 1)$, equals zero when $x = 0$ or $x = 1$, the domain is all real numbers except 0 and 1. [In interval notation, the domain is $(-\infty, 0) \cup (0, 1) \cup (1, \infty)$.]

9. Find the domain of the function $f(x) = \sqrt{5 - x}$.

Solution

The radicand, $5 - x$, cannot be negative. Thus,

$5 - x \geq 0$

$-x \geq -5$

$x \leq 5$

[The domain is the interval $(-\infty, 5]$.]

10. Use a graphing utility to graph $f(x) = -x^2 + 5x + 3$. Then estimate the intervals on which the function is increasing or decreasing.

Solution

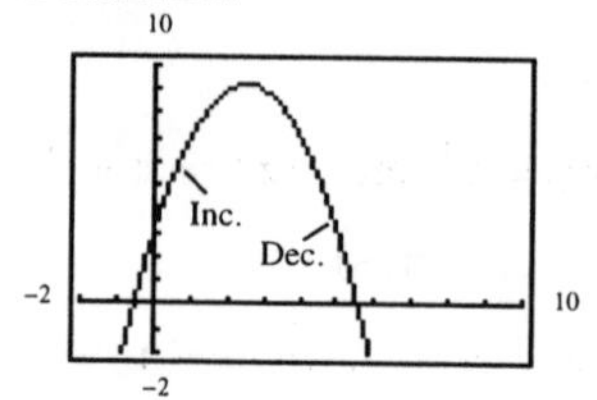

Increasing for $x < 2.5$

Decreasing for $x > 2.5$

11. Use a graphing utility to graph $f(x) = 2x^3 - 5x^2$. Then estimate the open intervals on which the function is increasing or decreasing.

Solution

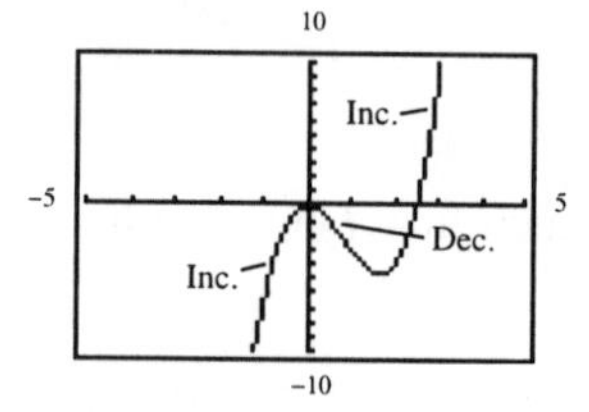

Increasing for $x < 0,\ x > \frac{5}{3}$

Decreasing for $0 < x < \frac{5}{3}$

12. Use a graphing utility to graph $f(x) = x^2 + 2$. Then estimate the open intervals on which the function is increasing or decreasing.

Solution

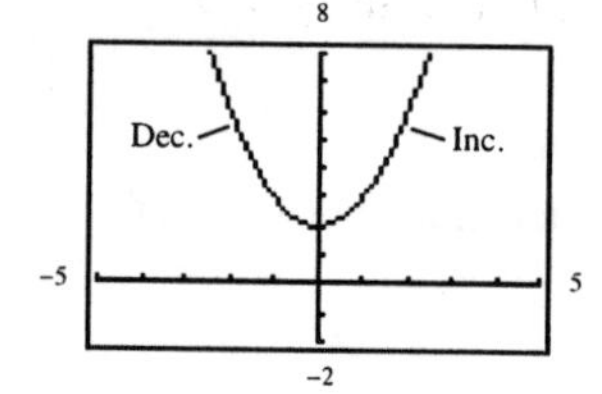

Increasing for $x > 0$

Decreasing for $x < 0$

13. Use a graphing utility to graph $f(x) = \sqrt{x + 1}$. Then estimate the open intervals on which the function is increasing or decreasing.

Solution

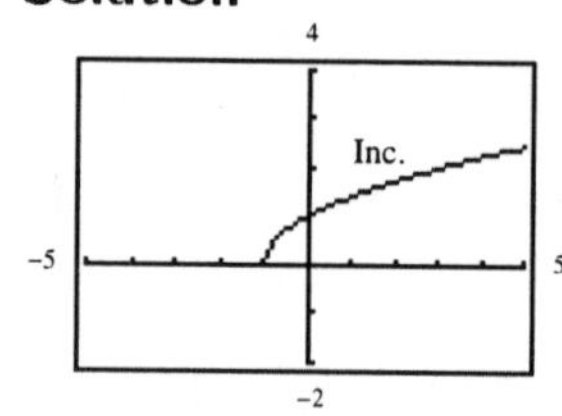

Increasing for $x > -1$

Note: The domain for this function is $[-1, \infty)$.

14. Compare the graph of $f(x) = \sqrt[3]{x}$ and $g(x) = \sqrt[3]{x-2}$.

Solution

The graph of $g(x)$ is obtained by shifting the graph of $f(x)$ two units to the right.

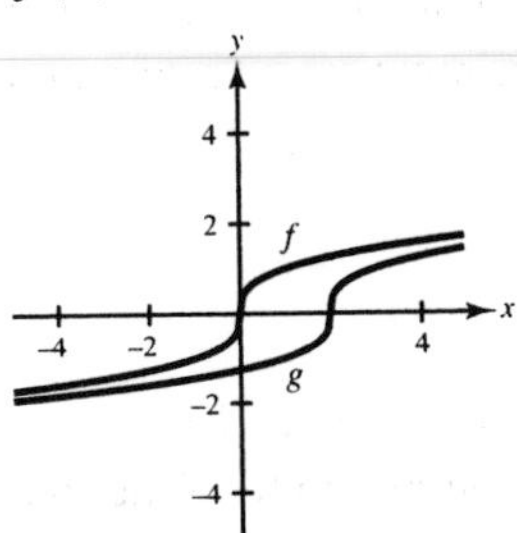

15. Compare the graph of $f(x) = \sqrt[3]{x}$ and $g(x) = \sqrt[3]{x+1}$.

Solution

The graph of $g(x)$ is obtained by shifting the graph of $f(x)$ one unit to the left.

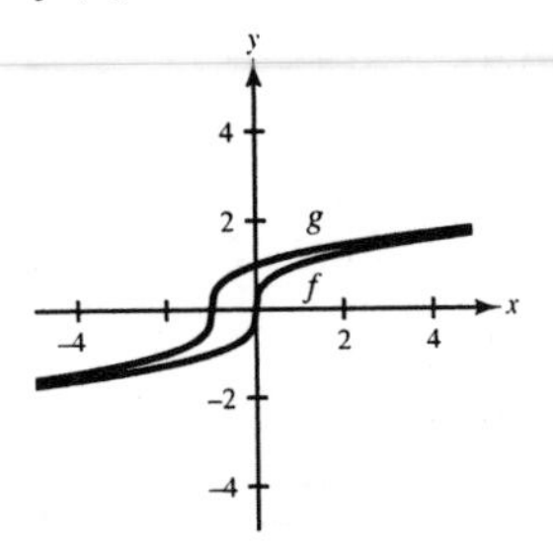

16. Find $(f \circ g)(x)$ for $f(x) = x^2$ and $g(x) = x - 1$.

Solution

$(f \circ g)(x) = f(g(x)) = f(x-1) = (x-1)^2$

17. Find $(f \circ g)(x)$ for $f(x) = \dfrac{1}{x}$ and $g(x) = x^3$.

Solution

$(f \circ g)(x) = f(g(x)) = f(x^3) = \dfrac{1}{x^3}$

18. Compare the graphs of f and f^{-1} for $f(x) = x^2,\ x \geq 0$.

Solution

$f(x) = x^2,\ x \geq 0$

$f^{-1}(x) = \sqrt{x}$

The graphs are reflections in the line $y = x$.

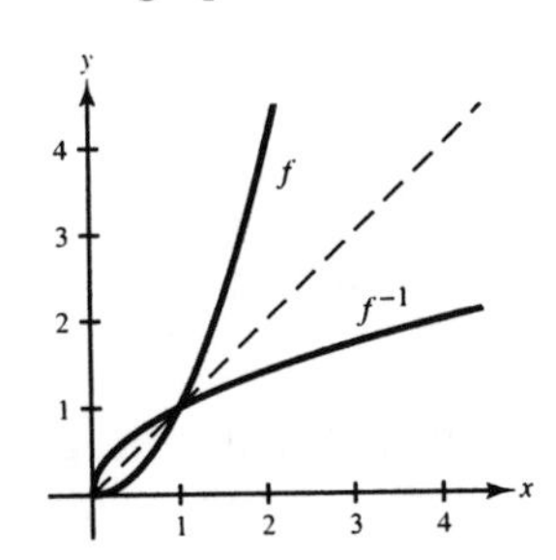

19. Compare the graphs of f and f^{-1} for $f(x) = \sqrt[3]{x-1}$.

Solution

$$f(x) = \sqrt[3]{x-1}$$
$$y = \sqrt[3]{x-1}$$
$$x = \sqrt[3]{y-1}$$
$$x^3 = y - 1$$
$$x^3 + 1 = y$$
$$f^{-1}(x) = x^3 + 1$$

The graphs are reflections in the line $y = x$.

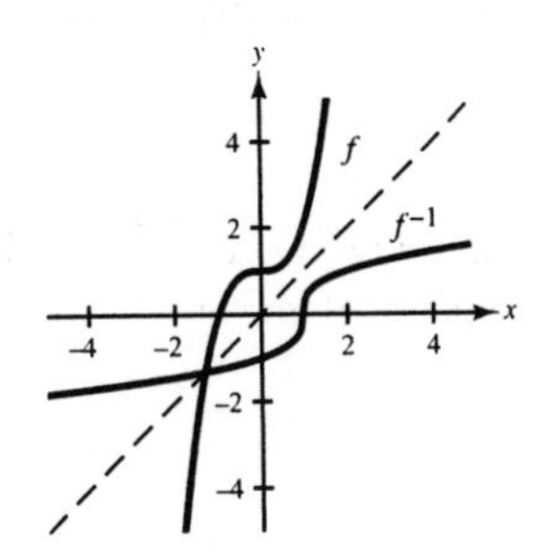

20. *Company Income* The income y for a company can be modeled by $y = (96.5 + 3.75t^2, \quad 4 \le t \le 14$, where $t = 4$ represents 1984. Solve this model for t and use the result to find the year in which the income was \$16,200.

Solution

$$y = (96.5 + 3.75t)^2, \quad 4 \le t \le 14$$

$$\sqrt{y} = 96.5 + 3.75t$$

$$\sqrt{y} - 96.5 = 3.75t$$

$$t = \frac{\sqrt{y} - 96.5}{3.75}$$

When $y = 16{,}200$, $t = \dfrac{\sqrt{16{,}200} - 96.5}{3.75} \approx 8.21$. This would occur during the year 1988.

3.5 Quadratic Functions

You should know the following facts about parabolas.

- $f(x) = ax^2 + bx + c, \ a \neq 0$, is a quadratic function, and its graph is a parabola.
- If $a > 0$, the parabola opens upward. If $a < 0$, the parabola opens downward.
- To find the x-intercepts (if any), solve $ax^2 + bx + c = 0$.
- The standard form of the equation of a parabola is $f(x) = a(x - h)^2 + k$, where $a \neq 0$.
  (a) The vertex is (h, k).
  (b) The axis is the vertical line $x = h$.
- The vertex is $(-b/2a, f(-b/2a))$.

SOLUTIONS TO SELECTED EXERCISES

1. Match $f(x) = (x - 3)^2$ with its graph. [The graphs are labeled (a), (b), (c), (d), (e), (f), (g), and (h) in the textbook.]

Solution

$f(x) = (x - 3)^2$

Vertex: $(3, 0)$

Opens upward

Matches (g)

5. Match $f(x) = 4 - (x - 1)^2$ with its graph. [The graphs are labeled (a), (b), (c), (d), (e), (f), (g), and (h) in the textbook.]

Solution

$f(x) = -(x - 1)^2 + 4$

Vertex: $(1, 4)$

Opens downward

Matches (b)

9. Find an equation for the parabola.
(See figure in the textbook.)

Solution

Vertex: $(2, 0)$

Point: $(0, 4)$

$y = a(x - 2)^2 + 0$

$4 = a(0 - 2)^2$

$a = 1$

$y = (x - 2)^2$

13. Find an equation for the parabola.
(See figure in the textbook.)

Solution

Vertex: $(-3,\ 3)$

Point: $(-2, 1)$

$y = a(x + 3)^2 + 3$

$1 = a(-2 + 3)^2 + 3$

$a = -2$

$y = -2(x + 3)^2 + 3$

17. Sketch the graph of $f(x) = 16 - x^2$.
Identify the vertex and intercepts.

Solution

$f(x) = -(x - 0)^2 + 16$

Vertex: $(0, 16)$

Intercepts: $(-4,\ 0)$, $(0,\ 16)$, $(4,\ 0)$

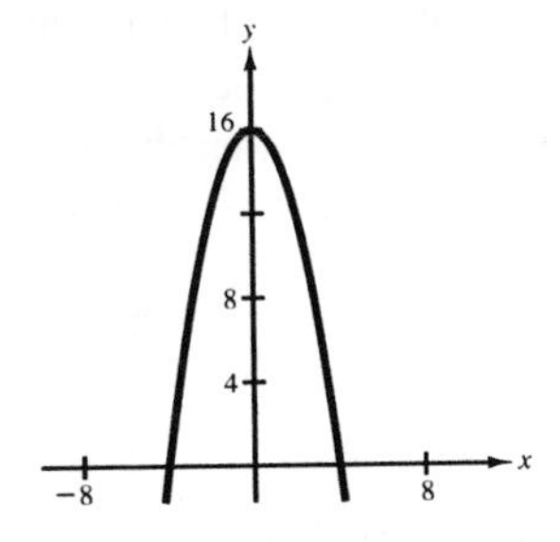

21. Sketch the graph of
$h(x) = x^2 - 8x + 16$.
Identify the vertex and intercepts.

Solution

$h(x) = (x - 4)^2 + 0$

Vertex: $(4, 0)$

Intercepts: $(0, 16)$, $(4, 0)$

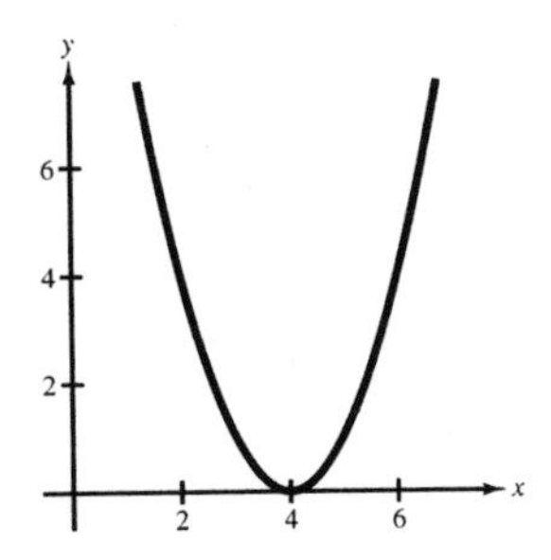

25. Sketch the graph of $f(x) = x^2 - x + \frac{5}{4}$. Identify the vertex and intercepts.

Solution

$f(x) = x^2 - x + \frac{5}{4}$

$f(x) = x^2 - x + \frac{1}{4} - \frac{1}{4} + \frac{5}{4}$

$f(x) = \left(x - \frac{1}{2}\right)^2 + 1$

Vertex: $\left(\frac{1}{2},\ 1\right)$

x-intercept: None since

$$0 = \left(x - \tfrac{1}{2}\right)^2 + 1$$

$$-1 = \left(x - \tfrac{1}{2}\right)^2$$

$\pm\sqrt{-1} = x - \frac{1}{2}$ has no real solution.

y-intercept: $\left(0,\ \frac{5}{4}\right)$

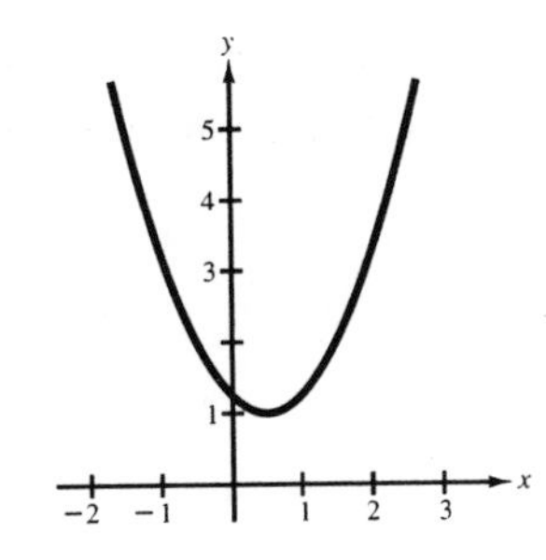

29. Sketch the graph of $h(x) = 4x^2 - 4x + 21$. Identify the vertex and intercepts.

Solution

$h(x) = 4x^2 - 4x + 21$

$h(x) = 4\left(x^2 - x + \frac{1}{4} - \frac{1}{4}\right) + 21$

$h(x) = 4\left(x^2 - x + \frac{1}{4}\right) - 1 + 21$

$h(x) = 4\left(x - \frac{1}{2}\right)^2 + 20$

Vertex: $\left(\frac{1}{2}, 20\right)$

x-intercept: None since

$$0 = 4\left(x - \tfrac{1}{2}\right)^2 + 20$$

$$-20 = 4\left(x - \tfrac{1}{2}\right)^2$$

$$-5 = \left(x - \tfrac{1}{2}\right)^2$$

$\pm\sqrt{-5} = x - \frac{1}{2}$ has no real solutions.

y-intercept: $(0, 21)$

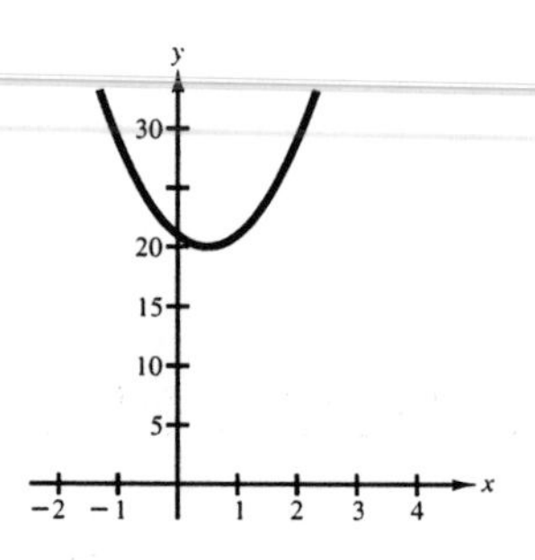

33. Find the quadratic function that has the vertex $(3, 4)$ and whose graph passes through the point $(1, 2)$.

Solution

$y = a(x - 3)^2 + 4$

$2 = a(1 - 3)^2 + 4$

$a = -\frac{1}{2}$

$y = -\frac{1}{2}(x - 3)^2 + 4$

37. Find two quadratic functions whose graphs have the x-intercepts $(-1, 0)$ and $(3, 0)$. Find one function that has a graph that opens upward and another that has a graph that opens downward. (Each exercise has many correct answers.)

Solution

$y = (x + 1)(x - 3) = x^2 - 2x - 3$

$y = -(x + 1)(x - 3) = -x^2 + 2x + 3$

41. Find two quadratic functions whose graphs have the x-intercepts $(-3, 0)$ and $\left(-\frac{1}{2}, 0\right)$. Find one function that has a graph that opens upward and another that has a graph that opens downward. (Each exercise has many correct answers.)

Solution

$y = (x + 3)(2x + 1) = 2x^2 + 7x + 3$

$y = -(x + 3)(2x + 1) = -2x^2 - 7x - 3$

45. *Maximum Area* A rancher has 200 feet of fencing with which to enclose two adjacent rectangular corrals. What dimensions will produce a maximum enclosed area? (See figure in the textbook.)

Solution

$4x + 3y = 200$

$$y = \frac{200 - 4x}{3}$$

$$A = (2x)y = 2x\left(\frac{200 - 4x}{3}\right) = \frac{400}{3}x - \frac{8}{3}x^2 = -\frac{8}{3}(x - 25)^2 + \frac{5000}{3}$$

The area will be maximum when $x = 25$ and $y = \dfrac{200 - 4x}{3} = \dfrac{100}{3}$ feet.

The dimensions are 50 ft $\times 33\frac{1}{3}$ ft.

49. *Minimum Cost* A manufacturer of lighting fixtures has daily production costs of

$$C = 800 - 10x + 0.25x^2$$

where C is the total cost in dollars and x is the number of units produced. How many fixtures should be produced each day to yield a minimum cost?

Solution

$$C = 0.25x^2 - 10x + 800$$
$$= 0.25(x^2 - 40x + 400) - 100 + 800$$
$$= 0.25(x - 20)^2 + 700$$

Vertex: $(20, 700)$

The cost is minimum when $x = 20$ fixtures.

53. *Maximum Height of a Diver* The path of a diver is given by

$$y = -\tfrac{4}{9}x^2 + \tfrac{24}{9}x + 10$$

where y is the height in feet and x is the horizontal distance from the end of the diving board in feet. Use a graphing utility and the trace feature to find the maximum height of the diver. (See figure in the textbook.)

Solution

Algebraically:

$y = -\frac{4}{9}x^2 + \frac{24}{9}x + 10 = -\frac{4}{9}(x - 3)^2 + 14$

Vertex: $(3, 14)$

Maximum when $y = 14$ feet

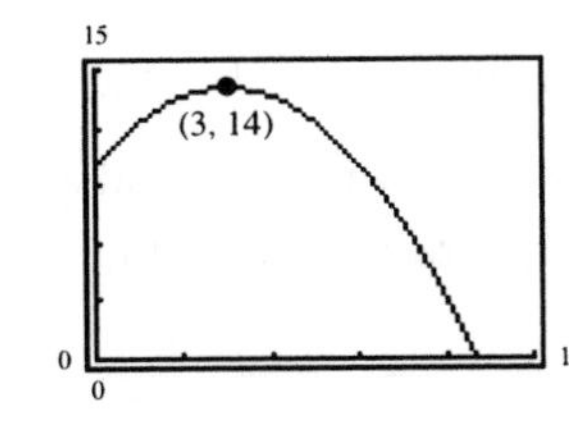

3.6 Polynomial Functions of Higher Degree

You should know the following basic principles about polynomials.

- $f(x) = a_n x^n + a_{n-1}x^{n-1} + \cdots + a_2x^2 + a_1x + a_0$ is a polynomial function of degree n.
- If n is odd and
  (a) $a_n > 0$, then
    1. $f(x) \to \infty$ as $x \to \infty$
    2. $f(x) \to -\infty$ as $x \to -\infty$

  (b) $a_n < 0$, then
    1. $f(x) \to -\infty$ as $x \to \infty$
    2. $f(x) \to \infty$ as $x \to -\infty$
- If n is even and
  (a) $a_n > 0$, then
    1. $f(x) \to \infty$ as $x \to \infty$
    2. $f(x) \to \infty$ as $x \to -\infty$

  (b) $a_n < 0$, then
    1. $f(x) \to -\infty$ as $x \to \infty$
    2. $f(x) \to -\infty$ as $x \to -\infty$
- The following are equivalent for a polynomial function.
  (a) $x = a$ is a zero of a function.
  (b) $x = a$ is a solution of the polynomial equation $f(x) = 0$.
  (c) $(x - a)$ is a factor of the polynomial.
  (d) $(a, 0)$ is an x-intercept of the graph of f.
- The graph of a polynomial function of degree n has, at most, $n - 1$ turning points.
- A polynomial of degree n has at most n zeros.
- If f is a polynomial function such that $a < b$ and $f(a) \neq f(b)$, then f takes on every value between $f(a)$ and $f(b)$ in the interval $[a, b]$.
- If you can find a value where a polynomial is positive and another value where it is negative, then there is at least one real zero between the values.

SOLUTIONS TO SELECTED EXERCISES

1. Match the polynomial function $f(x) = -3x + 5$ with its graph. [The graphs are labeled (a), (b), (c), (d), (e), (f), (g), and (h) in the textbook.]

Solution

$f(x) = -3x + 5$ is a line with negative slope. Matches (e).

5. Match the polynomial function $f(x) = 3x^3 - 9x^2 + 1$ with its graph. [The graphs are labeled (a), (b), (c), (d), (e), (f), (g), and (h) in the textbook.]

Solution

$f(x) = 3x^3 - 9x^2 + 1$

The degree is odd and the leading coefficient is positive. The graph rises to the right and falls to the left. The y-intercept is $(0, 1)$. The graph has, at most, two turning points, and three x-intercepts. Matches (f).

9. Determine the right-hand and left-hand behavior of the graph of $f(x) = 2x^2 - 3x + 1$.

Solution

The degree is even and the leading coefficient is positive. The graph rises to the left and right.

17. Determine the right-hand and left-hand behavior of the graph of $h(t) = -\frac{2}{3}(t^2 - 5t + 3)$.

Solution

The degree is even and the leading coefficient is negative. The graph falls to the left and right.

13. Determine the right-hand and left-hand behavior of the graph of $f(x) = 2x^5 - 5x + 7.5$.

Solution

The degree is odd and the leading coefficient is positive. The graph falls to the left and rises to the right.

21. Find all the real zeros of $h(t) = t^2 - 6t + 9$ algebraically. Then use a graphing utility to confirm your results.

Solution

$h(t) = t^2 - 6t + 9 = (t - 3)^2$

$t = 3$

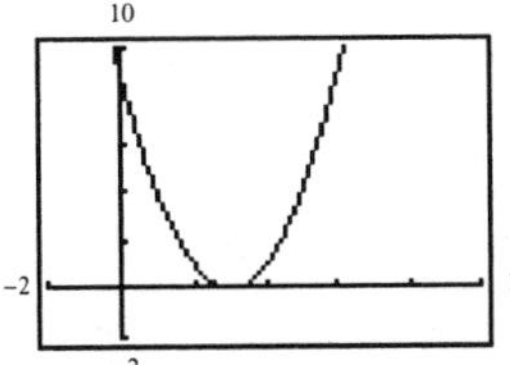

25. Find all the real zeros of
$f(x) = 3x^2 - 12x + 3$
algebraically. Then use a graphing utility to confirm your results.

Solution

$$f(x) = 3x^2 - 12x + 3$$
$$= 3(x^2 - 4x + 1)$$
$$x = \frac{4 \pm \sqrt{16 - 4}}{2} = 2 \pm \sqrt{3}$$

29. Find all the real zeros of
$g(t) = \frac{1}{2}t^4 - \frac{1}{2}$
algebraically. Then use a graphing utility to confirm your results.

Solution

$$g(t) = \frac{1}{2}t^4 - \frac{1}{2}$$
$$= \frac{1}{2}(t + 1)(t - 1)(t^2 + 1)$$
$$t = \pm 1$$

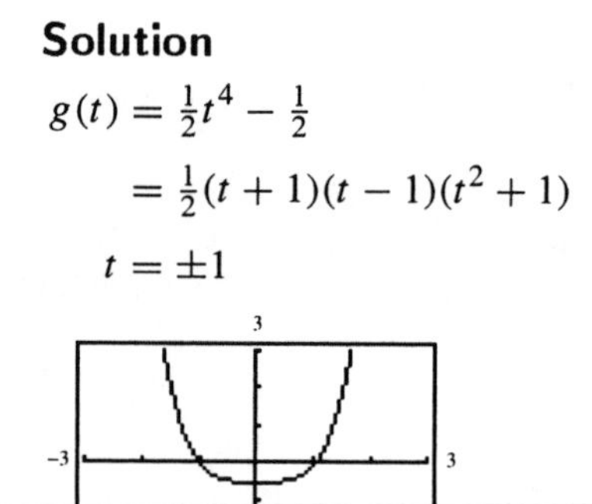

33. Find all the real zeros of $f(x) = 5x^4 + 15x^2 + 10$ algebraically. Then use a graphing utility to confirm your results.

Solution

$$f(x) = 5x^4 + 15x^2 + 10 = 5(x^4 + 3x^2 + 2)$$
$$= 5(x^2 + 2)(x^2 + 1)$$

No real zeros

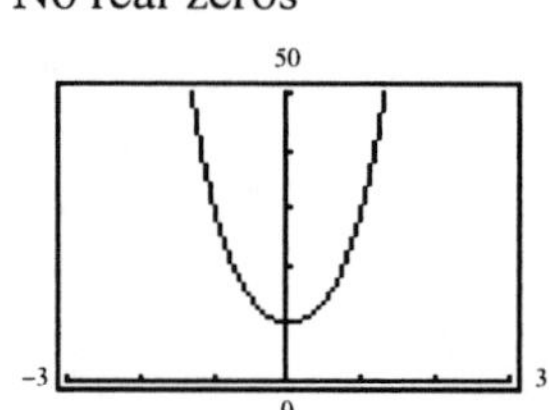

37. Use the graph of $y = x^3$ to sketch the graph of $f(x) = (x - 2)^3 - 2$.

Solution

Horizontal shift 2 units to the right

Vertical shift 2 units down

41. Use the graph of $y = x^4$ to sketch the graph of $f(x) = 4 - x^4$.

Solution

Reflection in the x-axis

Vertical shift 4 units upward

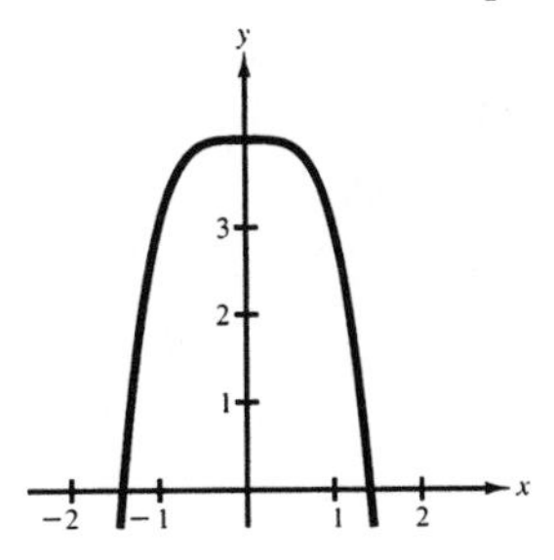

45. Analyze and sketch the graph of $f(t) = \frac{1}{4}(t^2 - 2t + 15)$ by hand. Then use a graphing utility to confirm your sketch.

Solution

$f(t) = \frac{1}{4}(t^2 - 2t + 15) = \frac{1}{4}(t - 1)^2 + \frac{7}{2}$

Vertex: $\left(1, \frac{7}{2}\right)$

Opens upward

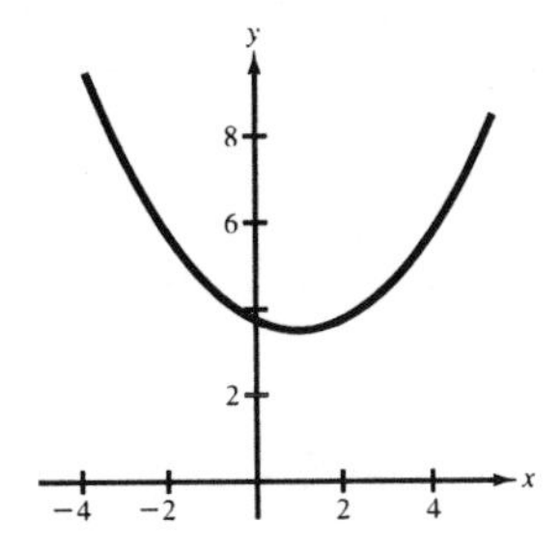

49. Analyze and sketch the graph of $f(x) = x^3 - 4x$ by hand. Then use a graphing utility to confirm your sketch.

Solution

$f(x) = x(x + 2)(x - 2)$

x-intercepts $(0, 0)$, $(\pm 2, 0)$

x	-3	-1	1	3
$f(x)$	-15	3	-3	15

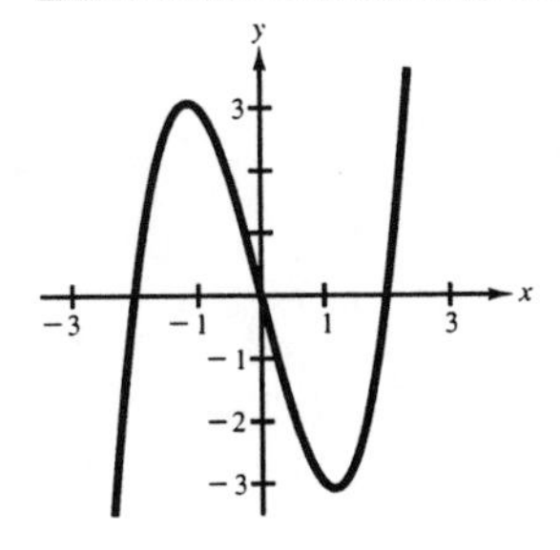

53. Analyze and sketch the graph of $f(x) = 1 - x^6$ by hand. Then use a graphing utility to confirm your sketch.

Solution

$f(x) = (1 + x^3)(1 - x^3)$

x-intercepts $(\pm 1, 0)$

y-intercept $(0, 1)$

x	2	-2
$f(x)$	-31	-31

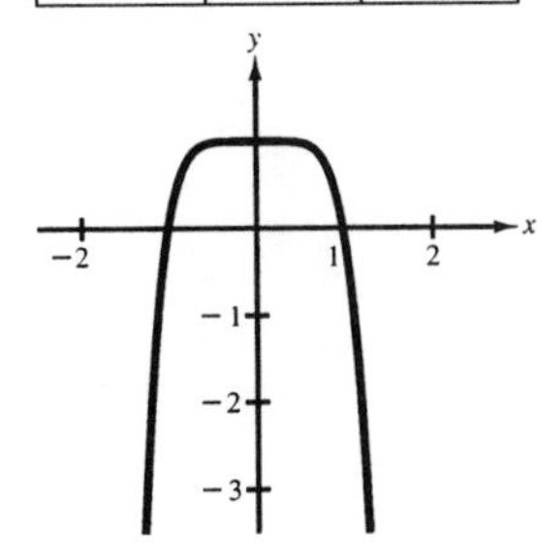

57. Follow the procedure given in Example 8 (Section 3.6) in the textbook to approximate the zero of $f(x) = x^4 - 10x^2 - 11$ in the interval $[3, 4]$. Give your approximation to the nearest tenth. (If you have a graphing utility, use it to help approximate the zero.)

Solution

$x \approx 3.3$

x	3.0	3.1	3.2	3.3	3.4	3.5	3.6	3.7	3.8	3.9	4.0
$f(x)$	−20	−14.7	−8.54	−1.31	7.03	16.6	27.4	39.5	53.1	68.2	85

61. *Advertising Expenses* The total revenue for a soft-drink company is related to its advertising expense by the function

$$R = \frac{1}{50{,}000}(-x^3 + 600x^2), \quad 0 \le x \le 400$$

where R is the total revenue in millions of dollars and x is the amount spent on advertising (in tens of thousands of dollars). Use the graph of this function to estimate the point on the graph at which the function is increasing most rapidly. (See figure in the textbook.) This point is called the **point of diminishing returns** because any expenses above this amount will yield less return per dollar invested in advertising.

Solution

The function is increasing on the interval $[0, 400]$. It is increasing most rapidly (the slope of the tangent line is largest) at $(200, 320)$ which corresponds to spending $2,000,000 on advertising.

3.7 Rational Functions

- You should know the following basic facts about rational functions.
  - (a) A function of the form $f(x) = P(x)/Q(x)$, $Q(x) \neq 0$, where $P(x)$ and $Q(x)$ are polynomials, is called a rational function.
  - (b) The domain of a rational function is the set of all real numbers except those which make the denominator zero.
  - (c) If $f(x) = P(x)/Q(x)$ is in reduced form, and a is a value such that $Q(a) = 0$, then the line $x = a$ is a vertical asymptote of the graph of f.
  - (d) The line $y = b$ is a horizontal asymptote of the graph of f if $f(x) \to b$ as $x \to \infty$ or $x \to -\infty$.
  - (e) If $f(x) = \dfrac{a_n x^n + \cdots + a_0}{b_m x^m + \cdots + b_0}$ is a rational function and
    - $(i) n < m$, then the x-axis is a horizontal asymptote.
    - $(ii) n = m$, then $y = \dfrac{a_n}{b_m}$ is a horizontal asymptote.
    - $(iii) n > m$, then the graph has no horizontal asymptote.
- Be able to graph rational functions.

SOLUTIONS TO SELECTED EXERCISES

1. Find the domain of the function.

$$f(x) = \frac{1}{x^2}$$

Solution

Since $x = 0$ makes the denominator zero, the domain is the set of all real numbers except 0.

5. Find the domain of the function.

$$f(x) = \frac{3x^2 + 1}{x^2 + 9}$$

Solution

Since the denominator does not equal zero for any real number, the domain is the set of all real numbers.

9. Match the function $f(x) = \dfrac{2}{x + 1}$ with its graph.
[The graphs are labeled (a), (b), (c), (d), (e), and (f) in the textbook.]

Solution

$$f(x) = \frac{2}{x + 1}$$

x-intercept: None

y-intercept: $(0, 2)$

Vertical asymptote: $x = -1$

Horizontal asymptote: x-axis

Matches (f)

13. Match the function $f(x) = \dfrac{x - 2}{x - 1}$ with its graph.
[The graphs are labeled (a), (b), (c), (d), (e), and (f) in the textbook.]

Solution

$$f(x) = \frac{x - 2}{x - 1}$$

x-intercept: $(2, 0)$

y-intercept: $(0, 2)$

Vertical asymptote: $x = 1$

Horizontal asymptote: $y = 1$

Matches (c)

15. Compare the graph of $f(x) = \frac{1}{x}$ with the graph of g.

$$g(x) = f(x) + 1 = \frac{1}{x} + 1 = \frac{1+x}{x}$$

Solution

Vertical shift one unit upward

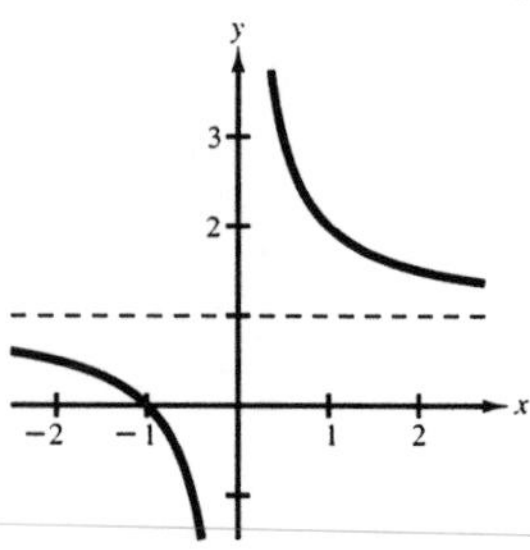

19. Compare the graph of $f(x) = \frac{4}{x^2}$ with the graph of g.

$$g(x) = f(x) - 2 = \frac{4}{x^2} - 2 = \frac{4 - 2x^2}{x^2}$$

Solution

Vertical shift two units downward

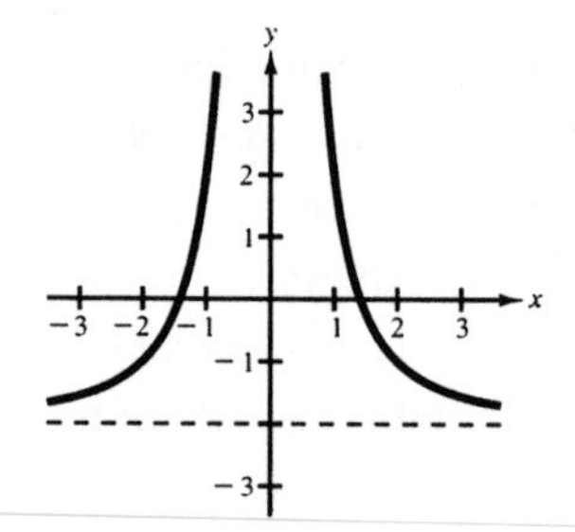

23. Compare the graph of $f(x) = \frac{8}{x^3}$ with the graph of g.

$$g(x) = f(x) + 1 = \frac{8}{x^3} + 1 = \frac{8 + x^3}{x^3}$$

Solution

Vertical shift one unit upward

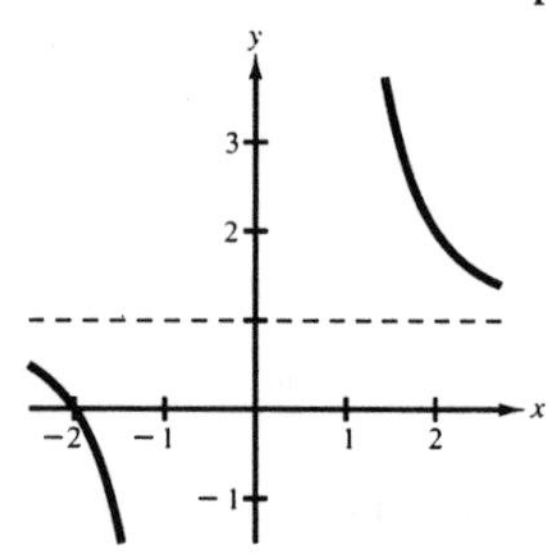

27. Sketch the graph of the function.

$$f(x) = \frac{1}{x+2}$$

As sketching aids, check for intercepts, vertical asymptotes, and horizontal asymptotes.

Solution

Intercept: $(0, \frac{1}{2})$
Vertical asymptote: $x = -2$
Horizontal asymptote: $y = 0$

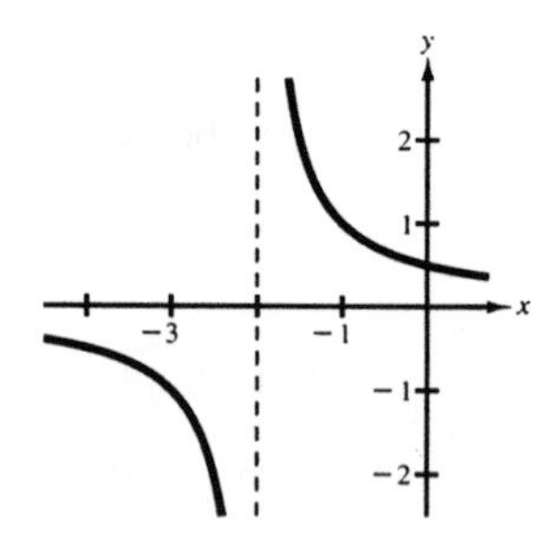

31. Sketch the graph of the following.

$$f(x) = \frac{x+1}{x+2}$$

As sketching aids, check for intercepts, vertical asymptotes, and horizontal asymptotes.

Solution

Intercepts: $(-1,\ 0)$, $\left(0,\ \frac{1}{2}\right)$
Vertical asymptote: $x = -2$
Horizontal asymptote: $y = 1$

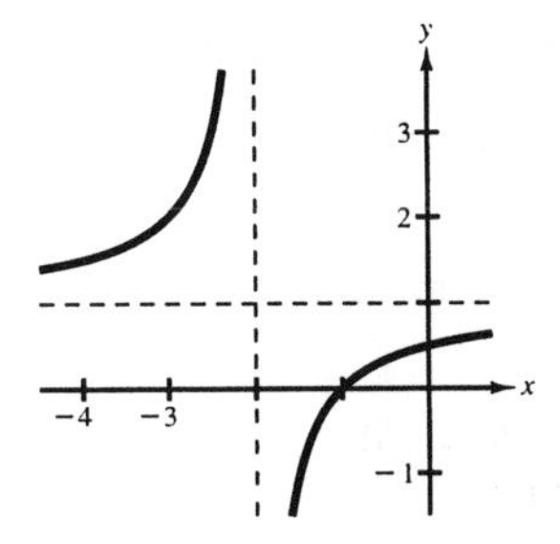

37. Sketch the graph of the following.

$$g(x) = \frac{1}{x+2} + 2 = \frac{2x+5}{x+2}$$

As sketching aids, check for intercepts, vertical asymptotes, and horizontal asymptotes.

Solution

Intercepts: $\left(-\frac{5}{2},\ 0\right)$, $\left(0,\ \frac{5}{2}\right)$
Vertical asymptote: $x = -2$
Horizontal asymptote: $y = 2$

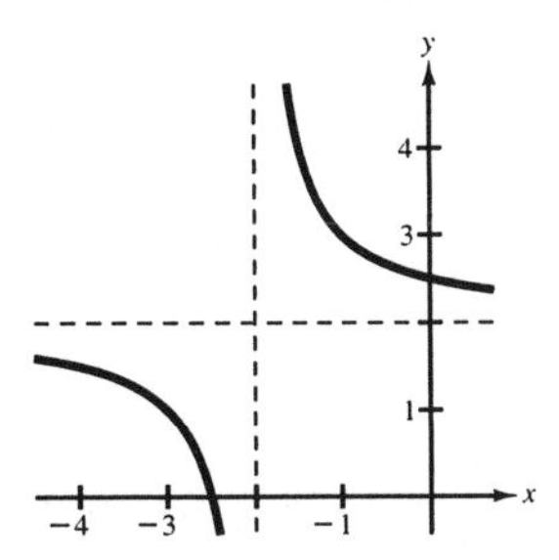

41. Sketch the graph of the following.

$$f(x) = \frac{x^2}{x^2+9}$$

As sketching aids, check for intercepts, vertical asymptotes, and horizontal asymptotes.

Solution

Intercept: $(0, 0)$
Horizontal asymptote: $y = 1$
y-axis symmetry

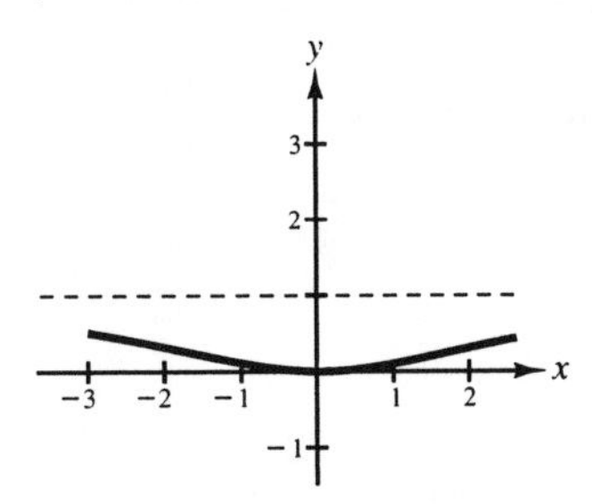

43. Sketch the graph of the following.

$$h(x) = \frac{x^2}{x^2-9} = \frac{x^2}{(x+3)(x-3)}$$

As sketching aids, check for intercepts, vertical asymptotes, and horizontal asymptotes.

Solution

Intercept: $(0, 0)$
Vertical asymptotes: $x = \pm 3$
Horizontal asymptote: $y = 1$
y-axis symmetry

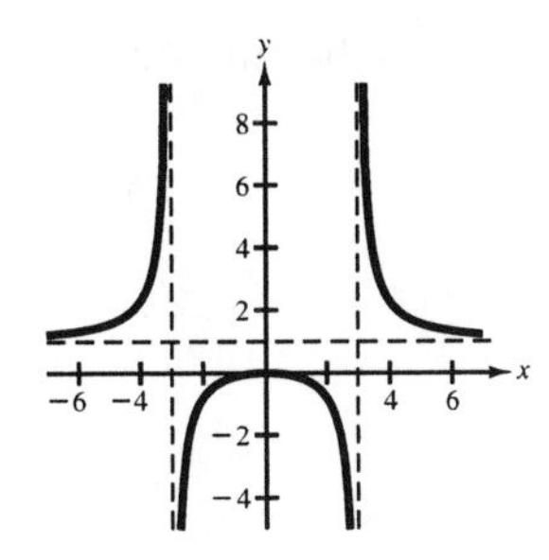

49. Sketch the graph of the following.

$$f(x) = \frac{3x}{x^2 - x - 2} = \frac{3x}{(x+1)(x-2)}$$

As sketching aids, check for intercepts, vertical asymptotes, and horizontal asymptotes.

Solution

Intercept: $(0, 0)$
Vertical asymptotes: $x = -1, 2$
Horizontal asymptote: $y = 0$

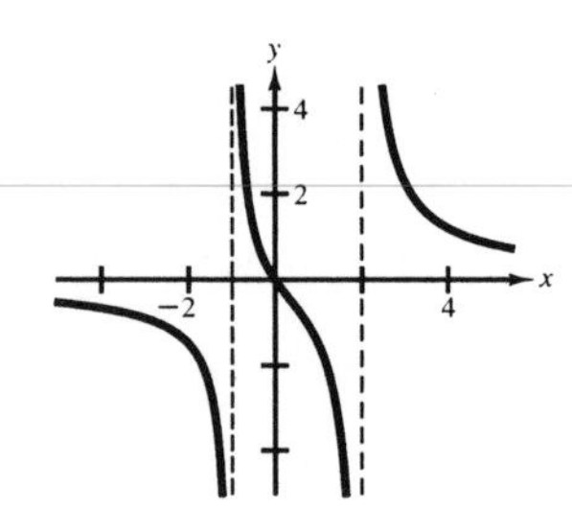

51. *Seizure of Illegal Drugs* The cost in millions of dollars for the federal government to seize p per cent of a certain illegal drug as it enters the country is given by

$$C = \frac{528p}{100 - p}, \quad 0 \leq p < 100.$$

(a) Find the cost of seizing 25%.

(b) Find the cost of seizing 50%.

(c) Find the cost of seizing 75%.

(d) According to this model, would it be possible to seize 100% of the drug?

Solution

(a) $C(25) = \dfrac{528(25)}{100 - 25} = \$176$ million

(b) $C(50) = \dfrac{528(50)}{100 - 50} = \$528$ million

(c) $C(75) = \dfrac{528(75)}{100 - 75} = \$1584$ million

(d) $C \Rightarrow \infty$ as $x \Rightarrow 100^-$

No, it would not be possible to seize 100% of the drug.

55. *Average Cost* The cost of producing x units of a product is $C = 150{,}000 + 0.25x$ and therefore the average cost per unit is

$$\overline{C} = \frac{C}{x} = \frac{150{,}000 + 0.25x}{x}, \quad 0 < x.$$

(a) Sketch the graph of this average cost function.

(b) Find the average cost of producing $x = 1000$, $x = 10{,}000$, and $x = 100{,}000$ units. What can you conclude?

Solution

(a)

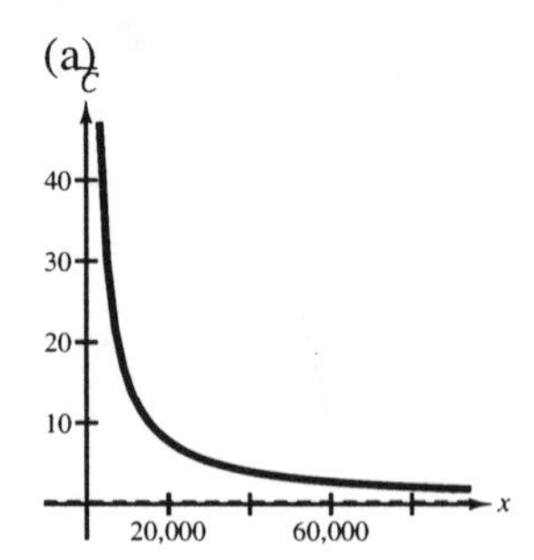

(b) $\overline{C}(1000) = \$150.25$

$\overline{C}(10{,}000) = \$15.25$

$\overline{C}(100{,}000) = \$1.75$

As the number of units increase, the average cost per unit decreases.

61. *Sale of Long-Playing Albums* The sales of long-playing albums in the United States from 1975 to 1994 can be approximated by the model

$$L = \frac{178.77 - 8.49t}{1 - 0.117t + 0.005t^2}, \quad 5 \le t \le 21$$

where L is the number of albums sold (in millions) and $t = 5$ represents 1975. Notice the drop in sales of albums after about 1980. How would you account for this drop in sales? (See figure in the textbook.) (*Source: Recording Industry Association of America.*)

Solution

The sale of cassettes and compact discs have replaced the sale of long-playing albums.

Review Exercises for Chapter 3

SOLUTIONS TO SELECTED EXERCISES

3. Evaluate $f(x) = |x| + 5$ and simplify the results.

(a) $f(0)$ (b) $f(-3)$ (c) $f\left(\frac{1}{2}\right)$ (d) $f(-x^2)$

Solution

(a) $f(0) = |0| + 5 = 5$

(b) $f(-3) = |-3| + 5 = 8$

(c) $f\left(\frac{1}{2}\right) = \left|\frac{1}{2}\right| + 5 = \frac{11}{2} = 5\frac{1}{2}$

(d) $f(-x^2) = |-x^2| + 5 = x^2 + 5$

9. Find the domain of $h(x) = \sqrt{x+9}$.

Solution

$x + 9 \ge 0$ when $x \ge -9$

Domain: $x \ge -9$ or the interval $[-9, \infty)$

11. Find the domain of the following function.

$$g(t) = \frac{\sqrt{t-3}}{t-5}$$

Solution

Domain: $t \ge 3$, $t \ne 5$ or $[3, 5) \cup (5, \infty)$

15. Determine if the set of ordered pairs $\{(1, -3), (2, -7), (3, -3)\}$ represents a function from A to B when $A = \{1, 2, 3\}$ and $B = \{-3, -4, -7\}$. Give a reason for your answer.

Solution

Since each element of A is paired with only one element of B, $\{(1, -3), (2, -7), (3, -3)\}$ represents a function from A to B.

21. Determine (a) the range and domain of $f(x) = x^3 - 4x^2$, (b) determine the intervals over which the function is increasing, decreasing, or constant, and (c) determine if the function is even, odd, or neither. (See figure in the textbook.)

Solution

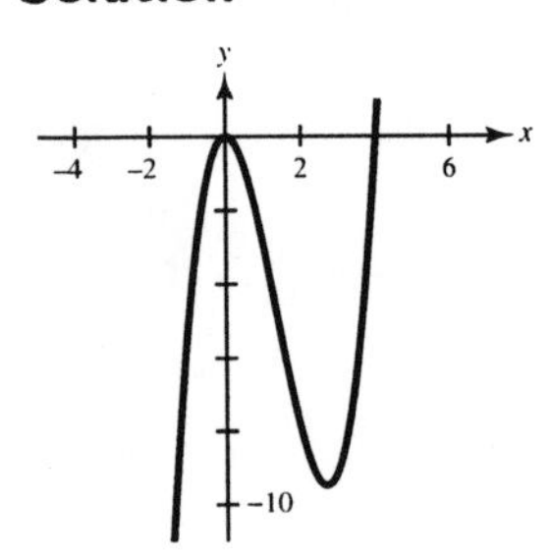

(a) Domain: All real numbers

Range: All real numbers

(b) Increasing for $x < 0,\ x > \frac{8}{3}$

Decreasing for $0 < x < \frac{8}{3}$

(c) $f(-x) = (-x)^3 - 4(-x)^2 = -x^3 - 4x^2$

Neither odd nor even.

25. Sketch the graph of $g(x) = \sqrt{x^2 - 16}$.

Solution

Domain: $(-\infty, -4] \bigcup [4, \infty)$

x-intercepts: $(\pm 4, 0)$

y-axis symmetry

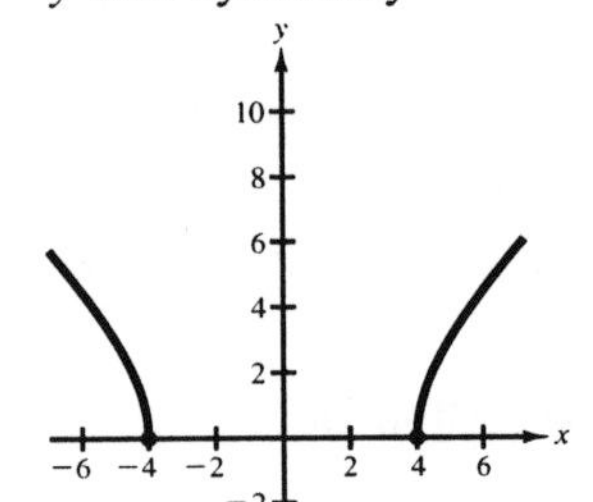

29. Sketch the graph of the following function.

$$g(x) = \begin{cases} x + 2, & x < 0 \\ 2, & x = 0 \\ x^2 + 2, & x > 0 \end{cases}$$

Solution

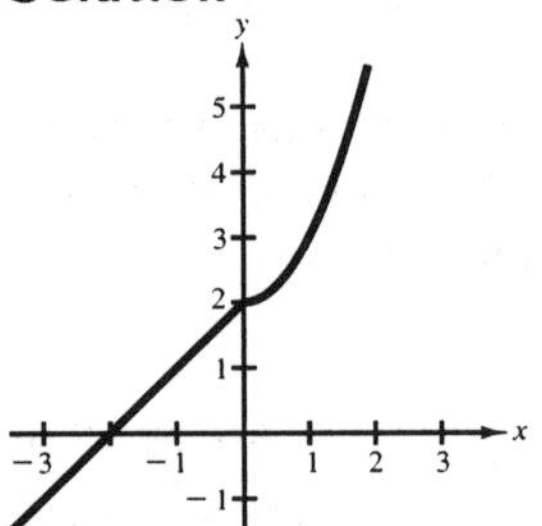

35. Use the graph of $f(x) = x^2$ to sketch the graph of $g(x) = -x^2 + 3$.

Solution

Reflection in the x-axis

Vertical shift 3 units upward

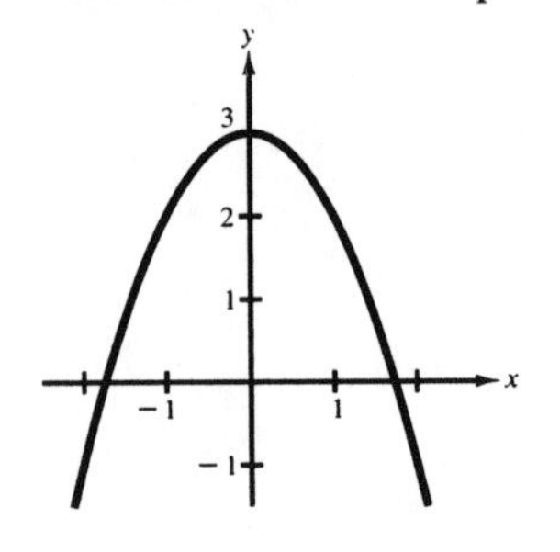

39. Use the graph of $f(x) = \sqrt[3]{x}$ to sketch the graph of $y = 2\sqrt[3]{x}$.

Solution

Vertically stretched by a factor of 2.

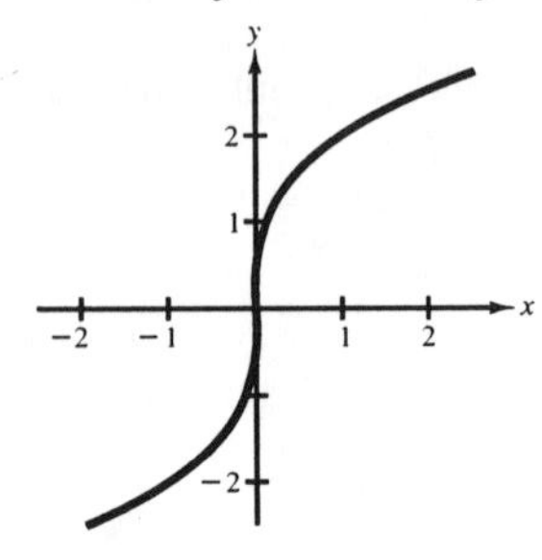

43. Given $f(x) = 3x + 2$ and $g(x) = x^2 - 4x$, find $(f+g)(x)$, $(f-g)(x)$, $(fg)(x)$, and $(f/g)(x)$. What is the domain of f/g?

Solution

$$(f+g)(x) = (3x+2) + (x^2 - 4x) = x^2 - x + 2$$

$$(f-g)(x) = (3x+2) - (x^2 - 4x) = -x^2 + 7x + 2$$

$$(fg)(x) = (3x+2)(x^2 - 4x) = 3x^3 - 10x^2 - 8x$$

$$\left(\frac{f}{g}\right)(x) = \frac{3x+2}{x^2-4x}, \; x \neq 0, \; x \neq 4$$

47. Evaluate $(f/g)(0)$ for $f(x) = x^2 - 1$ and $g(x) = x + 2$.

Solution

$$\left(\frac{f}{g}\right)(0) = \frac{f(0)}{g(0)} = \frac{-1}{2} = -\frac{1}{2}$$

51. Given $f(x) = 1/x$ and $g(x) = 3x + x^2$, find (a) $f \circ g$ and (b) $g \circ f$.

Solution

(a) $(f \circ g)(x) = f(g(x)) = f(3x + x^2) = \dfrac{1}{3x + x^2}, \; x \neq 0, \; x \neq -3$

(b) $(g \circ f)(x) = g(f(x)) = g\left(\dfrac{1}{x}\right) = 3\left(\dfrac{1}{x}\right) + \left(\dfrac{1}{x}\right)^2 = \dfrac{3}{x} + \dfrac{1}{x^2} = \dfrac{3x+1}{x^2}, \; x \neq 0$

53. *Cost* The daily cost of producting x units in a manufacturing process is given by the function $C(x) = 30x + 800$. The number of units produced in t hours is given by $x(t) = 150t$. Find and interpret $(C \circ x)(t)$.

Solution

$$(C \circ x)(t) = C(x(t)) = C(150t) = 30(150t) + 800 = 4500t + 800$$

This function represents the daily cost in terms of hours t of production.

55. Given $f(x) = \sqrt[3]{x-3}$ and $g(x) = x^3 + 3$, show that f and g are inverses of each other.

Solution

Algebraically

$$f(g(x)) = f(x^3 + 3) = \sqrt[3]{(x^3+3) - 3} = \sqrt[3]{x^3} = x$$

$$g(f(x)) = g(\sqrt[3]{x-3}) = (\sqrt[3]{x-3})^3 + 3 = (x-3) + 3 = x$$

Graphically

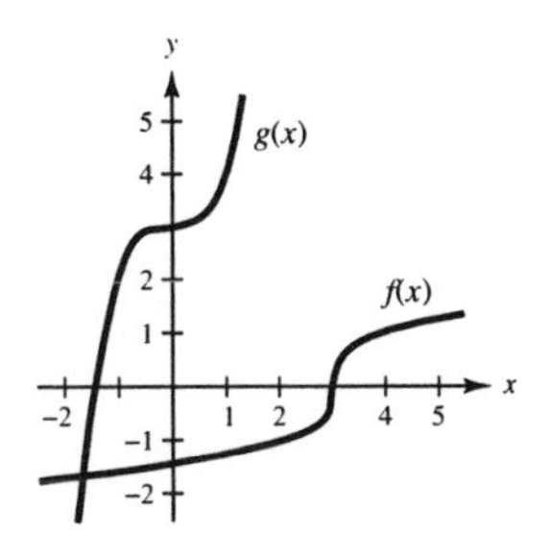

59. Determine whether $f(x) = x^4$ has an inverse. If it does, find the inverse and graph f and f^{-1} in the same coordinate plane.

Solution

$f(x) = x^4$

$y = x^4$

$x = y^4$

$y = \pm\sqrt[4]{x}$

Since y is not a function of x, f does **not** have an inverse. Also, the graph of f fails the horizontal line test.

63. Sketch the graph of $g(x) = -x^2 + 6x + 8$. Identify the vertex and intercepts.

Solution

$g(x) = -x^2 + 6x + 8 = -(x - 3)^2 + 17$

Vertex: $(3, 17)$

y-intercept: $(0, 8)$

Use the Quadratic Formula on $0 = -x^2 + 6x + 8$.

x-intercepts: $(3 \pm \sqrt{17},\ 0)$

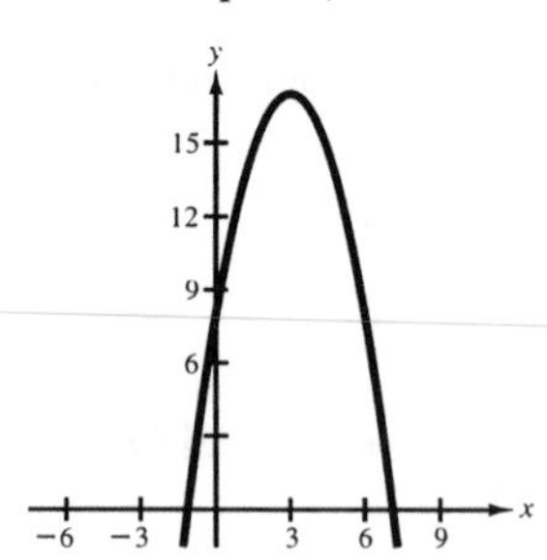

67. Find the quadratic function that has the vertex of $(2, 3)$ and whose graph passes through the point $(-1, 6)$.

Solution

$f(x) = a(x - 2)^2 + 3$

$6 = a(-1 - 2)^2 + 3$

$3 = 9a$

$\frac{1}{3} = a$

$f(x) = \frac{1}{3}(x - 2)^2 + 3 = \frac{1}{3}(x^2 - 4x + 4) + \frac{9}{3} = \frac{1}{3}(x^2 - 4x + 13)$

71. *Maximum Profit* The profit for a company is given by

$$P = -0.0001x^2 + 130x - 300{,}000$$

where x is the number of units produced.

(a) Use a graphing utility to graph the profit function.

(b) Graphically estimate the number of units that should be produced to yield a maximum profit.

(c) Explain how to confirm the result of part (b) algebraically.

Solution

(a)

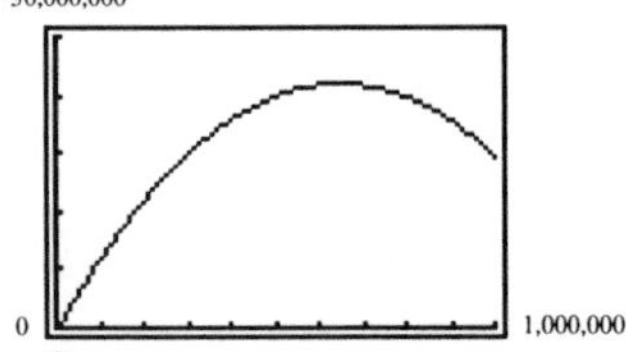

(b) $x = 650{,}000$ units

(c) $P = -0.0001x^2 + 130x - 300{,}000$

$= -0.0001(x - 650{,}000)^2 + 41{,}950{,}000$

The profit is maximum ($41,950,000) when $x = 650{,}000$ units.

75. Determine the right- and left-hand behavior of the graph of the function
$f(x) = \frac{3}{4}(x^4 + 3x^2 + 2)$.

Solution

The degree is even and the leading coefficient is positive. The graph rises to the left and right.

79. Find all the real zeros of the function
$f(x) = x^3 - 6x^2 - 3x + 18$.

Solution

$$f(x) = x^3 - 6x^2 - 3x + 18 = x^2(x - 6) - 3(x - 6)$$
$$= (x - 6)(x^2 - 3) = (x - 6)(x - \sqrt{3})(x + \sqrt{3})$$

The real zeros are $x = 6$ and $x = \pm\sqrt{3}$.

83. Find the domain of the following function.

$$f(x) = \frac{2x^2 + 5x - 3}{x^2 + 2}$$

Identify any horizontal or vertical asymptotes.

Solution

Domain: all real numbers

Horizontal asymptote: $y = \frac{2}{1} = 2$

Vertical asymptote: None since $x^2 + 2 \neq 0$ for any real x.

87. Sketch the graph of the following function.

$$f(x) = \frac{4}{(x - 1)^2}$$

As sketching aids, check for intercepts, symmetry, vertical asymptotes, and horizontal asymptotes.

Solution

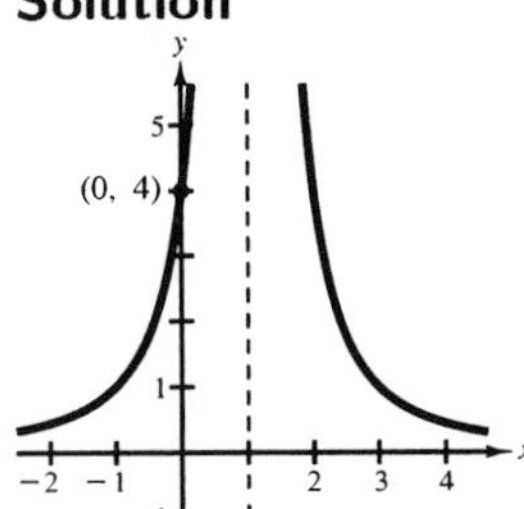

y-intercept: $(0, 4)$

Vertical asymptote: $x = 1$

Horizontal asymptote: $y = 0$

91. *Population of Fish* The Parks and Wildlife Commission introduces 50,000 game fish into a large lake. The population of the fish (in thousands) is

$$N = \frac{20(4+3t)}{1+0.05t} = \frac{60t+80}{0.05t+1}, \quad 0 \le t$$

where t is the time in years.

(a) Find the population when $t = 5,\ 10$, and 25.

(b) What is the limiting number of this population of fish in the lake as time progresses?

Solution

(a) $N(5) = 304$ thousand fish

$N(10) \approx 453.3$ thousand fish

$N(25) \approx 702.2$ thousand fish

(b) The population is limited by the horizontal asymptote.

$$N = \frac{60}{0.05} = 1200 \text{ thousand fish}$$

Test for Chapter 3

1. Find the domain of $f(x) = 2x^2 - 3x + 8$.

Solution

Domain: No restriction

All real numbers $(-\infty, \infty)$

2. Find the domain of $g(x) = \sqrt{x-7}$.

Solution

Domain: $x \ge 7$ or $[7, \infty)$

3. Find the domain of $g(t) = \dfrac{\sqrt{t-2}}{t-7}$.

Solution

Domain: $t \ge 2,\ t \ne 7$

$[2, 7) \cup (7, \infty)$

4. Find the domain of $h(x) = \dfrac{3}{x^2-4}$.

Solution

Domain: $x \ne \pm 2$

$(-\infty, -2) \cup (2, 2) \cup (2, \infty)$

5. Decide whether the following statement "the equation $2x - 3y = 5$ identifies y as a function of x" is true or false. Explain.

Solution

$$2x - 3y = 5$$
$$-3y = -2x + 5$$
$$y = \tfrac{2}{3}x - \tfrac{5}{3}$$

Each value of x yields only one y-value, so it is a function of x.

True.

6. Decide whether the following statement "the equation $y = \pm\sqrt{x^2-16}$ indicates y is a function of x" is true or false. Explain.

Solution

$y = \pm\sqrt{x^2-16}$

Some x-values yield two y-values.

i.e. $x = 25 \quad \Rightarrow \quad y = \pm 3$

Thus, y is **not** a function of x.

False.

7. Decide whether the following statement "if $A = \{3,\ 4,\ 5\}$ and $B = \{-1,\ -2,\ -3\}$, the set $C = \{(3,\ -9), (4,\ -2), (5,\ -3)\}$ represents a function form A to B" is true or false.

Solution

$A = \{3, 4, 5\},\ B = \{-1, -2, -3\}$

$C = \{(3, -9), (4, -2), (5, -3)\}$

Since -9 is not an element of B, C does **not** represent a function from A to B.

False.

8. Determine (a) the range and domain of the function $f(x) = x^2 + 2$, (b) determine the intervals over which the function is increasing, decreasing, or constant, and (c) determine if the function is even or odd.

Solution

(a) Domain: $(-\infty,\ \infty)$

Range: $[2,\ \infty)$

(b) Increasing on $(0,\ \infty)$

Decreasing on $(-\infty, 0)$

(c)

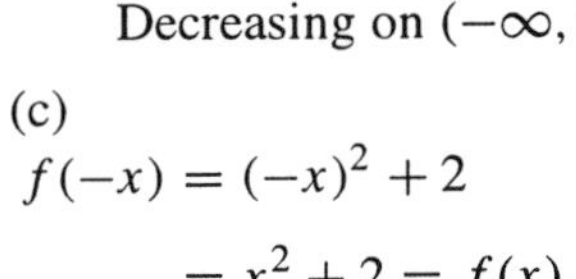

$f(-x) = (-x)^2 + 2$

$= x^2 + 2 = f(x)$

The function is even.

9. Determine (a) the range and domain of the function $g(x) = \sqrt{x^2 - 4}$ (b) determine the intervals over which the function is increasing, decreasing, or constant, and (c) determine if the function is even or odd.

Solution

(a) Domain: $(-\infty, -2] \cup [2,\ \infty)$

Range: $[0,\ \infty)$

(b) Increasing on $(2,\ \infty)$

Decreasing on $(-\infty, -2)$

(c) $f(-x) = \sqrt{(-x)^2 - 4}$

$= \sqrt{x^2 - 4} = f(x)$

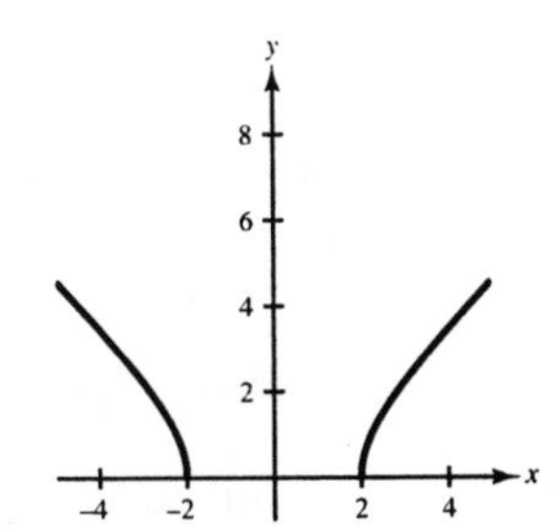

The function is even.

10. Sketch the graph of the following function.

$$f(x) = \frac{3}{x + 2}$$

Solution

x	-4	-3	-1	0
y	$-\frac{3}{2}$	-3	3	$\frac{3}{2}$

Vertical asymptote: $x = -2$

Horizontal asymptote: $y = 0$

y-intercept: $(0, \frac{3}{2})$

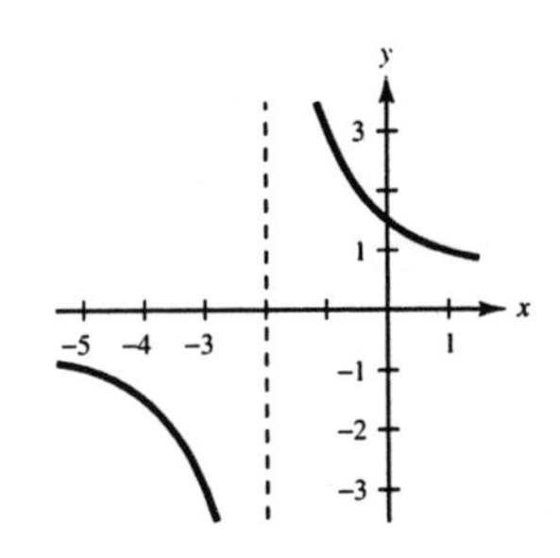

11. Sketch the graph of the following function.

$$g(x) = \begin{cases} x+1, & x<0 \\ 1, & x=0 \\ x^2+1, & x>0 \end{cases}$$

Solution

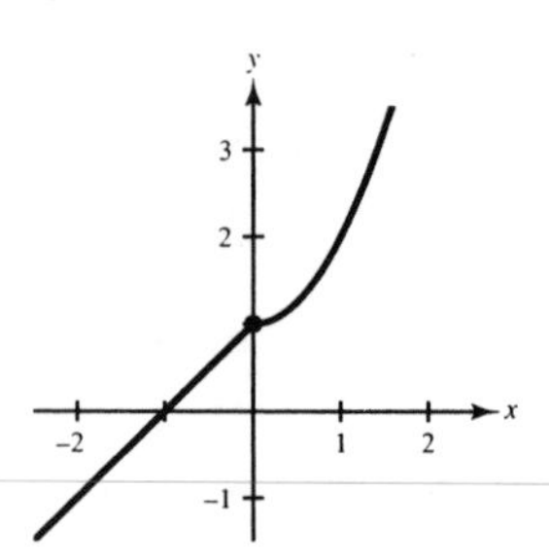

12. Sketch the graph of $h(x) = (x-3)^2 + 4$.

Solution

Parabola

Vertex: (3, 4)

x	2	1	4	5
y	5	8	5	8

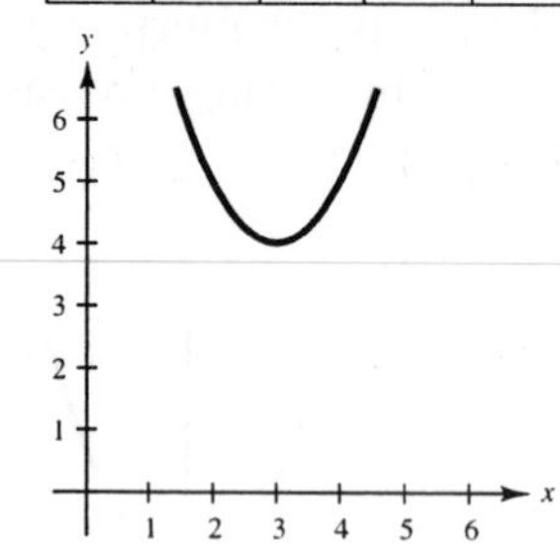

13. Find f^{-1} given $f(x) = x^3 - 5$.

Solution

$$f(x) = x^3 - 5$$
$$y = x^3 - 5$$
$$x = y^3 - 5$$
$$x + 5 = y^3$$
$$y = \sqrt[3]{x+5}$$
$$f^{-1}(x) = \sqrt[3]{x+5}$$

14. Given $f(x) = x^2 + 2$ and $g(x) = 2x - 1$, find $(f+g)(x)$.

Solution

$(f+g)(x) = f(x) + g(x) = (x^2+2) + (2x-1) = x^2 + 2x + 1$

15. Given $f(x) = x^2 + 2$ and $g(x) = 2x - 1$, find $(f-g)(x)$.

Solution

$(f-g)(x) = f(x) - g(x) = (x^2+2) - (2x-1) = x^2 - 2x + 3$

16. Given $f(x) = x^2 + 2$ and $g(x) = 2x - 1$, find $(fg)(x)$.

Solution

$(fg)(x) = f(x) \cdot g(x) = (x^2+2)(2x-1) = 2x^3 - x^2 + 4x - 2$

17. Given $f(x) = x^2 + 2$ and $g(x) = 2x - 1$, find $(f/g)(x)$.

Solution

$$\left(\frac{f}{g}\right)(x) = \frac{f(x)}{g(x)} = \frac{x^2+2}{2x-1}, \quad x \neq \frac{1}{2}$$

18. Given $f(x) = x^2 + 2$ and $g(x) = 2x - 1$, find $(f \circ g)(x)$.

Solution

$(f \circ g)(x) = f(g(x)) = f(2x-1) = (2x-1)^2 + 2 = 4x^2 - 4x + 1 + 2 = 4x^2 - 4x + 3$

19. Given $f(x) = x^2 + 2$ and $g(x) = 2x - 1$, find $g(x)$.

Solution

$(g(x) = g(f(x)) = g(x^2 + 2) = 2(x^2 + 2) - 1 = 2x^2 + 4 - 1 = 2x^2 + 3$

20. The profit for a company is given by

$$P = 300 + 10x - 0.002x^2$$

where x is the number of units produced. What production level will yield a maximum profit?

Solution

$$P = 300 + 10x - 0.002x^2 = -0.002(x^2 - 5000x) + 300$$

$$= -0.002(x^2 - 5000x + 6{,}250{,}000 - 6{,}250{,}000) + 300$$

$$= -0.002(x - 2500)^2 + 12{,}800$$

The profit is maximum (\$12,800) when $x = 2500$ units.

Practice Test for Chapter 3

1. Given $f(x) = x^2 - 2x + 1$, find $f(x - 3)$.

2. Given $f(x) = 4x - 11$, find $\dfrac{f(x) - f(3)}{x - 3}$.

3. Find the domain and range of $f(x) = \sqrt{36 - x^2}$.

4. Which equations determine y as a function of x?

(a) $6x - 5y + 4 = 0$ (b) $x^2 + y^2 = 9$ (c) $y^3 = x^2 + 6$

5. Sketch the graph of $f(x) = \begin{cases} 2x + 1 \text{ if } x \geq 0, \\ x^2 - x \text{ if } x \leq 0. \end{cases}$

6. Given $f(x) = x^2 - 2x + 16$ and $g(x) = 2x + 3$, find $f(g(x))$.

7. Given $f(x) = x^3 + 7$, find $f^{-1}(x)$.

8. Sketch the graph of $f(x) = x^2 - 6x + 5$ and identify the vertex and the intercepts.

9. Find the number of units x that produce a minimum cost C if $C = 0.01x^2 - 90x + 15,000$.

10. Find the quadratic function that has a maximum at (1, 7) and passes through the point (2, 5).

11. Find two quadratic functions that have x-intercepts (2, 0) and $\left(\frac{4}{3}, 0\right)$.

12. Use the Leading Coefficient Test to determine the right-hand and left-hand behavior of the graph of the polynomial function $f(x) = -3x^5 + 2x^3 - 17$.

13. Find all the real zeros of $f(x) = x^5 - 5x^3 + 4x$.

14. Find a polynomial function with 0, 3, and -2 as zeros.

15. Sketch $f(x) = x^3 - 12x$.

16. Sketch the graph of the following.

$$f(x) = \frac{x-1}{2x}$$

Label all intercepts and asymptotes.

17. Sketch the graph of the following.

$$f(x) = \frac{3x^2 - 4}{x}$$

Label all intercepts and asymptotes.

18. Find all the the asymptotes of the following.

$$f(x) = \frac{8x^2 - 9}{x^2 + 1}$$

19. Find all the asymptotes of the following.

$$f(x) = \frac{4x^2 - 2x + 7}{x - 1}$$

20. Sketch the graph of the following.

$$f(x) = \frac{x-5}{(x-5)^2}$$

CHAPTER FOUR
Zeros of Polynomial Functions

4.1 Polynomial Division and Synthetic Division

You should know the following basic techniques and principles of polynomial division.

- The Division Algorithm (Long Division of Polynomials)
- Synthetic Division
- $f(k)$ is equal to the remainder of $f(x)$ divided by $(x - k)$.
- $f(k) = 0$ if and only if $(x - k)$ is a factor of $f(x)$.
- If $f(k) = 0$, then $(k, 0)$ is an x-intercept of the graph of f.

SOLUTIONS TO SELECTED EXERCISES

5. Divide $x^4 + 5x^3 + 6x^2 - x - 2$ by $x + 2$, using long division.

Solution

$$\begin{array}{r|l}
 & x^3 + 3x^2 \qquad\qquad - 1 \\
x + 2 & x^4 + 5x^3 + 6x^2 - x - 2 \\
 & -(x^4 + 2x^3) \\
\hline
 & 3x^3 + 6x^2 \\
 & -(3x^3 + 6x^2) \\
\hline
 & -x - 2 \\
 & -(-x - 2) \\
\hline
 & 0
\end{array}$$

$$\frac{x^4 + 5x^3 + 6x^2 - x - 2}{x + 2} = x^3 + 3x^2 - 1$$

9. Divide $6x^3 + 10x^2 + x + 8$ by $2x^2 + 1$, using long division.

Solution

$$\begin{array}{r|l}
 & 3x + 5 \\
2x^2 + 0x + 1 & 6x^3 + 10x^2 + x + 8 \\
 & -(6x^3 + 0x^2 + 3x) \\
\hline
 & 10x^2 - 2x + 8 \\
 & -(10x^2 + 0x + 5) \\
\hline
 & -2x + 3
\end{array}$$

$$\frac{6x^3 + 10x^2 + x + 8}{2x^2 + 1} = 3x + 5 + \frac{-2x + 3}{2x^2 + 1} = 3x + 5 - \frac{2x - 3}{2x^2 + 1}$$

13. Divide $x^4 + 3x^2 + 1$ by $x^2 - 2x + 3$, using long division.

Solution

$$
\begin{array}{rrrrrrr}
 & & x^2 & +2x & +4 & & \\
\hline
x^2 - 2x + 3 \,\big) & x^4 + & 0x^3 + & 3x^2 & +0x & +1 & \\
 & -(x^4 - & 2x^3 + & 3x^2) & & & \\
\hline
 & & 2x^3 + & 0x^2 & +0x & & \\
 & & -(2x^3 - & 4x^2 & +6x) & & \\
\hline
 & & & 4x^2 & -6x & +1 & \\
 & & & -(4x^2 & -8x & +12) & \\
\hline
 & & & & 2x & -11 &
\end{array}
$$

$$\frac{x^4 + 3x^2 + 1}{x^2 - 2x + 3} = x^2 + 2x + 4 + \frac{2x - 11}{x^2 - 2x + 3}$$

17. Divide $4x^3 - 9x + 8x^2 - 18$ by $x + 2$, using synthetic division.

Solution

$$
\begin{array}{r|rrrr}
-2 & 4 & 8 & -9 & -18 \\
 & & -8 & 0 & 18 \\
\hline
 & 4 & 0 & -9 & 0
\end{array}
$$

Quotient: $4x^2 - 9$

21. Divide $5x^3 - 6x^2 + 8$ by $x - 4$, using synthetic division.

Solution

$$
\begin{array}{r|rrrr}
4 & 5 & -6 & 0 & 8 \\
 & & 20 & 56 & 224 \\
\hline
 & 5 & 14 & 56 & 232
\end{array}
$$

Quotient: $5x^2 + 14x + 56 + \dfrac{232}{x - 4}$

25. Divide $x^3 + 512$ by $x + 8$, using synthetic division.

Solution

$$
\begin{array}{r|rrrr}
-8 & 1 & 0 & 0 & 512 \\
 & & -8 & 64 & -512 \\
\hline
 & 1 & -8 & 64 & 0
\end{array}
$$

Quotient: $x^2 - 8x + 64$

29. Divide $5 - 3x + 2x^2 - x^3$ by $x + 1$, using snythetic division.

Solution

$$
\begin{array}{r|rrrr}
-1 & -1 & 2 & -3 & 5 \\
 & & 1 & -3 & 6 \\
\hline
 & -1 & 3 & -6 & 11
\end{array}
$$

Quotient: $-x^2 + 3x - 6 + \dfrac{11}{x + 1}$

33. Use synthetic division to show that x is a solution of the third-degree polynomial equation $x^3 - 7x + 6 = 0,\ x = 2$, and use the result to factor the the polynomial completely.

Solution

$$
\begin{array}{r|rrrr}
2 & 1 & 0 & -7 & 6 \\
 & & 2 & 4 & -6 \\
\hline
 & 1 & 2 & -3 & 0
\end{array}
$$

$$x^3 - 7x + 6 = (x - 2)(x^2 + 2x - 3) = (x - 2)(x + 3)(x - 1)$$

37. Use synthetic division to show that x is a solution of the third- degree polynomial equation $x^3 + 2x^2 - 3x - 6 = 0,\ x = \sqrt{3}$, and use the result to factor the polynomial completely.

Solution

$$\begin{array}{r|llll} \sqrt{3} & 1 & 2 & -3 & -6 \\ & & \sqrt{3} & 3+2\sqrt{3} & 6 \\ \hline & 1 & 2+\sqrt{3} & 2\sqrt{3} & 0 \end{array}$$

$$\begin{array}{r|lll} -\sqrt{3} & 1 & 2+\sqrt{3} & 2\sqrt{3} \\ & & -\sqrt{3} & -2\sqrt{3} \\ \hline & 1 & 2 & 0 \end{array}$$

$$x^3 + 2x^2 - 3x - 6 = (x - \sqrt{3})(x + \sqrt{3})(x + 2)$$

41. Express $f(x) = x^3 + 3x^2 - 2x - 14$ in the form $f(x) = (x - k)q(x) + r$ for the value of $k = \sqrt{2}$, and demonstrate that $f(k) = r$.

Solution

$$\begin{array}{r|llll} \sqrt{2} & 1 & 3 & -2 & -14 \\ & & \sqrt{2} & 2+3\sqrt{2} & 6 \\ \hline & 1 & 3+\sqrt{2} & 3\sqrt{2} & -8 \end{array}$$

$$f(x) = (x - \sqrt{2})[x^2 + (3 + \sqrt{2})x + 3\sqrt{2}] - 8$$

$$f(\sqrt{2}) = 2\sqrt{2} + 6 - 2\sqrt{2} - 14 = -8$$

45. Given $h(x) = 3x^3 + 5x^2 - 10x + 1$, use synthetic division to find the following function values.

(a) $h(3)$ (b) $h\left(\frac{1}{3}\right)$ (c) $h(-2)$ (d) $h(-5)$

Solution

(a)
$$\begin{array}{r|llll} 3 & 3 & 5 & -10 & 1 \\ & & 9 & 42 & 96 \\ \hline & 3 & 14 & 32 & 97 \end{array}$$

$h(3) = 97$

(b)
$$\begin{array}{r|llll} \frac{1}{3} & 3 & 5 & -10 & 1 \\ & & 1 & 2 & -\frac{8}{3} \\ \hline & 3 & 6 & -8 & -\frac{5}{3} \end{array}$$

$h\left(\frac{1}{3}\right) = -\frac{5}{3}$

(c)
$$\begin{array}{r|llll} -2 & 3 & 5 & -10 & 1 \\ & & -6 & 2 & 16 \\ \hline & 3 & -1 & -8 & 17 \end{array}$$

$h(-2) = 17$

(d)
$$\begin{array}{r|llll} -5 & 3 & 5 & -10 & 1 \\ & & -15 & 50 & -200 \\ \hline & 3 & -10 & 40 & -199 \end{array}$$

$h(-5) = -199$

51. Match $f(x) = x^3 + 5x^2 + 6x + 2$ with its graph and use the result to find all real solutions of $f(x) = 0$. [The graphs are labeled (a), (b), (c), and (d) in the textbook.]

Solution

Matches graph (a)

$x = -1$ is a real solution of $f(x) = 0$.

$$\begin{array}{r|rrrr} -1 & 1 & 5 & 6 & 2 \\ & & -1 & -4 & -2 \\ \hline & 1 & 4 & 2 & 0 \end{array}$$

$(x + 1)(x^2 + 4x + 2) = 0$

$x = -1$ or $x = -2 \pm \sqrt{2}$

(Use the Quadratic Formula.)

53. Simplify the following rational function.

$$\frac{4x^3 - 8x^2 + x + 3}{2x - 3}$$

Solution

$$\begin{array}{r|rrrr} \frac{3}{2} & 2 & -4 & \frac{1}{2} & \frac{3}{2} \\ & & 3 & -\frac{3}{2} & -\frac{3}{2} \\ \hline & 2 & -1 & -1 & 0 \end{array}$$

$$\frac{4x^3 - 8x^2 + x + 3}{2x - 3} = \frac{2x^3 - 4x^2 + \frac{1}{2}x + \frac{3}{2}}{x - \frac{3}{2}}$$

$$= 2x^2 - x - 1 = (2x + 1)(x - 1)$$

57. Simplify the following rational function.

$$\frac{x^4 + 6x^3 + 11x^2 + 6x}{x^2 + 3x + 2}$$

Solution

$$\begin{array}{r|rrrrr} -1 & 1 & 6 & 11 & 6 & 0 \\ & & -1 & -5 & -6 & 0 \\ \hline & 1 & 5 & 6 & 0 & 0 \end{array}$$

$$\begin{array}{r|rrrr} -2 & 1 & 5 & 6 & 0 \\ & & -2 & -6 & 0 \\ \hline & 1 & 3 & 0 & 0 \end{array}$$

$$\frac{x^4 + 6x^3 + 11x^2 + 6x}{(x + 1)(x + 2)} = x^2 + 3x = x(x + 3)$$

61. *Profit* A company that produces compact discs estimates that the profit for selling a particular disc is

$$P = -153.6x^3 + 5760x^2 - 100{,}000, \ 0 \le x \le 35$$

where P is the profit in dollars and x is the advertising expense (in ten thousands of dollars). For this disc, the advertising expense was \$300,000 ($x = 30$) and the profit was \$936,800.

(a) From the graph shown in the textbook, it appears that the company could have obtained the same profit by spending less on advertising. Use the graph to estimate another amount that the company could have spent that would produce the same profit.

(b) Use synthetic division with $x = 30$ to algebraically confirm the result of part (a).

Solution

(a) From the graph it appears that $x \approx \$190{,}000$ would produce the same profit.

(b) $-153.6x^3 + 5760x^2 - 100{,}000 = 936{,}800$

$0 = 153.6x^3 - 5760x^2 + 1{,}036{,}800$

$x = 30$ is a real solution.

$$\begin{array}{r|rrrr} 30 & 153.6 & -5760 & 0 & 1{,}036{,}800 \\ & & 4608 & -34{,}560 & -1{,}036{,}800 \\ \hline & 153.6 & -1152 & -34{,}560 & 0 \end{array}$$

$(x - 30)(153.6x^2 - 1152x - 34{,}560) = 0$

$$x = 30 \text{ or } x = \frac{1152 \pm \sqrt{22{,}560{,}768}}{307.2}$$

The other nonnegative real solution is:

$$x = \frac{1152 + \sqrt{22{,}560{,}768}}{307.2} \approx 19.2116 = \$192{,}116$$

4.2 Real Zeros of Polynomial Functions

- You should know Descartes's Rule of Signs.
  (a) The number of positive real zeros of f is either equal to the number of variations of sign of f or is less than that number by an even integer.
  (b) The number of negative real zeros of f is either equal to the number of variations in sign of $f(-x)$ or is less than that number by an even integer.
  (c) When there is only one variation in sign, there is exactly one positive (or negative) real zero.
- You should know the Rational Zero Test.
- You should know shortcuts for the Rational Zero Test.
  (a) Use a programmable calculator.
  (b) Sketch a graph.
  (c) After finding a root, use snythetic division to reduce the degree of the polynomial.
- You should be able to observe the last row obtained from synthetic division in order to determine upper or lower bounds.
  (a) If the test value is positive and all of the entries in the last row are positive or zero, then the test value is an upper bound.
  (b) If the test value is negative and the entries in the last row alternate from positive to negative, then the test value is a lower bound. (Zero entries count as positive or negative.)

SOLUTIONS TO SELECTED EXERCISES

5. Use Descartes's Rule of Signs to determine the possible number of positive and negative zeros of $g(x) = 2x^3 - 3x^2 - 3$.

Solution

$g(x) = 2x^3 - 3x^2 - 3$

Sign variations: 1, positive zeros: 1

$g(-x) = -2x^3 - 3x^2 - 3$

Sign variations: 0, negative zeros: 0

9. Use Descartes's Rule of Signs to determine the possible number of positive and negative zeros of $g(x) = 5x^5 + 10x$.

Solution

$g(x) = 5x^5 + 10x = 5x(x^4 + 2)$

Sign variations: positive zeros: 0

$g(-x) = -5x^5 - 10x$

Sign variations: 0, negative zeros: 0

13. Given $f(x) = -4x^3 + 15x^2 - 8x - 3$, use the Rational Zero Test to list all possible rational zeros of f. Then use a graphing utility to graph the function. Use the graph to help determine which of the possible rational zeros are actual zeros of the function.

Solution

Possible rational zeros: $\pm 1, \pm 3, \pm \frac{1}{2}, \pm \frac{3}{2}, \pm \frac{1}{4}, \pm \frac{3}{4}$

From the graph we see that $x = -\frac{1}{4}, 1, 3$ are the zeros of $f(x)$.

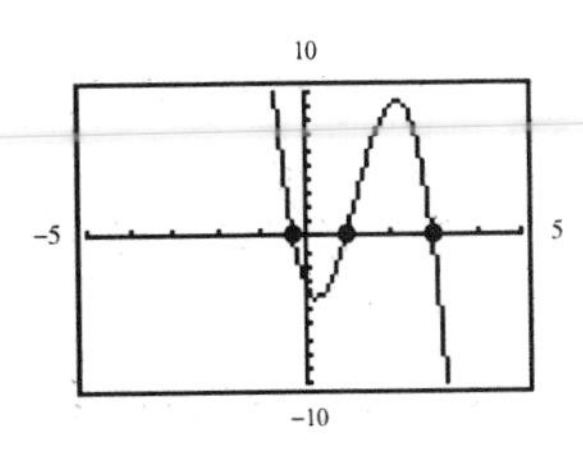

17. Given $f(x) = x^4 - 4x^3 + 15$, use synthetic division to determine whether the give x-value is an upper bound of the zeros of f, a lower bound of the zeros of f, or neither.

(a) $x = 4$ (b) $x = -1$ (c) $x = 3$

Solution

(a)
$$\begin{array}{r|rrrrr} 4 & 1 & -4 & 0 & 0 & 15 \\ & & 4 & 0 & 0 & 0 \\ \hline & 1 & 0 & 0 & 0 & 15 \end{array}$$

4 is an upper bound.

(b)
$$\begin{array}{r|rrrrr} -1 & 1 & -4 & 0 & 0 & 15 \\ & & -1 & 5 & -5 & 5 \\ \hline & 1 & -5 & 5 & -5 & 20 \end{array}$$

-1 is a lower bound.

(c)
$$\begin{array}{r|rrrrr} 3 & 1 & -4 & 0 & 0 & 15 \\ & & 3 & -3 & -9 & -27 \\ \hline & 1 & -1 & -3 & -9 & -12 \end{array}$$

3 is not an upper bound or a lower bound.

21. Find the real zeros of $f(x) = x^3 - 6x^2 + 11x - 6$.

Solution

Descartes's Rule of Signs: 1 or 3 positive zeros

Possible rational zeros: $\pm 1, \pm 2, \pm 3, \pm 6$

$$\begin{array}{r|rrrr} 1 & 1 & -6 & 11 & -6 \\ & & 1 & -5 & 6 \\ \hline & 1 & -5 & 6 & 0 \end{array}$$

$$f(x) = (x-1)(x^2 - 5x + 6)$$
$$= (x-1)(x-2)(x-3)$$

Real zeros: 1, 2, 3

25. Find the real zeros of $h(t) = t^3 + 12t^2 + 21t + 10$.

Solution

Descartes's Rule of Signs: 3 or 1 negative zeros

Possible rational zeros: $\pm 1, \pm 2, \pm 5, \pm 10$

$$\begin{array}{r|rrrr} -1 & 1 & 12 & 21 & 10 \\ & & -1 & -11 & -10 \\ \hline & 1 & 11 & 10 & 0 \end{array}$$

$$h(t) = (t+1)(t^2 + 11t + 10)$$
$$= (t+1)(t+1)(t+10)$$

Real zeros: -10, -1 (repeated)

29. Find the real zeros of $C(x) = 2x^3 + 3x^2 - 1$.

Solution

Descartes's Rule of Signs: 1 positive zero, 2 or no negative zeros

Possible rational zeros: $\pm\frac{1}{2}, \pm 1$

-1	2	3	0	-1
		-2	-1	1
	2	1	-1	0

$$C(x) = (x + 1)(2x^2 + x - 1)$$
$$= (x + 1)(x + 1)(2x - 1)$$

Real zeros: -1 (repeated), $\frac{1}{2}$

33. Find the real zeros of $f(y) = 4y^3 + 3y^2 + 8y + 6$.

Solution

Descartes's Rule of Signs: 3 or 1 negative zeros

Possible rational zeros: $\pm\frac{1}{4}, \pm\frac{1}{2}, \pm\frac{3}{4}, \pm 1, \pm\frac{3}{2}, \pm 2, \pm 3, \pm 6$

$-\frac{3}{4}$	4	3	8	6
		-3	0	-6
	4	0	8	0

$$f(y) = \left(y + \tfrac{3}{4}\right)(4)(y^2 + 2)$$
$$= (4y + 3)(y^2 + 2)$$

Real zero: $-\frac{3}{4}$

37. Find all real solutions of $z^4 - z^3 - 2z - 4 = 0$.

Solution

Descartes's Rule of Signs: 1 positive zero, 1 negative zero

Possible rational zeros: $\pm 1, \pm 2, \pm 4$

-1	1	-1	0	-2	-4
		-1	2	-2	4
	1	-2	2	-4	0

2	1	-2	2	-4
		2	0	4
	1	0	2	0

$$(z + 1)(z - 2)(z^2 + 2) = 0$$

Real zeros: $z = -1, 2$

41. Find all real solutions of $2x^4 - 11x^3 - 6x^2 + 64x + 32 = 0$.

Solution

Descartes's Rule of Signs: 2 or no positive zeros, 2 or no negative zeros

Possible rational zeros:

$\pm\frac{1}{2}, \pm 1, \pm 2, \pm 4, \pm 8, \pm 16, \pm 32$

-2	2	-11	-6	64	32
		-4	30	-48	-32
	2	-15	24	16	0

4	2	-15	24	16
		8	-28	-16
	2	-7	-4	0

$$(x - 2)(x + 4)(2x^2 - 7x - 4) = 0$$
$$(x - 2)(x - 4)(2x + 1)(x - 4) = 0$$

Real zeros: $x = -2, -\frac{1}{2}, 4$ (repeated)

43. Find all real solutions of $x^5 - 7x^4 + 10x^3 + 14x^2 - 24x = 0$.

Solution

$x(x^4 - 7x^3 + 10x^2 + 14x - 24) = 0$

Descartes's Rule of Signs: 3 or 1 positive solutions, 1 negative solution

Possible rational solutions: $\pm 1, \pm 2, \pm 3, \pm 4, \pm 6, \pm 8, \pm 12, \pm 24$

3	1	−7	10	14	−24
		3	−12	−6	24
	1	−4	−2	8	0

4	1	−4	−2	8
		4	0	−8
	1	0	−2	0

$x(x-3)(x-4)(x^2-2) = 0$

Real solutions: $x = 0, 3, 4$

Other real solutions: $x = -\sqrt{2}, \sqrt{2}$

47. Given $f(x) = 4x^3 + 7x^2 - 11x - 18$, (a) list all possible rational zeros of f, (b) sketch the graph of f so that some of the possible zeros in part (a) can be discarded, and then (c) determine all real zeros of f.

Solution

(a) $\pm 1, \pm 2, \pm 3, \pm 6, \pm 9, \pm 18, \pm \frac{1}{2}, \pm \frac{3}{2}, \pm \frac{9}{2}, \pm \frac{1}{4}, \pm \frac{3}{4}, \pm \frac{9}{4}$

(b)

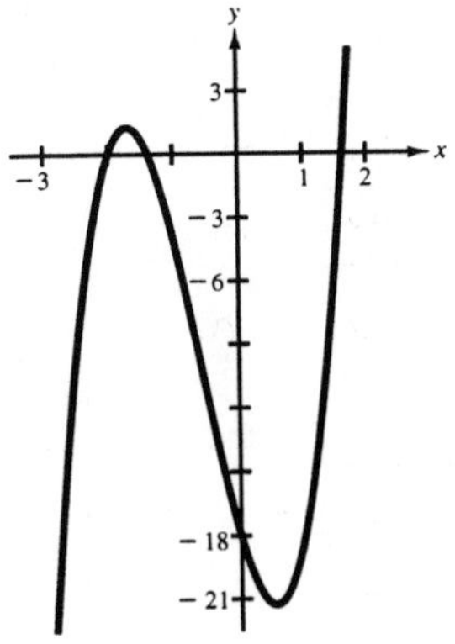

−2	4	7	−11	−18
		−8	2	18
	4	−1	−9	0

$f(x) = (x+2)(4x^2 - x - 9)$

(c) Real zeros: $-2, \dfrac{1 \pm \sqrt{145}}{8}$ (Quadratic Formula)

51. Find the rational zeros of $f(x) = x^3 - \frac{1}{4}x^2 - x + \frac{1}{4}$.

Solution

$f(x) = x^3 - \frac{1}{4}x^2 - x + \frac{1}{4} = \frac{1}{4}(4x^3 - x^2 - 4x + 1)$

Possible rational zeros: $\pm \frac{1}{4}, \pm \frac{1}{2}, \pm 1$

By testing, we see that the rational zeros are $x = -1, \frac{1}{4}, 1$

57. *Dimensions of a Box* An open box is to be made from a rectangular piece of material, 12 inches by 10 inches, by cutting equal squares from each corner and turning up the sides. Find the dimensions of the box, given that the volume is to be 96 cubic inches.

Solution

$V = 96 = (12 - 2x)(10 - 2x)(x),\ 0 < x < 5$

$$0 = 4x^3 - 44x^2 + 120x - 96$$

$$0 = 4(x^3 - 11x + 30x - 24)$$

$$0 = 4(x - 2)(x^2 - 9x + 12)$$

$$x = 2 \text{ or } x = \frac{9 - \sqrt{33}}{2} \approx 1.63$$

Box: 8 in. × 6 in. × 2 in. or 8.74 in. × 6.74 in. × 1.63 in.

59. *Dimensions of a Package* A rectangle package to be sent by a postal service can have a maximum combined length and girth (perimeter of a cross section) of 108 inches. Find the dimensions of the package, given that the volume is 11,664 cubic inches. (See figure in textbook.)

Solution

$4x + y = 108 \Rightarrow y = 108 - 4x$

$$V = x^2y = x^2(108 - 4x) = -4x^3 + 108x^2 = 11{,}664$$

$$4x^3 - 108x^2 + 11{,}664 = 0$$

$$x^3 - 27x^2 + 2916 = 0$$

$$(x + 9)(x - 18)^2 = 0$$

$$x = 18$$

$$y = 36$$

Box: 18 in. × 18 in. × 36 in.

4.3 Approximating Zeros of Polynomial Functions

- Be able to use the ZOOM and TRACE keys on your graphing utility to approximate the zeros of polynomial functions.
- Be able to use the SOLVER or ROOT keys on your graphing utility to approximate the zeros of polynomial functions.

SOLUTIONS TO SELECTED EXERCISES

3. Match $f(x) = 2x^3 - 6x^2 + 6x - 1$ with its graph. Then approximate the real zeros of the function to three decimal places. [The graphs are labeled (a), (b), (c), (d), (e), and (f) in the textbook.]

Solution

$f(x) = 2x^3 - 6x^2 + 6x - 1$ matches graph (d). The only real zero is at $x \approx 0.206$.

9. Use a graphing utility to approximate the real zeros of $f(x) = x^3 - 27x - 27$. Use an accuracy of 0.001.

Solution

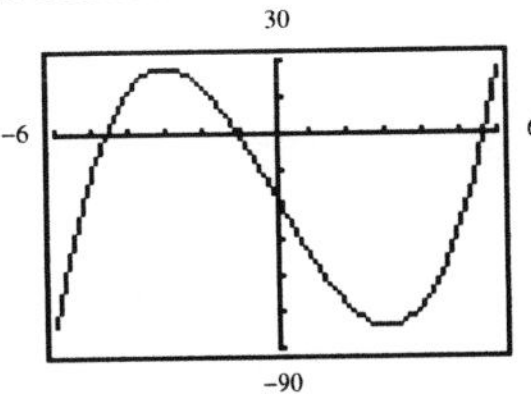

From the graph we see that there are three real zeros. Using ZOOM and TRACE we determine the zeros to be $x \approx -4.596$, $x \approx -1.042$, and $x \approx 5.638$.

13. Use a graphing utility to approximate the real zeros of $f(x) = -x^3 + 2x^2 + 4x + 5$. Use an accuracy of 0.001.

Solution

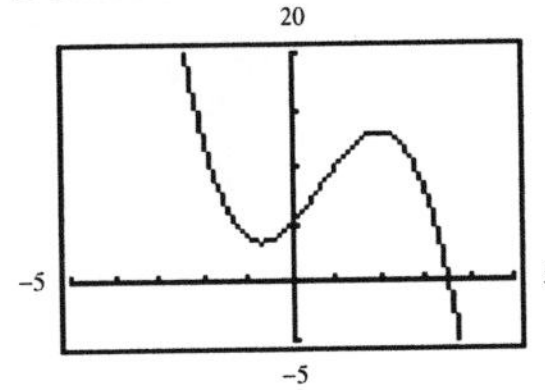

From the graph we see that there is only one real zero. Using ZOOM and TRACE we determine the zero to be $x \approx 3.533$.

17. Use a graphing utility to approximate the real zeros of $f(x) = -x^3 - 4x + 1$. Use an accuracy of 0.001.

Solution

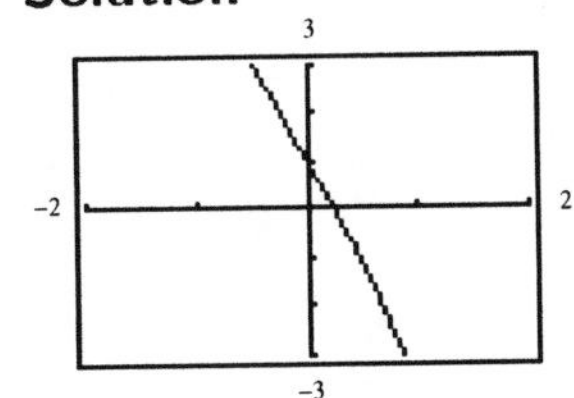

From the graph we see that there is only one real zero. Using ZOOM and TRACE we determine the zero to be $x \approx 0.246$.

21. *Cost of Dental Care* From 1970 to 1993, the cost of dental care in the United States rose according to the function

$$C = \sqrt{1.9t^3 + 13.4t^2 + 128.6t + 1536.8}, \ 0 \le t \le 23$$

where C is the dental care index and $t = 0$ represents 1970. In 1984 the dental care index was about 84. According to this model, when did the cost of dental care reach a level that was *twice* the 1984 cost? (*Source: U.S. Bureau of Labor Statistics.*)

Solution

$$\sqrt{1.9t^3 + 13.4t^2 + 128.6t + 1536.8} = 2(84)$$

$$1.9t^3 + 13.4t^2 + 128.6t + 1536.8 = 28{,}224$$

$$1.9t^3 + 13.4t^2 + 128.6t - 26{,}687.2 = 0$$

Using the methods of this section, we have $t \approx 21$ which corresponds with the year 1991.

25. *Advertising Costs* A company that produces portable cassette players estimates that the profit for selling a particular model is given by

$$P = -76x^3 + 4{,}830x^2 - 320{,}000, \ 0 \le x \le 60$$

where P is the profit in dollars and x is the advertising expense (in ten thousands of dollars). According to this model, how much money should the company spend to obtain a profit of $2,500,000?

Solution

$$2{,}500{,}000 = -76x^3 + 4{,}830x^2 - 320{,}000$$

$$76x^3 - 4{,}830x^2 + 2{,}820{,}000 = 0$$

Using the methods of this section, we have $x \approx 38.4$ or $x \approx 46.1$ which corresponds to $384,000 or $461,000 being spent on advertising.

Mid-Chapter Quiz for Chapter 4

1. Completely factor the polynomial $2x^3 - x^2 - 13x - 6$ given $x + 2$ as one factor.

Solution

$$\begin{array}{r|rrrr}
 & 2x^2 & -\ 5x & -\ 3 & \\ \hline
x+2 & 2x^3 & -\ x^2 & -\ 13x & -\ 6 \\
 & -\ 2x^3 & +\ 4x^2 & & \\ \hline
 & & -\ 5x^2 & -\ 13x & \\
 & & -(-\ 5x^2 & -\ 10x) & \\ \hline
 & & & -\ 3x & -\ 6 \\
 & & & -(-\ 3x & -\ 6) \\ \hline
 & & & & 0
\end{array}$$

Thus, $2x^3 - x^2 - 13x - 6 = (x + 2)(2x^2 - 5x - 3)$

$$= (x + 2)(2x + 1)(x - 3)$$

2. Completely factor the polynomial $3x^3 - 5x^2 - 58x + 40$ given $x - \frac{2}{3}$ as one factor.

Solution

$$
\begin{array}{r|l}
 & 3x^2 \quad - \quad 3x \quad - 60 \\
\hline
x - \frac{2}{3} & 3x^3 \quad - 5x^2 \quad - 58x \quad - 40 \\
 & 3x^3 \quad - 2x^2 \\
\cline{2-2}
 & \quad - 3x^2 \quad - 58x \\
 & - (- 3x^2 \ + \ 2x) \\
\cline{2-2}
 & \qquad - 60x \quad - 40 \\
 & \qquad - (- 60x \quad - 40) \\
\cline{2-2}
 & \qquad\qquad 0
\end{array}
$$

Thus,
$$
\begin{aligned}
3x^3 - 5x^2 - 58x - 40 &= \left(x - \tfrac{2}{3}\right)(3x^2 - 3x - 60) \\
&= \left(x - \tfrac{2}{3}\right)(3)(x^2 - x - 20) \\
&= (3x - 2)(x - 5)(x + 4)
\end{aligned}
$$

3. Use synthetic division to evaluate $f(2)$ for $f(x) = 3x^3 - 5x^2 + 9$.

Solution

$$
\begin{array}{r|rrrr}
2 & 3 & -5 & 0 & 9 \\
 & & 6 & 1 & 4 \\
\hline
 & 3 & 1 & 2 & 13
\end{array}
$$

Since $r = 13$, $f(2) = 13$.

4. Use synthetic division to evaluate $f(2)$ for $f(x) = 0.3x^4 - 1.8x^3 + 0.7x^2 - 3$

Solution

$$
\begin{array}{r|rrrrr}
2 & 0.3 & -1.8 & 0.7 & 0 & -3 \\
 & & 0.6 & -2.4 & -3.4 & -6.8 \\
\hline
 & 0.3 & -1.2 & -1.7 & -3.4 & -9.8
\end{array}
$$

Since $r = -9.8$, $f(2) = -9.8$.

5. Simplify. $\dfrac{2x^4 + 9x^3 - 32x^2 - 99x + 180}{x^2 + 2x - 15}$

Solution

$$
\begin{array}{r|l}
 & \qquad\qquad\qquad 2x^2 \ + \ 5x \ - \ 12 \\
\hline
x^2 + 2x - 15 & 2x^4 + 9x^3 \quad - 32x^2 \quad - 99x \quad + 180 \\
 & - (2x^4 + 4x^3 \quad - 30x^2) \\
\cline{2-2}
 & \qquad 5x^3 \quad - \ 2x^2 \quad - 99x \\
 & \qquad - (5x^3 \quad + 10x^2 \quad - 75x) \\
\cline{2-2}
 & \qquad\qquad - 12x^2 \quad - 24x \quad + 180 \\
 & \qquad\qquad - (- 12x^2 \quad - 24x \quad + 180) \\
\cline{2-2}
 & \qquad\qquad\qquad\qquad 0
\end{array}
$$

Thus,
$$
\begin{aligned}
\frac{2x^4 + 9x^3 - 32x^2 - 99x + 180}{x^2 + 2x - 15} &= 2x^2 + 5x - 12 \\
&= (2x - 3)(x + 4)
\end{aligned}
$$

6. The profit for a company is $P = -95x^3 + 5650x^2 - 250{,}000$, $0 \le x \le 55$, where x is the advertising expense (in ten thousands of dollars). What is the profit for an advertising expense of \$450,000? Use a graphing utility to approximate another advertising expense that would yield the same profit.

Solution

When the advertising expense is \$450,000,

$$P = -95(45)^3 + 5650(45)^2 - 250{,}000$$

$$= \$2{,}534{,}375$$

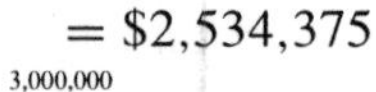

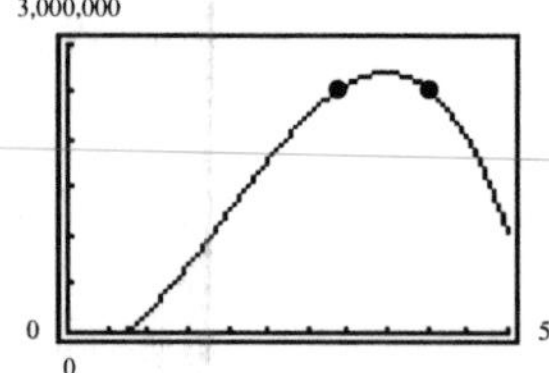

When $x \approx 33.764$, the advertising expense would yield the same profit.

7. Find all real zeros of $f(x) = -2x^3 - 7x^2 + 10x + 35$.

Solution

Possible rational zeros: $\pm 1, \pm 5, \pm 7, \pm 35, \pm\frac{1}{2}, \pm\frac{5}{2}, \pm\frac{7}{2}, \pm\frac{35}{2}$

By testing we see that $x = -\frac{7}{2}$ is a zero.

$-\frac{7}{2}$	-2	-7	10	35
		7	0	-35
	-2	0	10	0

Thus, $f(x) = \left(x + \frac{7}{2}\right)(-2x^2 + 10)$

$$= \left(x + \frac{7}{2}\right)(-2)(x^2 - 5)$$

$$= -(2x + 7)(x^2 - 5)$$

The real zeros of $f(x)$ are

$x = -\frac{7}{2}$ and $x = \pm\sqrt{5}$.

8. Find all the real zeros of $f(x) = 4x^4 - 37x^2 + 9$

Solution

$$f(x) = 4x^4 - 37x^2 + 9 = (4x^2 - 1)(x^2 - 9)$$

$$= (2x + 1)(2x - 1)(x + 3)(x - 3)$$

The real zeros of $f(x)$ are $x = \pm\frac{1}{2}$ and $x = \pm 3$.

9. Find all the real zeros of $f(x) = 3x^4 + 4x^3 - 3x - 4$.

Solution

$$f(x) = 3x^4 + 4x^3 - 3x - 4$$

$$= x^3(3x + 4) - (3x + 4)$$

$$= (3x + 4)(x^3 - 1)$$

The real zeros of $f(x)$ are $x = -\frac{4}{3}$ and $x = 1$.

10. Find all the real zeros of $f(x) = 2x^3 - 3x^2 + 2x - 3$.

Solution

$$f(x) = 2x^3 - 3x^2 + 2x - 3$$

$$= x^2(2x - 3) + (2x - 3)$$

$$= (2x - 3)(x^2 + 1)$$

The only real zero of $f(x)$ is $x = \frac{3}{2}$.

11. Find all the real zeros of $f(x) = x^4 - 5x^2 + 4$.

Solution

$$f(x) = x^4 - 5x^2 + 4 = (x^2 - 1)(x^2 - 4)$$

$$= (x + 1)(x - 1)(x + 2)(x - 2)$$

The real zeros of $f(x)$ are $x = \pm 1$, $x = \pm 2$

12. Find all the real zeros of $f(x) = 3x^3 - 4x^2 + 1$.

Solution

Possible rational zeros: $\pm 1,\ \pm\frac{1}{3}$

By testing we see that $x = 1$ is a zero.

$$\begin{array}{r|rrrr} 1 & 3 & -4 & 0 & 1 \\ & & 3 & -1 & -1 \\ \hline & 3 & -1 & -1 & 0 \end{array}$$

Thus, $f(x) = (x-1)(3x^2 - x - 1)$. By the Quadratic Formula, the zeros of $3x^2 - x - 1$ are

$$x = \frac{1 \pm \sqrt{13}}{6}.$$

The real zeros of $f(x)$ are $x = 1,\ x = \dfrac{1 \pm \sqrt{13}}{6}$.

13. Find all the real zeros of $f(x) = x^4 + x^3 + x^2 + 3x - 6$.

Solution

Possible rational zeros: $\pm 1,\ \pm 2,\ \pm 3,\ \pm 6$

By testing we see that $x = 1$ and $x = -2$ are zeros.

$$\begin{array}{r|rrrrr} 1 & 1 & 1 & 1 & 3 & -6 \\ & & 1 & 2 & 3 & 6 \\ \hline & 1 & 2 & 3 & 6 & 0 \end{array}$$

$$\begin{array}{r|rrrr} -2 & 1 & 2 & 3 & 6 \\ & & -2 & 0 & -6 \\ \hline & 1 & 0 & 3 & 0 \end{array}$$

Thus, $f(x) = (x-1)(x+2)(x^2+3)$. The only real zeros of $f(x)$ are $x = 1$ and $x = -2$.

14. Find all the real zeros of $f(x) = 5x^3 - 7x^2 + 2$.

Solution

Possible rational zeros: $\pm 1,\ \pm 2,\ \pm\frac{1}{5},\ \pm\frac{2}{5}$

By testing we see that $x = 1$ is a zero.

$$\begin{array}{r|rrrr} 1 & 5 & -7 & 0 & 2 \\ & & 5 & -2 & -2 \\ \hline & 5 & -2 & -2 & 0 \end{array}$$

Thus, $f(x) = (x-1)(5x^2 - 2x - 2)$. By the Quadratic Formula, the zeros of $5x^2 - 2x - 2$ are

$$x = \frac{2 \pm \sqrt{44}}{10} = \frac{1 \pm \sqrt{11}}{5}.$$

The real zeros of $f(x)$ are $x = 1,\ x = \dfrac{1 \pm \sqrt{11}}{5}$.

15. Use Descarte's Rule of Signs to determine the possible number of positive and negative zeros. Then use a graphing utility to determine the actual number of positive and negative zeros for $f(x) = -2x^3 - x^2 + 11x - 5$.

Solution

$f(x) = -2x^3 - x^2 + 11x - 5$ has two variations in signs.

$$f(-x) = -2(-x)^3 - (-x)^2 + 11(-x) - 5$$
$$= 2x^3 - x^2 - 11x - 5 \text{ has one variation in sign.}$$

Thus, $f(x)$ has either two or no positive real zeros and one negative real zero.

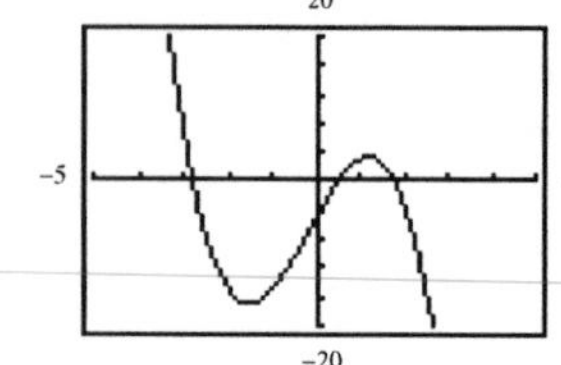

From the graph we see that $f(x)$ has two positive zeros and one negative zero.

16. Use Descarte's Rule of Signs to determine the possible number of positive and negative zeros. Then use a graphing utility to determine the actual number of positive and negative zeros for $f(x) = x^4 - 19x^2 + 48$.

Solution

$f(x) = x^4 - 19x^2 + 48$ has two variations in signs.

$f(-x) = (-x)^4 - 19(-x)^2 + 48 = f(x)$ has two variations in signs. Thus, $f(x)$ has either two or no positive real zeros and either two or no negative real zeros.

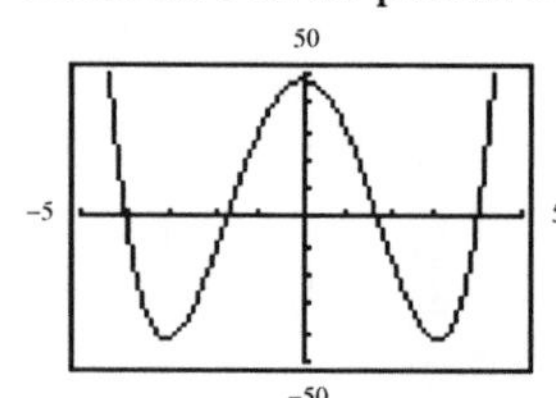

From the graph we see that $f(x)$ has two positive and two negative zeros.

In Exercises 17–20, an open box is to be made from a 12-inch by 14-inch piece of material by cutting equal squares from each corner and turning up the sides. The volume of the box is to be 160 cubic inches.

17. Write an equation to find the dimensions of the box.

Solution

Length: $l = 14 - 2x$

Width: $w = 12 - 2x$

Height: $h = x$

Volume: $V = l \cdot w \cdot h$

$$160 = (14 - 2x)(12 - 2x)(x)$$

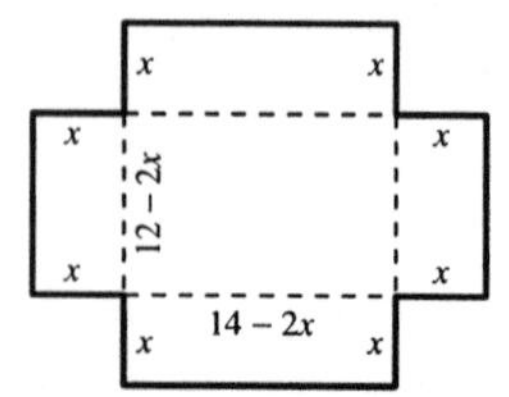

18. Find the dimensions of the box analytically.

Solution

$$160 = (14 - 2x)(12 - 2x)x$$

$$160 = 4x^3 - 52x^2 + 168x$$

$4x^3 - 52x^2 + 168x - 160 = 0$

$4(x^3 - 13x^2 + 42x - 40) = 0,\ 0 < x < 6$

$x^3 - 13x^2 + 42x - 40 = 0$

Possible rational zeros: $\pm1,\ \pm2,\ \pm4,\ \pm5,\ \pm8,\ \pm10,\ \pm20,\ \pm40$

By testing we see that $x = 2$ is a zero.

$$\begin{array}{r|rrrr} 2 & 1 & -13 & 42 & -40 \\ & & 2 & -22 & 40 \\ \hline & 1 & -11 & 20 & 0 \end{array}$$

Thus, $x^3 - 13x^2 + 42x - 40 = (x - 2)(x^2 - 11x + 20)$.
By the Quadratic Formula, the zeros of $x^2 - 11x + 20$ are

$$x = \frac{11 \pm \sqrt{41}}{2}$$

$x \approx 8.702$ and $x \approx 2.298$

From the domain, we know that $x \approx 8.70$ is too large. Thus, we have two possible solutions.

For $x = 2$: Length $= 14 - 2x = 10$ inches

Width $= 12 - 2x = 8$ inches

Height $= x = 2$ inches

For $x \approx 2.30$: Length $= 14 - 2x = 3 + \sqrt{41}$ in. ≈ 9.4 inches

Width $= 12 - 2x = 1 + \sqrt{41}$ in. ≈ 7.4 inches

Height $= x = \dfrac{11 - \sqrt{41}}{2}$ in. ≈ 2.3 inches

19. Use a graphing utility to find the dimensions of the box.

Solution

Graph $f(x) = x^3 - 13x^2 + 42x - 40$ and locate the x-intercepts where $0 < x < 6$.

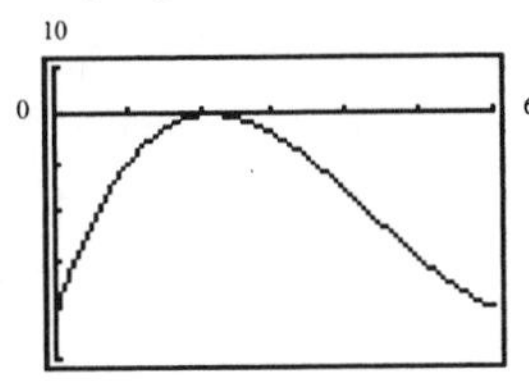

The two x-intercepts are at $x = 2$ and $x \approx 2.3$.

For $x = 2$, the dimensions are 10 in. $\times$ 8 in. $\times$ 2 in.

For $x \approx 2.298$, the dimensions are 9.404 in. $\times$ 7.404 in. $\times$ 2.298 in. (See Exercise 18.)

20. When solving this problem analytically or graphically, not all possible zeros are used. Explain.

Solution

Since the width is given by $12 - 2x$, x must be less than 6. Otherwise the width is zero or negative. Also, to make sense, x must be greater than zero. The feasible domain for this problem is $0 < x < 6$. For example, if we had used $\dfrac{11 + \sqrt{41}}{2} \approx 8.702$, we would have a length of $14 - 2x \approx -5.404$ and a width of $10 - 2x \approx -3.404$.

4.4 Complex Numbers

- Operations on Complex Numbers
  (a) Addition: $(a + bi) + (c + di) = (a + c) + (b + d)i$
  (b) Subtraction: $(a + bi) - (c + di) = (a - c) + (b - d)i$
  (c) Multiplication: $(a + bi)(c + di) = (ac - bd) + (ad + bc)i$
  (d) Division: $\dfrac{a + bi}{c + di} = \dfrac{a + bi}{c + di} \cdot \dfrac{c - di}{c - di} = \dfrac{ac + bd}{c^2 + d^2} + \dfrac{bc - ad}{c^2 + d^2}i$
- The complex conjugate of $a + bi$ is $a - bi$:

  $$(a + bi)(a - bi) = a^2 + b^2$$
- $\sqrt{-a} = \sqrt{a}i$ for $a > 0$.
- Be able to plot complex numbers in the complex plane.

SOLUTIONS TO SELECTED EXERCISES

5. Find the real numbers a and b so that the equation $(a - 1) + (b + 3)i = 5 + 8i$ is true.

Solution

$(a - 1) + (b + 3)i = 5 + 8i$ when $a - 1 = 5 \Rightarrow a = 6$ and $b + 3 = 8 \Rightarrow b = 5$.

9. Write $2 - \sqrt{-27}$ in standard form and find its complex conjugate.

Solution

$2 - \sqrt{-27} = 2 - \sqrt{27}\,i = 2 - 3\sqrt{3}\,i$. The complex conjugate is $2 + 3\sqrt{3}\,i$.

13. Write $-6i + i^2$ in standard form and find its complex conjugate.

Solution

$-6i + i^2 = -6i - 1 = -1 - 6i$. The complex conjugate is $-1 + 6i$.

17. Write 8 in standard form and find its complex conjugate.

Solution

8 is in standard form. The imaginary part is 0. The complex conjugate of 8 is 8.

21. Perform the operation $(8 - i) - (4 - i)$ and write the result in standard form.

Solution

$(8 - i) - (4 - i) = (8 - 4) + (-1 + 1)i = 4 + 0i = 4$

25. Perform the indicated operation and write the result in standard form.

$$-\left(\tfrac{3}{2} + \tfrac{5}{2}i\right) + \left(\tfrac{5}{3} + \tfrac{11}{3}i\right)$$

Solution

$-\left(\frac{3}{2} + \frac{5}{2}i\right) + \left(\frac{5}{3} + \frac{11}{3}i\right) = \left(-\frac{3}{2} + \frac{5}{3}\right) + \left(-\frac{5}{2} + \frac{11}{3}\right)i = \frac{1}{6} + \frac{7}{6}i$

29. Perform the operation $(\sqrt{-10})^2$ and write the result in standard form.

Solution

$(\sqrt{-10})^2 = (\sqrt{10}\,i)^2 = (\sqrt{10})^2(i^2) = -10$

33. Perform the operation $(4 + 5i)(4 - 5i)$ and write the result in standard form.

Solution

$(4 + 5i)(4 - 5i) = 16 - 25i^2 = 16 + 25 = 41$

39. Perform the operation $(\sqrt{14} + \sqrt{10}i)(\sqrt{14} - \sqrt{10}i)$ and write the result in standard form.

Solution

$(\sqrt{14} + \sqrt{10}\,i)(\sqrt{14} - \sqrt{10}\,i) = (\sqrt{14})^2 - (\sqrt{10})^2 i^2 = 14 + 10 = 24$

43. Perform the indicated operation and write the result in standard form.

$$\frac{4}{4 - 5i}$$

Solution

$$\frac{4}{4 - 5i} = \frac{4}{4 - 5i} \cdot \frac{4 + 5i}{4 + 5i} = \frac{16 + 20i}{16 + 25} = \frac{16}{41} + \frac{20}{41}i$$

47. Perform the indicated operation and write the result in standard form.

$$\frac{6 - 7i}{i}$$

Solution

$$\frac{6 - 7i}{i} = \frac{6 - 7i}{i} \cdot \frac{-i}{-i} = \frac{-6i + 7i^2}{1} = -7 - 6i$$

51. Perform the indicated operation and write the result in standard form.

$$\frac{5}{(1+i)^3}$$

Solution

$$\frac{5}{(1+i)^3} = \frac{5}{(1+i)^3} \cdot \frac{(1-i)^3}{(1-i)^3} = \frac{5[1^3 - 3(1)^2 i + 3(1)i^2 - i^3]}{(1+1)^3}$$

$$= \frac{5}{8}(1 - 3i - 3 + i) = \frac{5}{8}(-2 - 2i) = -\frac{5}{4} - \frac{5}{4}i$$

55. Perform the operation $(2+3i)^2 + (2-3i)^2$ and write the result in standard form.

Solution

$(2+3i)^2 + (2-3i)^2 = 4 + 12i + 9i^2 + 4 - 12i + 9i^2 = 8 - 18 = -10$

59. Solve $4x^2 + 16x + 17 = 0$.

Solution

By the Quadratic Formula, we have

$$x = \frac{-16 \pm \sqrt{16^2 - 4(4)(17)}}{2(4)} = \frac{-16 \pm \sqrt{-16}}{8} = \frac{-16 \pm 4i}{8} = -2 \pm \frac{1}{2}i$$

63. Solve $16t^2 - 4t + 3 = 0$.

Solution

By the Quadratic Formula, we have

$$t = \frac{-(-4) \pm \sqrt{(-4)^2 - 4(16)(3)}}{2(16)} = \frac{4 \pm \sqrt{-176}}{32} = \frac{4 \pm 4\sqrt{11}\,i}{32} = \frac{1}{8} \pm \frac{\sqrt{11}}{8}i$$

67. Plot the complex number 3.

Solution

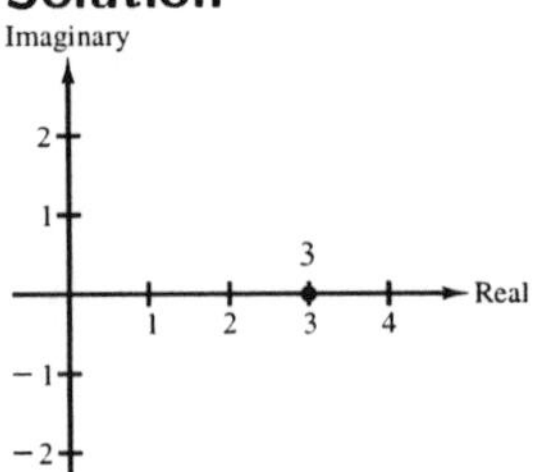

71. Decide whether $c = 0$ is in the Mandelbrot Set. Explain your reasoning.

Solution

The complex number 0 **is** in the Mandelbrot Set since for $c = 0$, the corresponding Mandelbrot sequence is $0, 0, 0, 0, 0, 0, \ldots$ which is bounded.

75. Decide whether $c = 1$ is in the Mandelbrot Set. Explain your reasoning.

Solution

The complex number 1 is **not** in the Mandelbrot Set since for $c = 1$, the corresponding Mandelbrot sequence is $1,\ 2,\ 5,\ 26,\ 677,\ 458,330$, which is unbounded.

4.5 The Fundamental Theorem of Algebra

- You should know that if f is a polynomial of degree $n > 0$, then f has exactly n zeros (roots) in the complex number system.
- You should know that if $a + bi,\ b \neq 0$, is a complex zero of a polynomial f, with real coefficients, then $a - bi$ is also a complex zero of f.
- You should know the difference between a factor that is irreducible over the rationals (such as $x^2 - 7$) and a factor that is irreducible over the reals (such as $x^2 + 9$).
- You should know that every polynomial of degree $n > 0$ with real coefficients can be written as the product of linear and irreducible (over the reals) quadratic factors with real coefficients.

SOLUTIONS TO SELECTED EXERCISES

3. Find all the zeros of $h(x) = x^2 - 4x + 1$ and write the polynomial as a product of linear factors.

Solution

Zeros: $x = \dfrac{4 \pm \sqrt{12}}{2} = 2 \pm \sqrt{3}$ (Quadratic Formula)

$h(x) = (x - 2 + \sqrt{3})(x - 2 - \sqrt{3})$

7. Find all the zeros of $f(z) = z^2 - 2z + 2$ and write the polynomial as a product of linear factors.

Solution

Zeros: $z = \dfrac{2 \pm \sqrt{4}\,i}{2} = 1 \pm i$ (Quadratic Formula)

$f(z) = (z - 1 + i)(z - 1 - i)$

11. Find all the zeros of $f(t) = t^3 - 3t^2 - 15t + 125$ and write the polynomial as a product of linear factors.

Solution

Possible rational zeros: $\pm 1, \pm 5, \pm 25, \pm 125$

$$\text{Zeros: } -5 \;\; \begin{array}{|rrrr} 1 & -3 & -15 & 125 \\ & -5 & 40 & -125 \\ \hline 1 & -8 & 25 & 0 \end{array}$$

($t = -5$ is a zero)

$t = \dfrac{8 \pm \sqrt{36}\,i}{2} = 4 \pm 3i$ (Quadratic Formula on $t^2 - 8t + 25 = 0$)

$f(t) = (t + 5)(t - 4 + 3i)(t - 4 - 3i)$

15. Find all the zeros of $f(x) = 16x^3 - 20x^2 - 4x + 15$ and write the polynomial as a product of linear factors.

Solution

Possible rational zeros: $\pm 1, \pm 3, \pm 5, \pm 15, \pm\frac{1}{2}, \pm\frac{3}{2}, \pm\frac{5}{2}, \pm\frac{15}{2}, \pm\frac{1}{4}, \pm\frac{3}{4}, \pm\frac{5}{4}, \pm\frac{15}{4}, \pm\frac{1}{8}, \pm\frac{3}{8}, \pm\frac{5}{8}, \pm\frac{15}{8}, \pm\frac{1}{16}, \pm\frac{3}{16}, \pm\frac{5}{16}, \pm\frac{15}{16}$

$$\text{Zeros: } -\tfrac{3}{4} \;\; \begin{array}{|rrrr} 16 & -20 & -4 & 15 \\ & -12 & 24 & -15 \\ \hline 16 & -32 & 20 & 0 \end{array}$$

($x = -\frac{3}{4}$ is a zero.)

$x = \dfrac{32 \pm \sqrt{256}\,i}{32} = 1 \pm \dfrac{1}{2}i$ (Quadratic Formula on $16x^2 - 32x + 20 = 0$)

$f(x) = (4x + 3)(2x - 2 + i)(2x - 2 - i)$

19. Find all the zeros of $f(x) = 5x^3 - 9x^2 + 28x + 6$ and write the polynomial as a product of linear factors.

Solution

Possible rational zeros: $\pm 1, \pm 2, \pm 3, \pm 6, \pm\frac{1}{5}, \pm\frac{2}{5}, \pm\frac{3}{5}, \frac{6}{5}$

$$\text{Zeros: } -\tfrac{1}{5} \;\; \begin{array}{|rrrr} 5 & -9 & 28 & 6 \\ & -1 & 2 & -6 \\ \hline 5 & -10 & 30 & 0 \end{array}$$

($x = -\frac{1}{5}$ is a zero.)

$x = \dfrac{10 \pm \sqrt{500}\,i}{10} = 1 \pm \sqrt{5}\,i$ (Quadratic Formula on $5x^2 - 10x + 30 = 0$)

$f(x) = (5x + 1)(x - 1 + \sqrt{5}\,i)(x - 1 - \sqrt{5}\,i)$

23. Find all the zeros of $f(x) = x^4 + 10x^2 + 9$ and write the polynomial as a product of linear factors.

Solution

$f(x) = x^4 + 10x^2 + 9 = (x^2 + 9)(x^2 + 1)$

Zeros: $x = \pm i,\ \pm 3i$

$f(x) = (x + i)(x - i)(x + 3i)(x - 3i)$

27. Find a polynomial with integer coefficients that has the zeros $1,\ 5i,\ -5i$.

Solution

$f(x) = (x - 1)(x - 5i)(x + 5i) = (x - 1)(x^2 + 25) = x^3 - x^2 + 25x - 25$

31. Find a polynomial with integer coefficients that has the zeros $i,\ -i,\ 6i,\ -6i$.

Solution

$f(x) = (x - i)(x + i)(x - 6i)(x + 6i) = (x^2 + 1)(x^2 + 36) = x^4 + 37x^2 + 36$

35. Find a polynomial with integer coefficients that has the zeros $\frac{3}{4},\ -2,\ -\frac{1}{2} + i$.

Solution

Since $-\frac{1}{2} + i$ is a zero, so is $-\frac{1}{2} - i$.

$$f(x) = (4x - 3)(x + 2)(2x + 1 - 2i)(2x + 1 + 2i)$$
$$= (4x^2 + 5x - 6)(4x^2 + 4x + 5) = 16x^4 + 36x^3 + 16x^2 + x - 30$$

39. Write $f(x) = x^4 - 4x^3 + 5x^2 - 2x - 6$ (a) as the product of factors that are irreducible over the *rationals*, (b) as the product of linear and quadratic factors that are irreducible over the *reals*, and (c) in completely factored form. (*Hint*: One factor is $x^2 - 2x - 2$.)

Solution

$$\begin{array}{r}
x^2 - 2x + 3 \\
x^2 - 2x - 2\,\overline{)\,x^4 - 4x^3 + 5x^2 - 2x - 6} \\
\underline{x^4 - 2x^3 - 2x^2 \qquad\qquad} \\
-2x^3 + 7x^2 - 2x \qquad \\
\underline{-(-2x^3 + 4x^2 + 4x)} \qquad \\
3x^2 - 6x - 6 \\
\underline{-(3x^2 - 6x - 6)} \\
0
\end{array}$$

(a) $f(x) = (x^2 - 2x - 2)(x^2 - 2x + 3)$

(b) $f(x) = (x - 1 - \sqrt{3})(x - 1 + \sqrt{3})(x^2 - 2x + 3)$ (Use the Quadratic Formula on $x^2 - 2x - 2 = 0$.)

(c) $f(x) = (x - 1 - \sqrt{3})(x - 1 + \sqrt{3})(x - 1 - \sqrt{2}\,i)(x - 1 + \sqrt{2}\,i)$ (Use the Quadratic Formula on $x^2 - 2x + 3 = 0$.)

43. Find all the zeros of $f(x) = 2x^4 - x^3 + 7x^2 - 4x - 4$, given that $2i$ is a zero.

Solution

Since $2i$ is a zero, so is $-2i$.

$$
\begin{array}{r|rrrrr}
2i & 2 & -1 & 7 & -4 & -4 \\
 & & 4i & -8-2i & 4-2i & 4 \\
\hline
-2i & 2 & -1+4i & -1-2i & -2i & 0 \\
 & & -4i & 2i & 2i & \\
\hline
 & 2 & -1 & -1 & 0 &
\end{array}
$$

$2x^2 - x - 1 = (2x+1)(x-1) = 0$ when $x = -\frac{1}{2}$ or $x = 1$.

Zeros: $-\frac{1}{2},\ 1,\ \pm 2i$

47. Find all the zeros of $f(x) = x^4 + 3x^3 - 5x^2 - 21x + 22$, given that $-3 + \sqrt{2i}$, is a zero.

Solution

Since $-3 + \sqrt{2}\,i$ is a zero, so is $-3 - \sqrt{2}\,i$.

$$
\begin{array}{r|rrrrr}
3+\sqrt{2}\,i & 1 & 3 & -5 & -21 & 22 \\
 & & -3+\sqrt{2}\,i & -2-3\sqrt{2}\,i & 27+2\sqrt{2}\,i & -22 \\
\hline
-3-\sqrt{2}\,i & 1 & \sqrt{2}\,i & -7-3\sqrt{2}\,i & 6+2\sqrt{2}\,i & 0 \\
 & & -3-\sqrt{2}\,i & 9+3\sqrt{2}\,i & -6-2\sqrt{2}\,i & \\
\hline
 & 1 & -3 & 2 & 0 &
\end{array}
$$

$x^2 - 3x + 2 = (x-1)(x-2) = 0$ when $x = 1$ or $x = 2$.

Zeros: $1,\ 2,\ -3 \pm \sqrt{2}\,i$

51. *Profit* The demand and cost equations for a product are $p = 140 - 0.0001x$ and $C = 80x + 150{,}000$, where p is the unit price, C is the total cost, and x is the number of units produced. The total profit obtained by producing and selling x units is given by

$$P = R - C = xp - C.$$

Determine a price p that would yield a profit of \$9 million. Use a graphing utility to explain why this is not possible.

Solution

Analytically

$$P = R - C = xp - C = x(140 - 0.0001x) - (80x + 150{,}000)$$

$$= -0.0001x^2 + 60x - 150{,}000 = 9{,}000{,}000$$

Thus, $0 = 0.0001x^2 - 60x + 9{,}150{,}000$

$$x = \frac{60 \pm \sqrt{-60}}{0.0002} = 300{,}000 \pm 10{,}000\sqrt{15}\,i.$$

Since the zeros are both complex, it is not possible to determine a price p that would yield a profit of 9 million dollars.

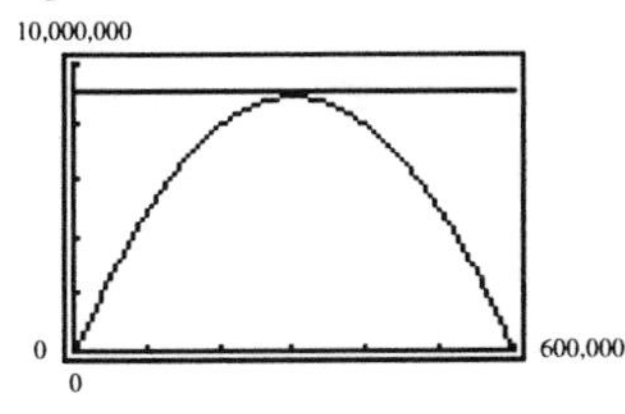

The graph has no x-intercepts so there are no real zeros of $f(x)$.

55. *Think About It* Another student claims that the polynomial $f(x) = x^4 - 7x^2 + 12$ may be factored over the rational numbers as $f(x) = (x - \sqrt{3})(x + \sqrt{3})(x - 2)(x + 2)$. Do you agree with this claim? Explain your answer.

Solution

$$f(x) = x^4 - 7x^2 + 12$$

$$= (x^2 - 3)(x^2 - 4)$$

$= (x^2 - 3)(x - 2)(x + 2)$ which is irreducible over the **rationals**.

$= (x - \sqrt{3})(x + \sqrt{3})(x - 2)(x + 2)$ which is irreducible over the **reals**.

The student's claim is not true because the student's factorization is over the **real** numbers, not over the **rational** numbers.

Review Exercises for Chapter 4

SOLUTIONS TO SELECTED EXERCISES

1. Divide $12x^2 + 5x - 2$ by $4x - 1$, by long division.

Solution

$$\begin{array}{r} 3x + 2 \\ 4x - 1 \overline{) \; 12x^2 + 5x - 2} \\ -(12x^2 - 3x) \\ \hline 8x - 2 \\ -(8x - 2) \\ \hline 0 \end{array}$$

$$\frac{12x^2 + 5x - 2}{4x - 1} = 3x + 2$$

3. Divide $2x^3 - x^2 - 22x + 21$ by $x - 1$, by synthetic division.

Solution

$$\begin{array}{c|cccc} 1 & 2 & -1 & -22 & 21 \\ & & 2 & 1 & -2 \\ \hline & 2 & 1 & -21 & 0 \end{array}$$

$$\frac{2x^3 - x^2 - 22x + 21}{x - 1} = 2x^2 + x - 21$$

7. Given $f(x) = 5x^3 - 10x + 7$, use synthetic division to evaluate the function.

(a) $f(1)$ (b) $f(-3)$

Solution

(a)
$$\begin{array}{c|cccc} 1 & 5 & 0 & -10 & 7 \\ & & 5 & 5 & -5 \\ \hline & 5 & 5 & -5 & 2 \end{array}$$

$f(1) = 2$

(b)
$$\begin{array}{c|cccc} -3 & 5 & 0 & -10 & 7 \\ & & -15 & 45 & -105 \\ \hline & 5 & -15 & 35 & -98 \end{array}$$

$f(-3) = -98$

13. Given $f(x) = -4x^3 + 8x^2 - 3x + 15$, use the Rational Zero Test to list all possible rational zeros of f. Verify using a graphing utility.

Solution

Possible rational zeros: $\pm 1, \pm 3, \pm 5, \pm 15,$

$\pm \frac{1}{2}, \pm \frac{3}{2}, \pm \frac{5}{2}, \pm \frac{15}{2}, \pm \frac{1}{4}, \pm \frac{3}{4}, \pm \frac{5}{4}, \pm \frac{15}{4}$

The only **real** root of $f(x)$ is $x \approx 2.357$ which is **not** one of the possible rational zeros.

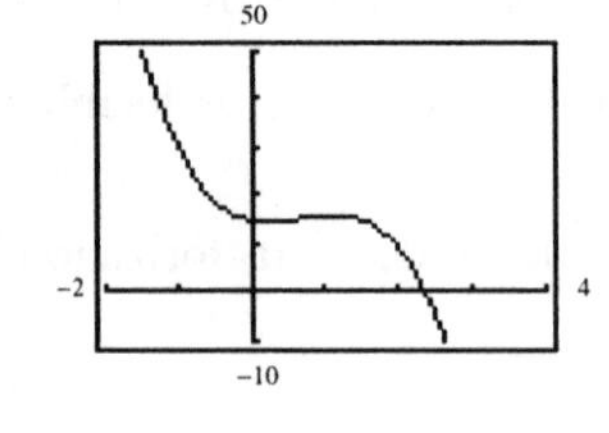

15. *Dimensions of a Room* A rectangular room has a volume of $x^3 + 13x^2 + 50x + 56$ cubic feet. The height of the room is $x + 2$. Find the number of square feet of floor space in the room.

Solution

$$\begin{array}{r|rrrr} -2 & 1 & 13 & 50 & 56 \\ & & -2 & -22 & -56 \\ \hline & 1 & 11 & 28 & 0 \end{array}$$

$$\frac{x^3 + 13x^2 + 50x + 56}{x + 2} = x^2 + 11x + 28$$

$$= (x + 4)(x + 7) \text{ square feet of floor space.}$$

19. Find all the real zeros of $h(x) = 3x^4 - 27x^2 + 60$.

Solution

$$h(x) = 3(x^4 - 9x^2 + 20) = 3(x^2 - 4)(x^2 - 5)$$

$$= 3(x + 2)(x - 2)(x + \sqrt{5})(x - \sqrt{5})$$

Zeros: $\pm 2, \ \pm\sqrt{5}$

21. Find all the real zeros of $C(x) = 3x^4 + 3x^3 - 7x^2 - x + 2$.

Solution

Possible rational zeros: $\pm 1, \ \pm 2, \ \pm\frac{1}{3}, \ \pm\frac{2}{3}$

$$\begin{array}{r|rrrrr} 1 & 3 & 3 & -7 & -1 & 2 \\ & & 3 & 6 & -1 & -2 \\ \hline -2 & 3 & 6 & -1 & -2 & 0 \\ & & -6 & 0 & 2 & \\ \hline & 3 & 0 & -1 & 0 & \end{array}$$

$$C(x) = 3x^4 + 3x^3 - 7x^2 - x + 2 = (x - 1)(x + 2)(3x^2 - 1)$$

Zeros: $1, \ -2, \ \pm\dfrac{\sqrt{3}}{3}$

25. Match the cubic equation $f(x) = x^3 - 3$ with the number of rational and irrational zeros given below.

(a) Rational zeros: 1
Irrational zeros: 2

(b) Rational zeros: 0
Irrantional zeros: 1

(c) Rational zeros: 3
Irrational zeros: 0

(d) Rational zeros: 1
Irrational zeros: 0

Solution

$$f(x) = x^3 - 3 = (x - \sqrt[3]{3})(x^2 + \sqrt[3]{3}x + \sqrt[3]{9})$$

Rational zeros: 0

Irrational zeros: 1

Matches (b)

27. Using a graphing utility, approximate the real zeros of $f(x) = 5x^3 - 11x - 3$. Use an accuracy of 0.001.

Solution

Using your graphing utility and the ZOOM and TRACE keys, we determine that the zeros are $x \approx -1.321$, $x \approx -0.283$, and $x \approx 1.604$.

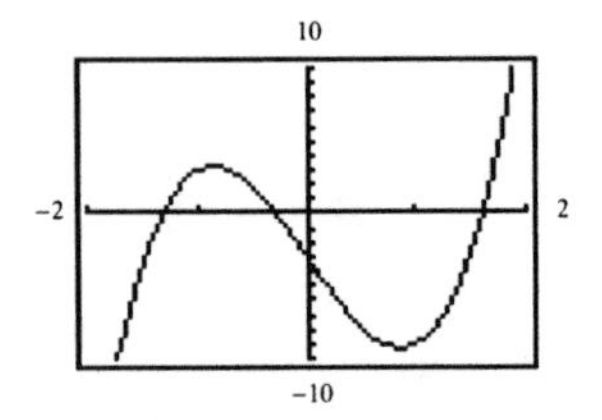

33. Write $\sqrt{-24}$ in standard form and find its complex conjugate.

Solution

$\sqrt{-24} = 0 + 2\sqrt{6}i = 2\sqrt{6}i$

Conjugate: $0 - 2\sqrt{6}i = -2\sqrt{6}i$

37. Perform the operation $(3 + \sqrt{-20})(5 + \sqrt{-10})$ and write the result in standard form.

Solution

$$\begin{aligned}(3 + \sqrt{-20})(5 + \sqrt{-10}) &= (3 + 2\sqrt{5}i)(5 + \sqrt{10}i)\\ &= 15 + 3\sqrt{10}i + 10\sqrt{5}i + 2\sqrt{50}i^2\\ &= (15 - 10\sqrt{2}) + (3\sqrt{10} + 10\sqrt{5})i\end{aligned}$$

41. Perform the operation $-2i(4 - 5i)$ and write the result in standard form.

Solution

$-2i(4 - 5i) = -8i + 10i^2 = -10 - 8i$

45. Perform the following operation and write the result in standard form.

$$\frac{3+i}{3-i}$$

Solution

$$\frac{3+i}{3-i} = \frac{3+i}{3-i} \cdot \frac{3+i}{3+i} = \frac{9+6i+i^2}{9+1} = \frac{8+6i}{10} = \frac{4}{5} + \frac{3}{5}i$$

49. Perform the operation $(3 + 2i)^2 + (3 - 2i)^2$ and write the result in standard form.

Solution

$$\begin{aligned}(3 + 2i)^2 + (3 - 2i)^2 &= 9 + 12i + 4i^2 + 9 - 12i + 4i^2\\ &= 18 + 8i^2 = 18 - 8 = 10\end{aligned}$$

53. Solve $4x^2 + 11x + 3 = 0$.

Solution

$4x^2 + 11x + 3 = 0$

$a = 4 \quad b = 11 \quad c = 3$ (Use the Quadratic Formula)

$$x = \frac{-11 \pm \sqrt{(11)^2 - 4(4)(3)}}{2(4)} = \frac{-11 \pm \sqrt{73}}{8}$$

57. Write $f(x) = x^4 - 625$ as a product of linear factors.

Solution

$f(x) = x^4 - 625 = (x^2 - 25)(x^2 + 25) = (x - 5)(x + 5)(x - 5i)(x + 5i)$

61. Write $g(x) = 4x^3 - 8x^2 + 9x - 18$ as a product of linear factors.

Solution

$g(x) = 4x^3 - 8x^2 + 9x - 18 = 4x^2(x - 2) + 9(x - 2)$

$= (x - 2)(4x^2 + 9) = (x - 2)(2x + 3i)(2x - 3i)$

65. Write $f(x) = x^4 + 5x^2 - 24$ (a) as the product of factors that are irreducible over the rationals, (b) as the product of linear and quadratic factors irreducible over the reals, and (c) in completely factored form.

Solution

(a) $f(x) = x^4 + 5x^2 - 24$

$= (x^2 - 3)(x^2 + 8)$ over the rationals

(b) $= (x + \sqrt{3})(x - \sqrt{3})(x^2 + 8)$ over the reals

(c) $= (x + \sqrt{3})(x - \sqrt{3})(x + \sqrt{8}\,i)(x - \sqrt{8}\,i)$

$= (x + \sqrt{3})(x - \sqrt{3})(x + 2\sqrt{2}\,i)(x - 2\sqrt{2}\,i)$ completely factored

67. Find all the zeros of $f(x) = x^3 - 2x^2 + 9x - 18$, given that $-3i$ is a zero.

Solution

Since $-3i$ is a zero, so is $3i$.

$-3i$	1	-2	9	-18
		$-3i$	$-9 + 6i$	18
$3i$	1	$-2 - 3i$	$6i$	0
		$3i$	$-6i$	
	1	-2	0	

$f(x) = x^3 - 2x^2 + 9x - 18 = (x + 3i)(x - 3i)(x - 2)$

Zeros: $\pm 3i,\ 2$

69. Find all the zeros of $f(x) = x^4 + 7x^3 + 24x^2 + 58x + 40$ given that $-1 + 3i$ is a zero.

Solution

Since $-1 + 3i$ is a zero, so is $-1 - 3i$.

$-1 + 3i$	1	7	24	58	40
		$-1 + 3i$	$-15 + 15i$	$-54 + 12i$	-40
$-1 - 3i$	1	$6 + 3i$	$9 + 15i$	$4 + 12i$	0
		$-1 - 3i$	$-5 - 15i$	$-4 - 12i$	
	1	5	4	0	

$$\begin{aligned} f(x) &= (x + 1 - 3i)(x + 1 + 3i)(x^2 + 5x + 4) \\ &= (x + 1 - 3i)(x + 1 + 3i)(x + 1)(x + 4) \end{aligned}$$

Zeros: $-1 \pm 3i,\ -1,\ -4$

Test for Chapter 4

1. Use synthetic division to show that $x = \frac{3}{2}$ is a solution of $f(x) = 12x^3 + 8x^2 - 49x + 15$ and then use the result to completely factor the polynomial.

Solution

3/2	12	8	-49	15
		18	39	-15
	12	26	-10	0

$$\begin{aligned} f(x) &= \left(x - \tfrac{3}{2}\right)(12x^2 + 26x - 10) \\ &= \left(x - \tfrac{3}{2}\right)(2)(6x^2 + 13x - 5) \\ &= (2x - 3)(3x - 1)(2x + 5) \end{aligned}$$

2. Simplify.

$$\frac{x^4 + 4x^3 - 19x^2 - 106x - 120}{x^2 - 3x - 10}$$

Solution

$$\begin{array}{r} x^2 + 7x + 12 \\ x^2 - 3x - 10 \;\big)\; x^4 + 4x^3 - 19x^2 - 106x - 120 \\ x^4 - 3x^3 - 10x^2 \\ \hline 7x^3 - 9x^2 - 106x \\ 7x^3 - 21x^2 - 70x \\ \hline 12x^2 - 36x - 120 \\ 12x^2 - 36x - 120 \\ \hline 0 \end{array}$$

$$\frac{x^4 + 4x^3 - 19x^2 - 106x - 120}{x^2 - 3x - 10} = x^2 + 7x + 12 = (x+3)(x+4)$$

3. List all possible rational zeros of

$$f(x) = 5x^4 - 3x^3 + 2x^2 + 11x + 12.$$

Solution

Possible Rational Zeros: $\pm 1,\ \pm 2,\ \pm 3,\ \pm 4,\ \pm 6,\ \pm 12,\ \pm\frac{1}{5},\ \pm\frac{2}{5},\ \pm\frac{3}{5},\ \pm\frac{4}{5},\ \pm\frac{6}{5},\ \pm\frac{12}{5}$

4. Use Descartes's Rule of Signs to determine the possible number of positive and negative zeros of $h(x) = -3x^5 + 2x^4 - 4x^3 + 3x^2 - 7$.

Solution

$h(x)$ has 4 variations in signs.

$h(x)$ has either 4, 2, or 0 positive real zeros.

$h(-x) = 3x^5 + 2x^4 + 4x^3 + 3x^2 - 7$ (One variation in sign)

$h(x)$ has one negative real zero.

5. Perform the operation $(12 + 3i) + (4 - 6i)$ and write the result in standard form.

Solution

$(12 + 3i) + (4 - 6i) = (12 + 4) + (3i - 6i) = 16 - 3i$

6. Perform the operation $(10 - 2i) - (3 + 7i)$ and write the result in standard form.

Solution

$(10 - 2i) - (3 + 7i) = (10 - 3) + (-2i - 7i) = 7 - 9i$

7. Perform the operation $(5 + \sqrt{-12})(3 - \sqrt{-12})$ and write the result in standard form.

Solution

$$(5 + \sqrt{-12})(3 - \sqrt{-12}) = (5 + 2i\sqrt{3})(3 - 2i\sqrt{3}) = 15 - 10i\sqrt{3} + 6i\sqrt{3} - 4i^2(3)$$

$$= 15 - 4i\sqrt{3} + 12 = 27 - 4\sqrt{3}i$$

8. Perform the operation $(4+3i)(2-5i)$ and write the result in standard form.

Solution

$(4+3i)(2-5i) = 8 - 20i + 6i - 15i^2 = 8 - 14i + 15 = 23 - 14i$

9. Perform the operation $(2-3i)^2$ and write the result in standard form.

Solution

$(2-3i)^2 = 4 - 12i + 9i^2 = 4 - 12i - 9 = -5 - 12i$

10. Perform the operation $(5+2i)^2$ and write the result in standard form.

Solution

$(5+2i)^2 = 25 + 20i + 4i^2 = 21 + 20i$

11. Perform the following operation and write the result in standard form.

$$\frac{1+i}{1-i}$$

Solution

$$\frac{1+i}{1-i} = \frac{1+i}{1-i} \cdot \frac{1+i}{1+i} = \frac{1+i+i+i^2}{1+1} = \frac{2i}{2} = i$$

12. Perform the following operation and write the result in standard form.

$$\frac{5-2i}{i}$$

Solution

$$\frac{5-2i}{i} = \frac{5-2i}{i} \cdot \frac{-i}{-i} = \frac{-5i+2i^2}{-i^2} = \frac{-2-5i}{1} = -2-5i$$

13. Find all the real zeros of $f(x) = x^5 - 5x^3 + 4x$.

Solution

$$\begin{aligned} f(x) &= x^5 - 5x^3 + 4x = x(x^4 - 5x^2 + 4) \\ &= x(x^2-1)(x^2-4) = x(x+1)(x-1)(x+2)(x-2) \end{aligned}$$

Real Zeros: $0, \pm 1, \pm 2$

14. Find all the real zeros of $g(x) = x^4 + 2x^3 - 9x^2 - 2x + 8$.

Solution

Possible Rational Zeros: $\pm 1, \pm 2, \pm 4, \pm 8$

1	1	2	−9	−2	8
		1	3	−6	−8
	1	3	−6	−8	0

−1	1	3	−6	−8
		−1	−2	8
	1	2	−8	0

$g(x) = (x-1)(x+1)(x^2+2x-8) = (x-1)(x+1)(x+4)(x-2)$

Real Zeros: $\pm 1, -4, 2$

15. Solve the quadratic equation $x^2 + 5x + 7 = 0$.

Solution

$$x = \frac{-5 \pm \sqrt{(5)^2 - 4(1)(7)}}{2(1)} = \frac{-5 \pm \sqrt{-3}}{2} = \frac{-5 \pm \sqrt{3}i}{2} = -\frac{5}{2} \pm \frac{\sqrt{3}}{2}i$$

16. Solve the quadratic equation $2x^2 - 5x + 11 = 0$.

Solution

$$x = \frac{-(-5) \pm \sqrt{(-5)^2 - 4(2)(11)}}{2(2)} = \frac{5 \pm \sqrt{-63}}{4} = \frac{5 \pm 3\sqrt{7}i}{4} = \frac{5}{4} \pm \frac{3\sqrt{7}}{4}i$$

17. Find all the zeros $f(x) = x^3 + 2x^2 + 5x + 10$, given that $\sqrt{5}\,i$ is a zero.

Solution

Since $\sqrt{5}i$ is a zero, so is $-\sqrt{5}i$.

$\sqrt{5}i$	1	2	5	10
		$\sqrt{5}i$	$-5+2\sqrt{5}i$	-10
$-\sqrt{5}i$	1	$2+\sqrt{5}i$	$2\sqrt{5}i$	0
		$-\sqrt{5}i$	$-2\sqrt{5}i$	
	1	2	0	

$f(x) = (x - \sqrt{5}i)(x + \sqrt{5}i)(x + 2)$

Zeros: $\pm\sqrt{5}i, -2$

18. Find a polynomial with integer coefficients that has $2, 5, 3i$, and $-3i$ as zeros.

Solution

$$P(x) = (x-2)(x-5)(x-3i)(x+3i) = (x-2)(x-5)(x^2+9)$$
$$= (x^2 - 7x + 10)(x^2 + 9) = x^4 - 7x^3 + 19x^2 - 63x + 90$$

19. Given that $-2 + \sqrt{3}\,i$ is a zero of $f(x) = x^4 + 4x^3 + 8x^2 + 4x + 7$, name another zero of $f(x)$.

Solution

Since $-2 + \sqrt{3}\,i$ is a zero of $f(x) = x^4 + 4x^3 + 8x^2 + 4x + 7$, so is its conjugate $-2 - \sqrt{3}\,i$.

20. A company estimates that the profit for selling its product is

$$P = -11x^3 + 900x^2 - 50{,}000, \quad 0 \le x$$

where P is the profit (in dollars) and x is the advertising expense (in ten thousands of dollars). How much should the company spend in advertising to obtain a profit of \$222,000? Explain your reasoning.

Solution

$-11x^3 + 900x^2 - 50{,}000 = 222{,}000$

$0 = 11x^3 - 900x^2 + 272{,}000, \quad 0 \le x$

From the calculator, we have $x = 20$ or $x \approx 77.7251$. Using the **smaller** value, the company should spend \$200,000 on advertising.

Practice Test for Chapter 4

1. Divide $3x^4 - 7x^2 + 2x - 10$ by $x - 3$ using long division.

2. Divide $x^3 - 11$ by $x^2 + 2x - 1$.

3. Use synthetic division to divide $3x^5 + 13x^4 + 12x - 1$ by $x + 5$.

4. Use synthetic division to find $f(-6)$ when $f(x) = 7x^3 + 40x^2 - 12x + 15$.

5. Find the real zeros of $f(x) = x^3 - 19x - 30$.

6. Find the real zeros of $f(x) = x^4 + x^3 - 8x^2 - 9x - 9$.

7. List all the possible rational zeros of the function $f(x) = 6x^3 - 5x^2 + 4x - 15$.

8. Find the rational zeros of $f(x) = x^3 - \frac{20}{3}x^2 + 9x - \frac{10}{3}$.

9. Write $f(x) = x^4 + x^3 + 3x^2 + 5x - 10$ as a product of linear factors.

10. Using your graphing utility, find the zero of the function $f(x) = x^3 + 2x - 1$ in the interval $[0, 1]$, accurate to within 0.001 units.

In Exercises 11–16, perform the indicated operations and write the result in standard form.

11. $(6 + 5i) + (-2 + 4i)$

12. $(11 - 5i) - (-2 + i)$

13. $(3 + 4i)(5 - 6i)$

14. $(5 + \sqrt{-8})(7 - \sqrt{-8})$

15. $(3 - 10i)^2$

16. $\dfrac{1 + 4i}{2 - 7i}$

17. Find a polynomial with real coefficients that has 2, $3 + i$, and $3 - i$ as zeros.

18. Use synthetic division to show that $3i$ is a zero of $f(x) = x^3 + 4x^2 + 9x + 36$.

19. Find all the zeros of $f(x) = x^3 - 8x^2 + 25x - 200$ given that $5i$ is a zero.

20. Find all the zeros of $f(x) = (x^2 + 6x - 3)(x^2 + x + 11)$.

CHAPTER FIVE
Exponential and Logarithmic Functions

5.1 Exponential Functions

- You should know that a function of the form $f(x) = a^x$, where $a > 0,\ a \neq 1$, is called an exponential function with base a. x can be any real number.
- You should be able to graph exponential functions.
- You should know formulas for compound interest.

  (a) For n compoundings per year: $A = P\left(1 + \frac{r}{n}\right)^{nt}$.

  (b) For continuous compoundings: $A = Pe^{rt}$.

SOLUTIONS TO SELECTED EXERCISES

5. Use a calculator to evaluate $\sqrt[4]{763}$. Round the result to three decimal places.

Solution

$\sqrt[4]{763} \approx 5.256$

9. Use a calculator to evaluate $100^{\sqrt{2}}$. Round the result to three decimal places.

Solution

$100^{\sqrt{2}} \approx 673.639$

13. Use a calculator to evaluate $e^{-3/4}$. Round the result to three decimal places.

Solution

$e^{-3/4} \approx 0.472$

17. Match $f(x) = 3^{-x}$ with its graph. [The graphs are labeled (a), (b), (c), (d), (e), (f), (g), and (h) in the textbook.]

Solution

$f(x) = 3^{-x}$ rises to the left.

Asymptote: $y = 0$

Intercept: $(0, 1)$

Graph (b)

21. Match $f(x) = -3^{x-2}$ with its graph. [The graphs are labeled (a), (b), (c), (d), (e), (f), (g), and (h) in the textbook.]

Solution

$f(x) = -3^{x-2}$ falls to the right.

Asymptote: $y = 0$

Intercept: $\left(0, -\frac{1}{9}\right)$

Point: $(2, -1)$

Graph (f)

25. Sketch the graph of $f(x) = 5^{-x}$.

Solution

$f(x) = \left(\frac{1}{5}\right)^x = 5^{-x}$

Asymptote: $y = 0$

Intercept: $(0, 1)$

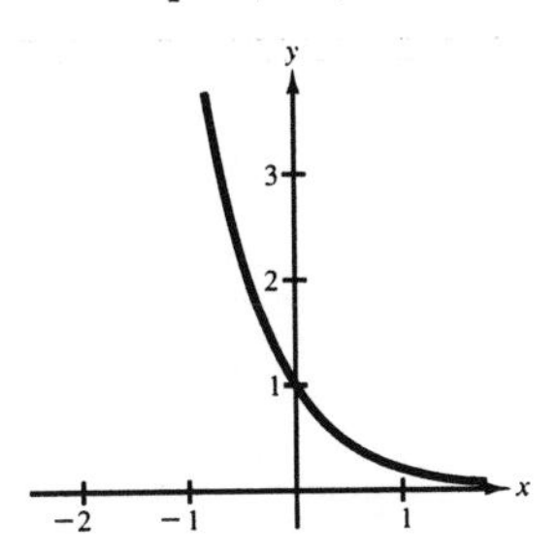

29. Sketch the graph of $g(x) = 5^{-x} - 3$.

Solution

This is the graph of Exercise 25 shifted down 3 units.

Asymptote: $y = -3$

Intercept: $(0, -2)$

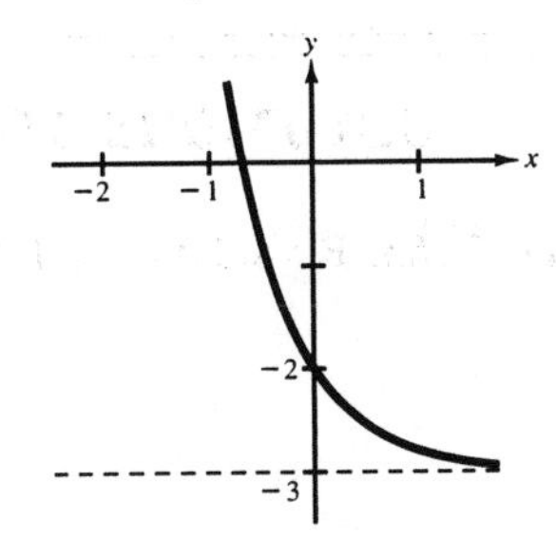

31. Sketch the graph of $y = 2^{-x^2}$.

Solution

Asymptote: $y = 0$

Intercept: $(0, 1)$

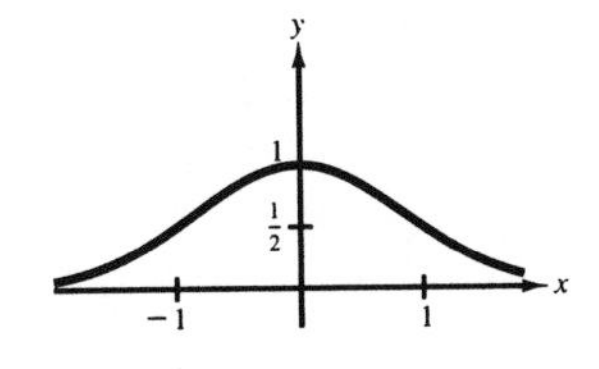

37. Sketch the graph of $f(x) = e^{2x}$.

Solution

Asymptote: $y = 0$

Intercept: $(0, 1)$

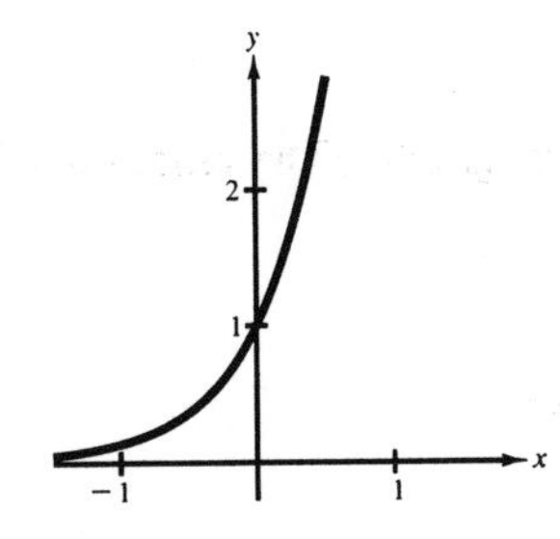

41. *Compound Interest* Determine the balance A for $P = \$2,500$ invested at the rate $r = 12\%$ for $t = 10$ years and compounded n times per year. Complete the table shown in the textbook.

Solution

Compounded n times per year: $A = 2500\left(1 + \frac{0.12}{n}\right)^{10n}$

Compounded continuously: $A = 2500e^{0.12(10)}$

n	1	2	4	12	365	Continuous compounding
A	\$7,764.62	\$8,017.84	\$8,155.09	\$8,250.97	\$8,298.66	\$8,300.29

45. *Compound Interest* Determine the amount of money P that should be invested at rate $r = 9\%$, compounded continuously, to produce a final balance of $A = \$100,000$ in t years. Complete the table shown in the textbook.

Solution

$100,000 = Pe^{0.09t}$

$P = 100,000e^{-0.09t}$

t	1	10	20	30	40	50
P	\$91,393.12	\$40,656.97	\$16,529.89	\$6,720.55	\$2,732.37	\$1,110.90

49. *Trust Fund* You deposited \$25,000 in a trust fund that pays 8.75% interest, compounded continuously, on the day that your grandchild was born. What would the balance of this account be on your grandchild's 25th birthday?

Solution

$A = 25,000e^{(0.0875)(25)} \approx \$222,822.57$

55. *Radioactive Decay* Five pounds of plutonium (Pu^{230}) is released in a nuclear accident. The amount of plutonium that is present after t years is given by $P = 5e^{-0.00002845t}$.

(a) Use a graphing utility to graph this function over the interval from $t = 0$ to $t = 100{,}000$ years.

(b) How much will remain after 100,000 years?

(c) Use the graph to estimate the half-life of Pu^{230}. Explain your reasoning.

Solution

(a)

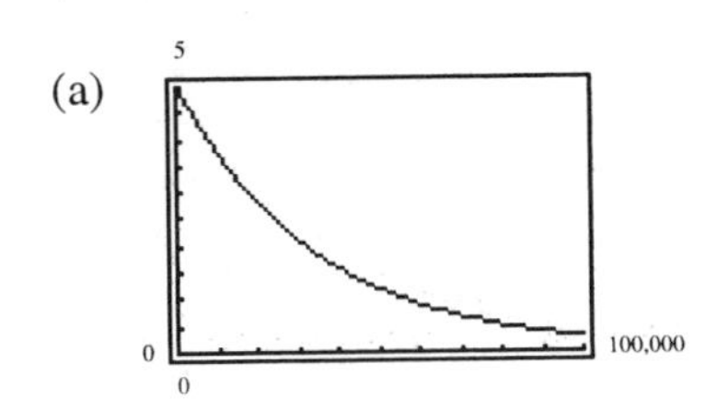

(b) $P(100{,}000) \approx 0.291$ lb.

(c) From the graph it appears that $P = 2.5$ when $t \approx 24{,}364$ years.

59. *Age at First Marriage* From 1970 to 1992, the average age of an American woman at her first marriage could be approximated by the model

$$A = 20.40 + \frac{1}{0.2163 + 2.351e^{-0.1824t}}, \quad 0 \leq t \leq 22$$

where A represents the average age and $t = 0$ represents 1970.

(a) Use the graph in the textbook to estimate the average age of an American woman at her first marriage in 1970, 1980, and 1992. See the figure in the textbook.

(b) Algebraically confirm the results of part (a). (*Source: U.S. Center for Health Statistics.*)

Solution

(a) 1970: $A \approx 20.8$ years old
1980: $A \approx 22$ years old
1992: $A \approx 24.3$ years old

(b) $A(0) \approx 20.7895$ years old
$A(10) \approx 22.0787$ years old
$A(22) \approx 24.2638$ years old

5.2 Logarithmic Functions

- You should know that a function of the form $f(x) = \log_a x$, where $a > 0,\ a \neq 1$, and $x > 0$, is called a logarithm of x to base a.
- You should be able to convert from logarithmic form to exponential form and vice versa. $y = \log_a x$ if and only if $x = a^y$.
- You should know the following properties of logarithms.
  (a) $\log_a 1 = 0$
  (b) $\log_a a = 1$
  (c) $\log_a a^x = x$
  (d) $\log_a x = \log_a y \Rightarrow x = y$
- You should know the definition of the natural logarithmic function.
  $\log_e x = \ln x,\ x > 0$
- You should know the properties of the natural logarithmic function.
  (a) $\ln 1 = 0$
  (b) $\ln e = 1$
  (c) $\ln e^x = x$
  (d) $\ln x = \ln y \Rightarrow x = y$
- You should be able to graph logarithmic functions.
- You should be able to find the domain of logarithmic functions.

SOLUTIONS TO SELECTED EXERCISES

3. Evaluate $\log_5 \left(\frac{1}{25}\right)$.

Solution

$\log_5 \left(\frac{1}{25}\right) = \log_5 5^{-2} = -2$

since $5^{-2} = \frac{1}{25}$.

5. Evaluate $\log_{16} 4$.

Solution

$\log_{16} 4 = \log_{16} 16^{1/2} = \frac{1}{2}$

since $16^{1/2} = 4$.

11. Evaluate $\ln e^3$.

Solution

$\ln e^3 = 3$

15. Evaluate $\log_a a^2$.

Solution

$\log_a a^2 = 2$

19. Use the definition of a logarithm to write $81^{1/4} = 3$ in logarithmic form. For instance, the logarithmic form of $2^3 = 8$ is $\log_2 8 = 3$.

Solution

$81^{1/4} = 3$

$\log_{81} 3 = \frac{1}{4}$

23. Use the definition of a logarithm to write $e^3 = 20.0855\ldots$ in logarithmic form. For instance, the logarithmic form of $2^3 = 8$ is $\log_2 8 = 3$.

Solution

$$e^3 = 20.0855\ldots$$

$\ln(20.0855\ldots) = 3$

27. Use a calculator to evaluate $\log_{10} 345$. Round your answers to three decimal places.

Solution

$\log_{10} 345 \approx 2.538$

31. Use a calculator to evaluate $\ln 18.42$. Round your answers to three decimal places.

Solution

$\ln 18.42 \approx 2.913$

35. Sketch $f(x) = 3^x$ and $g(x) = \log_3 x$ on the same coordinate plane.

Solution

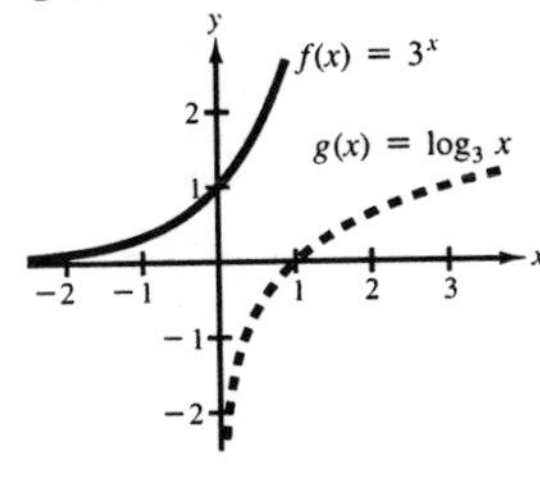

x	$f(x)$
-2	$\frac{1}{9}$
-1	$\frac{1}{3}$
0	1
1	3
2	9

x	$g(x)$
$\frac{1}{9}$	-2
$\frac{1}{3}$	-1
1	0
3	1
9	2

39. Match $f(x) = \ln x + 2$ to its graph. [The graphs are labeled (a), (b), (c), (d), (e), and (f) in the textbook.]

Solution

$f(x) = \ln x + 2$ rises to the right.

Asymptote: $x = 0$

Point on graph: $(1, 2)$

Graph (d)

43. Match $f(x) = \ln(1 - x)$ to its graph. [The graphs are labeled (a), (b), (c), (d), (e), and (f) in the textbook.]

Solution

$f(x) = \ln(1 - x)$ rises to the left.

Asymptote: $x = 1$

Intercept: $(0, 0)$

Graph (f)

47. Find the domain, vertical asymptote, and x-intercept of $h(x) = \log_4(x - 3)$, and sketch its graph.

Solution

Domain: $(3, \infty)$

Asymptote: $x = 3$

Intercept: $(4, 0)$

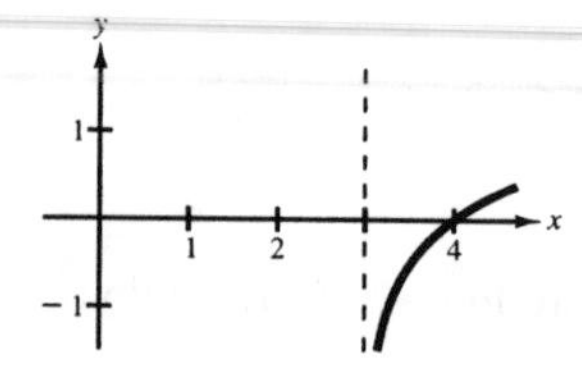

51. Find the domain, vertical asymptote, and x-intercept of $g(x) = \ln(-x)$, and sketch its graph.

Solution

Domain: $(-\infty, 0)$

Asymptote: $x = 0$

Intercept: $(-1, 0)$

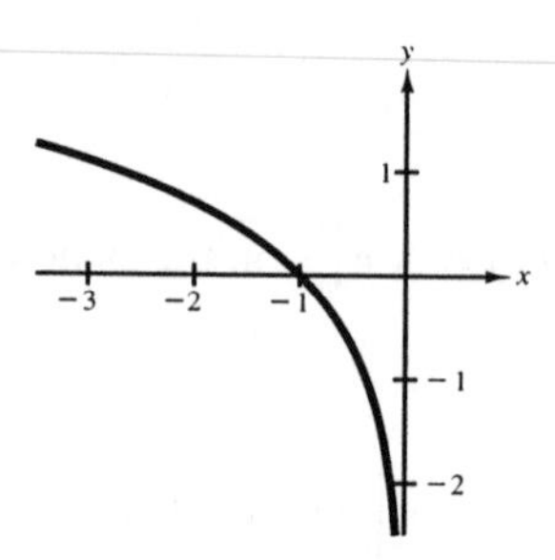

55. *Investment Time* A principal P invested at $9\frac{1}{2}\%$ and compounded continuously increases to an amount that is K times the original principal after t years, where t is given by

$$t = \frac{\ln K}{0.095}.$$

(a) Complete the table in the textbook and (b) use the completed table to graph this function.

Solution

(a)

K	1	2	3	4	6	8	10	12
t	0	7.3	11.6	14.6	18.9	21.9	24.2	26.2

(b)

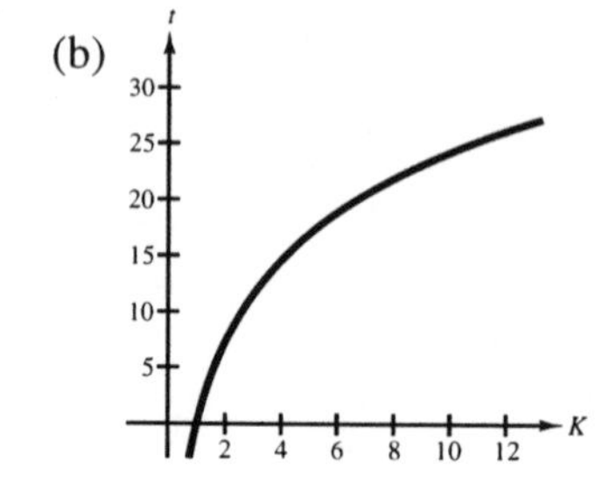

59. *Median Age of United States Population* Use the following model

$$A = 18.10 - 0.016t + 5.121 \ln t, \quad 10 \le t \le 80$$

which approximates the median age of the United States population from 1980 to 2050. In the model A is the median age and $t = 10$ represents 1980. Use the model to approximate the median age in the United States in 1980.

Solution

$A(10) \approx 29.73$ years

65. *Monthly Payment* Use the following model

$$t = \frac{5.315}{-6.7968 + \ln x}, \quad 1000 < x$$

which approximates the length of a home mortgage (of $120,000 at 10%) in terms of the monthly payment. In the model, t is the length of the mortgage in years and x is the monthly payment in dollars. Approximate the total amount paid over the term of the mortgage for a monthly payment of $1,167.41.

Solution

$A = (20)(12)(1167.41) = \$280,178.40$

5.3 Properties of Logarithms

- You should know the change of base formula.

$$\log_a x = \frac{\log_b x}{\log_b a}$$

- You should know the following properties of logarithms.

(a) $\log_a(uv) = \log_a u + \log_a v$ $\qquad \ln(uv) = \ln u + \ln v$

(b) $\log_a(u/v) = \log_a u - \log_a v$ $\qquad \ln(u/v) = \ln u - \ln v$

(c) $\log_a u^n = n \log_a u$ $\qquad \ln u^n = n \ln u$

- You should be able to rewrite logarithmic expressions.

SOLUTIONS TO SELECTED EXERCISES

3. Write $\log_2 x$ as a multiple of a common logarithm. For instance, $\log_2 3 = (1/\log_{10} 2)\log_{10} 3$.

Solution

$$\log_2 x = \frac{\log_{10} x}{\log_{10} 2}$$

7. Write $\log_2 x$ as a multiple of a natural logarithm. For instance, $\log_2 3 = (1/\ln 2)\ln 3$.

Solution

$$\log_2 x = \frac{\ln x}{\ln 2}$$

11. Evaluate $\log_{1/2} 4$. Round to three decimal places.

Solution

$$\log_{1/2} 4 = \frac{\log_{10} 4}{\log_{10}(1/2)} = \frac{\ln 4}{\ln(1/2)} = -2$$

15. Evaluate $\log_{15} 1,250$. Round to three decimal places.

Solution

$$\log_{15} 1250 = \frac{\log_{10} 1,250}{\log_{10} 15} = \frac{\ln 1,250}{\ln 15} \approx 2.633$$

19. Use the properties of logarithms to write $\log_{10} \frac{5}{x}$ as a sum, difference, and/or multiple of logarithms.

Solution

$$\log_{10} \frac{5}{x} = \log_{10} 5 - \log_{10} x$$

23. Use the properties of logarithms to write $\ln \sqrt{z}$ as a sum, difference, and/or multiple of logarithms.

Solution

$$\ln \sqrt{z} = \ln z^{1/2} = \tfrac{1}{2} \ln z$$

27. Use the properties of logarithms to write $\ln \sqrt{a-1}$ as a sum, difference, and/or multiple of logarithms.

Solution

$$\ln \sqrt{a-1} = \tfrac{1}{2} \ln(a-1)$$

31. Use the properties of logarithms to write $\log_b \frac{x^2}{y^3}$ as a sum, difference, and/or multiple of logarithms.

Solution

$$\log_b \frac{x^2}{y^3} = \log_b x^2 - \log_b y^3 = 2\log_b x - 3\log_b y$$

35. Use the properties of logarithms to write $\ln \frac{\sqrt{y}}{z^5}$ as a sum, difference, and/or multiple of logarithms.

Solution

$$\ln \frac{\sqrt{y}}{z^5} = \ln \sqrt{y} - \ln z^5 = \frac{1}{2} \ln y - 5 \ln z$$

39. Write $\log_{10} z - \log_{10} y$ as the logarithm of a single quantity.

Solution

$$\log_{10} z - \log_{10} y = \log_{10} \frac{z}{y}$$

43. Write $\ln x - 3\ln(x+1)$ as the logarithm of a single quantity.

Solution

$$\ln x - 3\ln(x+1) = \ln x - \ln(x+1)^3 = \ln \frac{x}{(x+1)^3}$$

49. Write $2\ln 3 - \frac{1}{2}\ln(x^2+1)$ as the logarithm of a single quantity.

Solution

$$2\ln 3 - \frac{1}{2}\ln(x^2+1) = \ln 3^2 - \ln\sqrt{x^2+1} = \ln \frac{3^2}{\sqrt{x^2+1}}$$

51. Write $\ln x - \ln(x+2) - \ln(x-2)$ as the logarithm of a single quantity.

Solution

$$\begin{aligned}\ln x - \ln(x+2) - \ln(x-2) &= \ln x - [\ln(x+2) + \ln(x-2)]\\ &= \ln x - \ln(x+2)(x-2)\\ &= \ln\left[\frac{x}{(x+2)(x-2)}\right]\\ &= \ln\left(\frac{x}{x^2-4}\right)\end{aligned}$$

55. Approximate $\log_b \left(\frac{3}{2}\right)$ using the properties of logarithms, given $\log_b 2 \approx 0.3562$, $\log_b 3 \approx 0.5646$, and $\log_b 5 \approx 0.8271$.

Solution

$$\log_b \tfrac{3}{2} = \log_b 3 - \log_b 2 \approx 0.5646 - 0.3562 = 0.2084$$

61. Approximate $\log_b 40$ using the properties of logarithms, given $\log_b 2 \approx 0.3562$, $\log_b 3 \approx 0.5646$, and $\log_b 5 \approx 0.8271$.

Solution

$$\log_b 40 = \log_b(2^3 \cdot 5) = 3\log_b 2 + \log_b 5 \approx 3(0.3562) + 0.8271 = 1.8957$$

65. Approximate $\log_b \sqrt{5b}$ using the properties of logarithms, given $\log_b 2 \approx 0.3562$, $\log_b 3 \approx 0.5646$, and $\log_b 5 \approx 0.8271$.

Solution

$$\log_b \sqrt{5b} = \tfrac{1}{2}[\log_b 5 + \log_b b] \approx \tfrac{1}{2}(0.8271 + 1) \approx 0.9136$$

69. Find the *exact* value of $\log_4 16^{1.2}$.

Solution

$$\log_4 16^{1.2} = 1.2(\log_4 16) = 1.2(2) = 2.4$$

73. Use the properties of logarithms to simplify $\log_4 8$.

Solution

$\log_4 8 = \log_4 4^{3/2} = \frac{3}{2}$

77. Use the properties of logarithms to simplify $\log_5 \left(\frac{1}{250}\right)$.

Solution

$$\log_5 \frac{1}{250} = \log_5 \left(\frac{1}{2} \cdot \frac{1}{125}\right)$$
$$= \log_5 \frac{1}{2} + \log_5 \frac{1}{125} = \log_5 2^{-1} + \log_5 5^{-3} = -\log_5 2 - 3$$

81. *Curve Fitting* Use the procedure demonstrated in Example 9 (Section 5.3 in the textbook) to find an equation that relates x and y. Explain the steps used to find the equation.

Solution

x	1	2	3	4	5	6
y	1.000	1.189	1.316	1.414	1.495	1.565
$\ln x$	0	0.693	1.099	1.386	1.609	1.792
$\ln y$	0	0.173	0.275	0.346	0.402	0.448

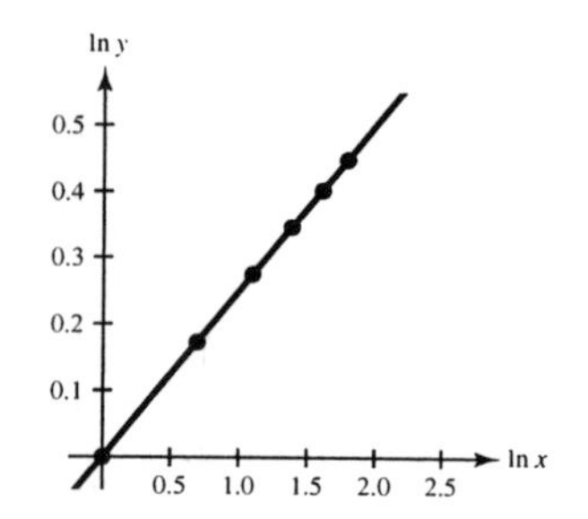

The slope of the line is $\frac{1}{4}$.

Thus, $\ln y = \frac{1}{4} \ln x$

$\ln y = \ln \sqrt[4]{x}$

$y = \sqrt[4]{x}.$

85. *Graphical Reasoning* Use a graphing utility to graph $f(x) = \ln\left(\frac{x}{3}\right)$ and $g(x) = \ln x - \ln 3$ in the same viewing rectangle. What do you observe about the two graphs? What property of logarithms is being demonstrated graphically?

Solution

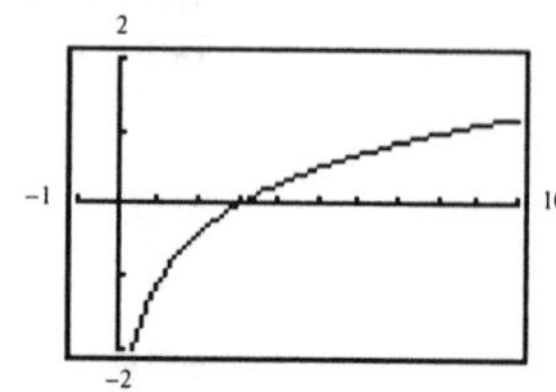

The graphs of $f(x) = \ln\left(\frac{x}{3}\right)$ and $g(x) = \ln x - \ln 3$ are the same. This illustrates the property of logarithms that says $\log_a\left(\frac{u}{v}\right) = \log_a u - \log_a v$.

89. *Nail Length* The approximate lengths and diameters (in inches) of common nails are given in the table in the textbook. Find a mathematical model that relates the diameter y of a common nail to its length x.

Solution

Length x	1	2	3	4	5	6
Diameter y	0.070	0.111	0.146	0.176	0.204	0.231
$\ln x$	0	0.693	1.099	1.386	1.609	1.792
$\ln y$	−2.659	−2.198	−1.924	−1.737	−1.590	−1.465

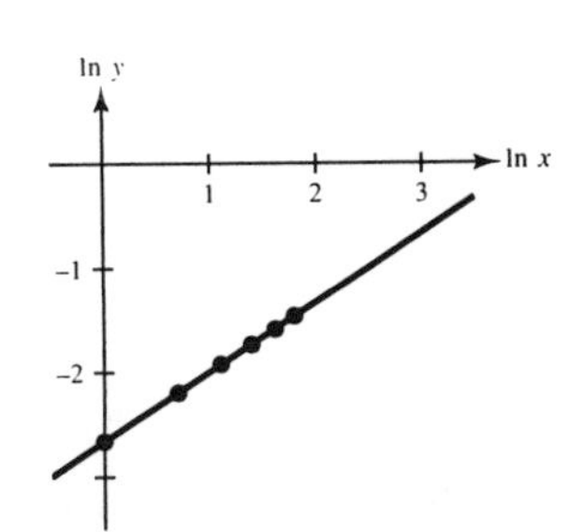

The slope of the line is $\frac{2}{3}$.

$$\text{Thus,} \quad \ln y = \frac{2}{3}\ln x - 2.659$$

$$\ln y - \ln x^{2/3} = -2.659$$

$$\ln \frac{y}{x^{2/3}} = -2.659$$

$$\frac{y}{x^{2/3}} = e^{-2.659}$$

$$y \approx 0.07x^{2/3}$$

Mid-Chapter Quiz for Chapter 5

1. Sketch the graph of $f(x) = 3^x$.

Solution

Asymptote: $y = 0$

Intercept: $(0, 1)$

x	−2	−1	0	1	2
y	$\frac{1}{9}$	$\frac{1}{3}$	1	3	9

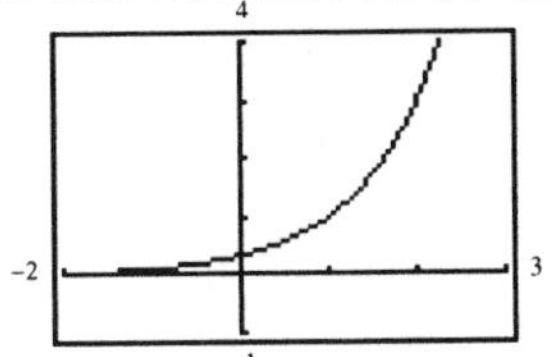

2. Sketch the graph of $g(x) = 2^{-x}$.

Solution

Asymptote: $y = 0$

Intercept: $(0, 1)$

x	−2	−1	0	1	2
y	4	2	1	$\frac{1}{2}$	$\frac{1}{4}$

3. Sketch the graph of $h(x) = \log_2 x$.

Solution

$y = \log_2 x \Rightarrow x = 2^y$

Asymptote: $x = 0$

Intercept: $(1, 0)$

x	$\frac{1}{4}$	$\frac{1}{2}$	1	2	4
y	-2	-1	0	1	2

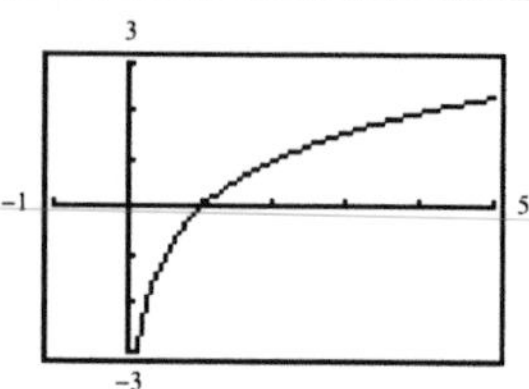

4. Sketch the graph of $h(x) = \log_3(x - 1)$.

Solution

$y = \log_3(x - 1) \Rightarrow x - 1 = 3^y$

$x = 3^y + 1$

Asymptote: $x = 1$

Intercept: $(2, 0)$

x	$1\frac{1}{9}$	$1\frac{1}{3}$	2	4	10
y	-2	-1	0	1	2

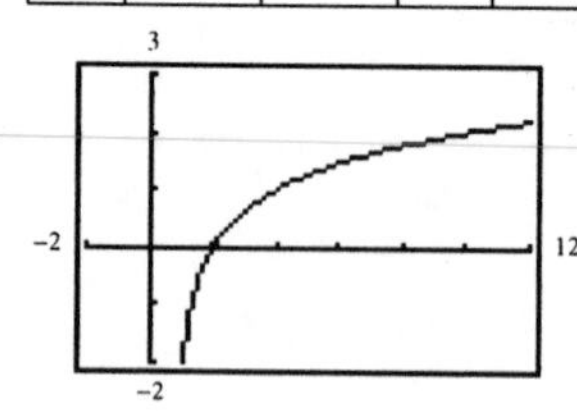

5. Find the balance compounded monthly and continuously.

$P = \$2{,}000,\ r = 8\%,\ t = 10$ years

Solution

Monthly: $A = 2{,}000\left(1 + \frac{0.08}{12}\right)^{(12)(10)} \approx \$4{,}439.28$

Continuously: $A = 2{,}000e^{(0.08)(10)} \approx \$4{,}451.08$

6. Find the balance compounded monthly and continuously.

$P = \$5{,}000,\ r = 9\%,\ t = 20$ years

Solution

Monthly: $A = 5{,}000\left(1 + \dfrac{0.09}{12}\right)^{(12)(20)} \approx \$30{,}045.76$

Continuously: $A = 5{,}000e^{(0.09)(20)} \approx \$30{,}248.24$

7. *Bacteria Growth* A bacteria population is modeled by

$P(t) = 100e^{0.3012t}$ where t is the time in hours.

Find

(a) $P(0)$, (b) $P(6)$, and (c) $P(12)$.

Solution

(a) $P(0) = 100e^0 = 100$ bacteria

(b) $P(6) = 100e^{(0.3012)(6)} \approx 609$ bacteria

(c) $P(12) = 100e^{(0.3012)(12)} \approx 3713$ bacteria

8. *Demand Function* Use the demand function $p = 600 - 0.4e^{0.003x}$ to find the price for a demand of $x = 1000$ units.

Solution

$p(1000) = 600 - 0.4e^{0.003(1000)} \approx \$591.97$

9. Evaluate $\log_{10} 100$.

Solution

$\log_{10} 100 = 2$

since $10^2 = 100$.

10. Evaluate $\ln e^4$.

Solution

$\ln e^4 = 4$

11. Evaluate $\log_4 \frac{1}{16}$.

Solution

$\log_4 \frac{1}{16} = \log_4 4^{-2} = -2$

12. Evaluate $\ln 1$.

Solution

$\ln 1 = 0$ since $e^0 = 1$.

13. Sketch the graph of $f(x) = 3^x$ and $g(x) = \log_3 x$ on the same coordinate plane. Discuss the special relationship between $f(x)$ and $g(x)$ that is demonstrated by their graphs.

Solution

$y = 3^x$

x	-2	-1	0	1	2
y	$\frac{1}{9}$	$\frac{1}{3}$	1	3	9

$y = \log_3 x \Rightarrow x = 3^y$

x	$\frac{1}{9}$	$\frac{1}{3}$	1	3	9
y	-2	-1	0	1	2

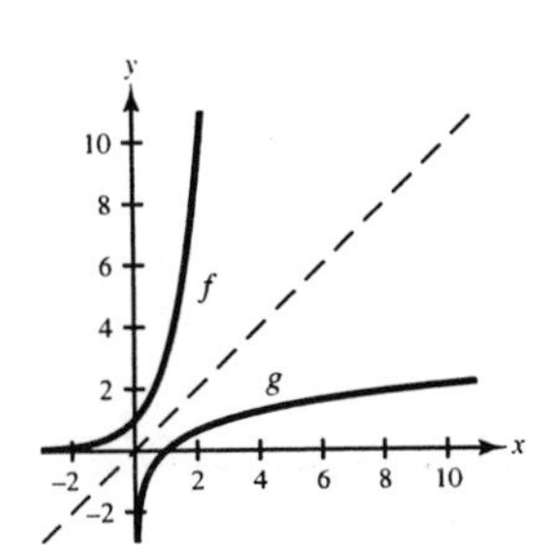

$f(x)$ and $g(x)$ are inverses of one another. They are reflected about the line $y = x$.

14. Expand the logarithmic expression $\log \sqrt[3]{\frac{xy}{z}}$.

Solution

$$\log \sqrt[3]{\frac{xy}{z}} = \frac{1}{3} \log \frac{xy}{z} = \frac{1}{3}(\log xy - \log z)$$

$$= \frac{1}{3}(\log x + \log y - \log z)$$

15. Expand the logarithmic expression $\ln\left(\frac{x^2+3}{x^3}\right)$.

Solution

$$\ln\left(\frac{x^2+3}{x^3}\right) = \ln(x^2+3) - \ln x^3 = \ln(x^2+3) - 3\ln x.$$

16. Condense the logarithmic expression $\ln x + \ln y - \ln 3$.

Solution

$$\ln x + \ln y - \ln 3 = \ln(xy) - \ln 3 = \ln\left(\frac{xy}{3}\right)$$

17. Condense the logarithmic expression $-3\log_{10} 4x$.

Solution

$$-3\log_{10} 4x = \log_{10}(4x)^{-3}$$

$$= \log_{10} \frac{1}{(4x)^3}$$

$$= \log_{10}\left(\frac{1}{64x^3}\right)$$

18. Find the exact value of $\log_7 \sqrt{343}$.

Solution

$\log_7 \sqrt{343} = \frac{1}{2}\log_7 343 = \frac{1}{2}\log_7 7^3 = \frac{1}{2}(3) = \frac{3}{2}$

19. Find the exact value of $\ln \sqrt[5]{e^6}$.

Solution

$\ln \sqrt[5]{e^6} = \ln e^{6/5} = \frac{6}{5}$

20. Find an equation that relates x and y.

x	1	2	3	4	6	8
y	1	1.260	1.442	1.587	1.817	2

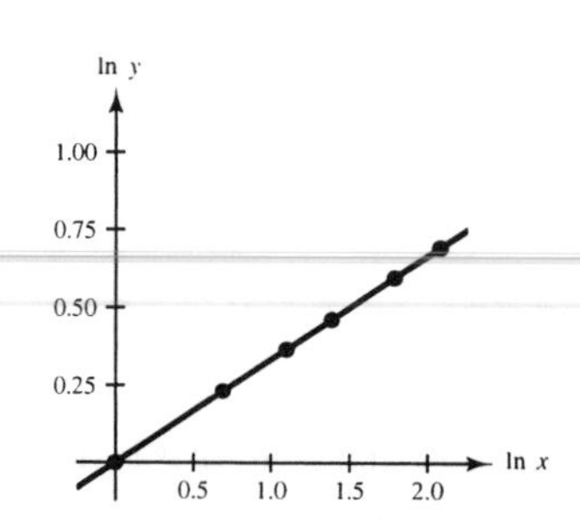

Solution

x	1	2	3	4	6	8
y	1	1.260	1.442	1.587	1.817	2
$\ln x$	0	0.693	1.099	1.386	1.792	2.079
$\ln y$	0	0.231	0.366	0.462	0.597	0.693

The slope of the line is approximately $\frac{1}{3}$

$\ln y = \frac{1}{3} \ln x$

$\ln y = \ln \sqrt[3]{x}$

$y = \sqrt[3]{x}$

or $y^3 = x$

5.4 Solving Exponential and Logarithmic Equations

- You should be able to solve exponential and logarithmic equations.
- To solve an exponential equation, take the logarithm of both sides.
- To solve a logarithmic equation, rewrite it in exponential form.
- Know the inverse properties.

  (1) $\log_a a^x = x$ $\quad \ln e^x = x$

  (2) $a^{\log_a x} = x$ $\quad e^{\ln x} = x$

SOLUTIONS TO SELECTED EXERCISES

3. Given $7^x = \frac{1}{49}$, solve for x.

Solution

$7^x = \frac{1}{49}$

$7^x = 7^{-2}$

$x = -2$

7. Given $\log_4 x = 3$, solve for x.

Solution

$\log_4 x = 3$

$x = 4^3$

$x = 64$

11. Apply the inverse properties of $\ln x$ and e^x to simplify $\ln e^{x^2}$.

Solution

$\ln e^{x^2} = x^2$

15. Apply the inverse properties of $\ln x$ and e^x to simplify $e^{\ln x^2}$.

Solution

$e^{\ln x^2} = x^2$

19. Solve $e^x = 18$.

Solution

$$e^x = 18$$
$$x = \ln 18 \approx 2.890$$

23. Solve $e^x - 5 = 10$.

Solution

$$e^x - 5 = 10$$
$$e^x = 15$$
$$x = \ln 15 \approx 2.708$$

27. Solve $7 - 2e^x = 5$.

Solution

$$7 - 2e^x = 5$$
$$-2e^x = -2$$
$$e^x = 1$$
$$x = \ln 1 = 0$$

31. Solve $e^{-x} = 28$.

Solution

$$e^{-x} = 28$$
$$-x = \ln 28$$
$$x = -\ln 28 \approx -3.332$$

35. Solve $500e^{0.06t} = 1,500$.

Solution

$$500e^{0.06t} = 1500$$
$$e^{0.06t} = 3$$
$$0.06t = \ln 3$$
$$t = \frac{\ln 3}{0.06} \approx 18.310$$

39. Solve $25e^{2x+1} = 962$.

Solution

$$25e^{2x+1} = 962$$
$$e^{2x+1} = 38.48$$
$$2x + 1 = \ln 38.48$$
$$x = \frac{-1 + \ln 38.48}{2} \approx 1.325$$

43. Solve $\ln x = 5$.

Solution

$$\ln x = 5$$
$$x = e^5 \approx 148.413$$

47. Solve $\ln 2x = 1$.

Solution

$$\ln 2x = 1$$
$$2x = e^1$$
$$x = \frac{e}{2} \approx 1.359$$

51. Solve $2\ln 4x = 0$.

Solution

$$2\ln 4x = 0$$
$$\ln 4x = 0$$
$$4x = 1$$
$$x = \tfrac{1}{4}$$

55. Solve $\ln 2x^2 = 5$.

Solution

$$\ln 2x^2 = 5$$
$$2x^2 = e^5$$
$$x^2 = \frac{e^5}{2}$$
$$x = \pm\sqrt{\frac{e^5}{2}} \approx \pm 8.614$$

59. Solve $\ln x + \ln(x-2) = 1$.

Solution

$$\ln x + \ln(x-2) = 1$$
$$\ln[x(x-2)] = 1$$
$$x(x-2) = e^1$$
$$x^2 - 2x - e = 0$$
$$x = \frac{2 \pm \sqrt{4+4e}}{2} = \frac{2 \pm 2\sqrt{1+e}}{2}$$

Using the positive value for x, we have $x = 1 + \sqrt{1+e} \approx 2.928$.

63. *Compound Interest* Find the time required for a \$1,000 investment to triple at interest rate $r = 0.0675$, compounded continuously.

Solution

$$3000 = 1000e^{0.0675t}$$
$$3 = e^{0.0675t}$$
$$0.0675t = \ln 3$$
$$t = \frac{\ln 3}{0.0675} \approx 16.28 \text{ years}$$

67. *Forest Yield* The yield V (in millions of cubic feet per acre) for a forest at age t years is given by

$$V = 6.7e^{-48.1/t}, \quad 0 \le t$$

(a) Use a graphing utility to find the time necessary to have a yield of 1.3 million cubic feet.

(b) Use a graphing utility to find the time necessary to have a yield of 2 million cubic feet.

Solution

(a) $t \approx 29.3$ years

(b) $t \approx 39.8$ years

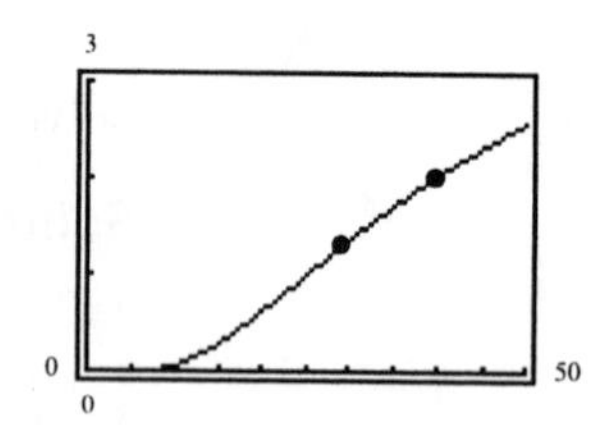

71. *Average Heights* The percentage of American males (between 18 and 24 years old) who are less than x inches tall is given by

$$p = \frac{1}{1 + e^{-0.6114(x-69.71)}}, \quad 60 \le x \le 77$$

where p is the percentage (in decimal form) and x is the height in inches.

(a) What is the median height for American males between 18 and 24 years old? (In other words, for what value of x is p equal to 0.5?) (See figure in the textbook.)

(b) Write a paragraph describing the height model.

(*Source: U.S. National Center for Health Statistics.*)

Solution

(a)

$$p = \frac{1}{1 + e^{-0.6114(x-69.71)}}, \quad 60 \le x \le 77$$

$$0.5 = \frac{1}{1 + e^{-0.6114(x-69.71)}}$$

$$0.5 + 0.5e^{-0.6114(x-69.71)} = 1$$

$$e^{-0.6114(x-69.71)} = 1$$

$$-0.6114(x - 69.71) = \ln 1 = 0$$

$$x = 69.71 \text{ inches in height}$$

(b) The model is an increasing function with horizontal asymptotes at $p = 0$ and $p = 1$.

75. *Native Prairie Grasses* The number of native prairie grasses per acre A is approximated by the model

$$A = 10.5 \cdot 10^{0.04x}, \quad 0 \le x \le 24$$

where x is the number of months since the acre was plowed. Use this model to approximate the number of months since the field was plowed in a test plot if $A = 280$.

Solution

$$280 = 10.5 \cdot 10^{0.04x}$$

$$\frac{280}{10.5} = 10^{0.04x}$$

$$\log(280/10.5) = 0.04x$$

$$\frac{\log(280/10.5)}{0.04} = x$$

$$x \approx 35.649 \text{ months}$$

5.5 Exponential and Logarithmic Models

- Know these basic types of mathematical models.
  (a) Exponential growth: $y = ae^{bx},\ b > 0$
  (b) Exponential decay: $y = ae^{-bx},\ b > 0$
  (c) Gaussian model: $y = ae^{-(x-b)^2/c}$
  (d) Logistics growth model: $y = \dfrac{a}{1 + be^{-(x-c)/d}}$
  (e) Logarithmic models: $y = \ln(a + bx)$
  $y = \log_{10}(a + bx)$

SOLUTIONS TO SELECTED EXERCISES

7. *Compound Interest* Find the initial investment and the time it takes to double for a savings account in which the annual percentage rate is 11% compounded continuously and the amount after 10 years is \$19,205.

Solution

$$19{,}205 = Pe^{(0.11)(10)}$$

$$\frac{19{,}205}{e^{1.1}} = P$$

$$P \approx \$6{,}392.79$$

Time to double: $\dfrac{\ln 2}{0.11} \approx 6.30$ years.

13. *Radioactive Decay* Given the radioactive isotope C^{14} which has a half-life of 5,730 years, find the initial quantity given the amount that remains after 1000 years is 2 grams.

Solution

$$y = ae^{bx}$$

$$\frac{1}{2}a = ae^{b(5.730)}$$

$$\frac{1}{2} = e^{5.730b}$$

$$b = \frac{\ln(1/2)}{5730} \approx -0.000120968$$

$$y = ae^{[\ln(1/2)/5730]x}$$

$$2 = ae^{[\ln(1/2)/5.740](1000)}$$

$a \approx 2.26$ grams for the initial quantity

17. Classify $y = 3e^{0.5t}$ as exponential growth or exponential decay.

Solution

Since $b = 0.5 > 0$, the model $y = 3e^{0.5t}$ represents exponential growth.

21. Find the constant k such that the exponential function $y = Ce^{kt}$ passes through (0, 1) and (4, 10). (See figure in the textbook.)

Solution

$y = e^{kt}$

$10 = e^{4k}$

$k = \frac{1}{4} \ln 10 \approx 0.5756$

$y = e^{0.5756t}$

25. *Population* The population P of a city is given by $P = 105{,}300e^{0.015t}$ where t is the time in years, with $t = 0$ corresponding to 1990. Sketch the graph of this equation. According to this model, in what year will the city have a population of 150,000?

Solution

$$P = 105{,}300e^{0.015t}$$

$$150{,}000 = 105{,}300e^{0.015t}$$

$$\ln \tfrac{1500}{1053} = 0.015t$$

$$t \approx 23.59 \Rightarrow 1990 + 24 = 2014$$

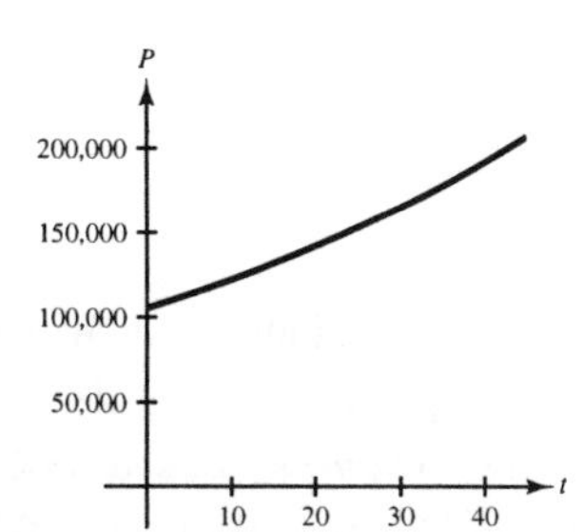

31. *Radioactive Decay* What percent of a present amount of radioactive radium (Ra^{226}) will remain after 100 years? Use the fact that radioactive radium has a half-life of 1,620 years.

Solution

$$y = Ce^{kt}$$

$$\frac{1}{2}C = Ce^{(1620)k}$$

$$\ln \frac{1}{2} = 1620k$$

$$k = \frac{\ln(1/2)}{1620}$$

When $t = 100$, we have $y = Ce^{[\ln(1/2)/1620](100)} \approx 0.958C = 95.8\%C$.

After 100 years, approximately 95.8% of the radioactive radium will remain.

35. *Number of Golfers* From 1970 to 1985 the number of golfers in the United States was increasing at the rate of 2.956% per year. Then in 1985 the growth pattern changed. (*Source: National Golf Foundation.*) The number of golfers from 1970 to 1991 was approximately

$$y = \begin{cases} 11{,}245e^{0.02956t}, & 0 \le t \le 15 \\ 8677.6 + 786.8t - e^{23.03-t}, & 15 \le t \le 21 \end{cases}$$

where y is the number of golfers (in thousands) and $t = 0$ represents 1970.

(a) Use a graphing ultility to graph this model.

(b) Describe the change in the growth pattern that occurred in 1985.

Solution

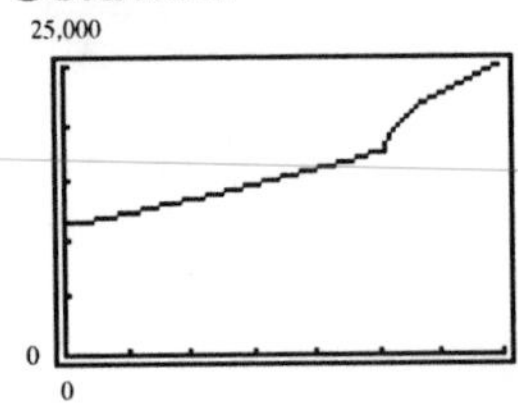

(b) The number of golfers began to increase at a faster rate in 1985.

37. *Odometer Readings* In 1990 the probability distribution of odometer readings for 1987 automobiles was approximately $p = 0.1337e^{-(x-32.2)^2/35.6}$ where x is the odometer reading in thousands of miles. (*Source: Based on Statistics from the U.S. Highway Administration.*) What was the average odometer reading for a 1987 automobile?

Solution

This is a Gaussian model with a mean of $b = 32.2$, which corresponds to 32,200 miles.

41. *Stocking a Lake with Fish* A certain lake was stocked with 500 fish, and the fish population increased according to the logistics curve

$$P = \frac{10{,}000}{1 + 19e^{-t/5}}, \quad 0 \le t$$

where t is measured in months.

(a) Use a graphing utility to graph this model.

(b) Find the fish population after five months.

(c) After how many months will the fish population be 2000?

Solution

(a)

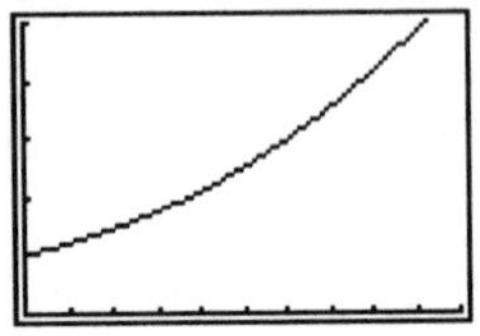

(b) $P(5) = \dfrac{10{,}000}{1 + 19e^{-1}} \approx 1252$ fish

(c)
$$2000 = \frac{10{,}000}{1 + 19e^{-t/5}}$$
$$1 + 19e^{-t/5} = 5$$
$$e^{-t/5} = \frac{4}{19}$$
$$t = -5\ln\left(\frac{4}{19}\right)$$
$$\approx 7.8 \text{ months}$$

43. *Endangered Species* A conservation group releases 100 animals of an endangered species into a game preserve. The organization believes that the preserve has a carrying capacity of 1,000 animals and that the growth of the herd will be modeled by the logistics curve

$$p(t) = \frac{1000}{1 + 9e^{-kt}}, \quad 0 \le t$$

where t is measured in years.

(a) Find k if the herd size is 134 after two years.

(b) Find the population after five years.

Solution

(a)
$$134 = \frac{1000}{1 + 9e^{-2k}}$$
$$134(1 + 9e^{-2k}) = 1000$$
$$e^{-2k} = \frac{866}{9(134)}$$
$$k = -\frac{1}{2}\ln\frac{433}{603} \approx 0.1656$$

(b)
$$p(t) = \frac{1000}{1 + 9e^{-0.1656t}}$$
$$p(5) = \frac{1000}{1 + 9e^{-0.1656(5)}}$$
$$\approx 203 \text{ animals}$$

47. *Earthquake Magnitudes* Use the Richter Scale in Example 6 (Section 5.5 in the textbook) for measuring the magnitude of earthquakes. Find the magnitude R of an earthquake of intensity I (let $I_0 = 1$), if (a) $I = 80{,}500{,}000$, (b) $I = 48{,}275{,}000$, (c) $I = 40{,}000{,}000$, and (d) $I = 20{,}000{,}000$.

Solution

If $I_0 = 1$, $R = \log I$

(a) $R = \log(80{,}500{,}000)$

$R \approx 7.91$

(b) $R = \log(48{,}275{,}000)$

$R \approx 7.68$

(c) $R = \log(40{,}000{,}000)$

$R \approx 7.60$

(d) $R = \log(20{,}000{,}000)$

$R \approx 7.30$

53. *Estimating the Time of Death* At 8:30 A.M., a coroner was called to the home of a person who had died during the night. In order to estimate the time of death, the coroner took the person's temperature twice. At 9:00 A.M. the temperature was 85.7° and at 9:30 A.M. the temperature was 82.8°. From these two temperature readings, the coroner was able to determine that the time elapsed since death and the body temperature were related by the formula

$$t = -2.5 \ln \frac{T - 70}{98.6 - 70}$$

where t is the time in hours that have elapsed since the person died and T is the temperature (in degrees Fahrenheit) of the person's body. Assume the person had a normal body temperature of 98.6° at death, and the room temperature was a constant 70°. (This formula is derived from a general cooling principle called Newton's Law of Cooling.) Use this formula to estimate the time of death of the person.

Solution

$$t = -2.5 \ln \frac{T - 70}{98.6 - 70}$$

$$t = -2.5 \ln \frac{85.7 - 70}{98.6 - 70} \approx 1.5 \text{ hours}$$

The time of death was approximately 9:00 A.M. − 1.5 hours = 7:30 A.M.

Review Exercises for Chapter 5

SOLUTIONS TO SELECTED EXERCISES

3. Match $f(x) = -2^x$ with its graph. [The graphs are labeled (a), (b), (c), (d), (e), (f), (g), and (h) in the textbook.]

Solution

Intercept: $(0, -1)$

Horizontal asymptote: x-axis

Decreasing on $(-\infty, \infty)$

Matches graph(a)

7. Match $f(x) = -\log_2 x$ with its graph. [The graphs are labeled (a), (b), (c), (d), (e), (f), (g), and (h) in the textbook.]

Solution

Intercept: $(1, 0)$

Vertical asymptote: $x = 0$

Decreasing on $(-\infty, \infty)$

Matches graph (h)

11. Sketch the graph of $y = \left(\frac{1}{2}\right)^x$.

Solution

x	-2	-1	0	1	2
y	4	2	1	$\frac{1}{2}$	$\frac{1}{4}$

15. Sketch the graph of $y = 4^{-x^2}$.

Solution

x	0	± 1	± 2
y	1	$\frac{1}{4}$	$\frac{1}{256}$

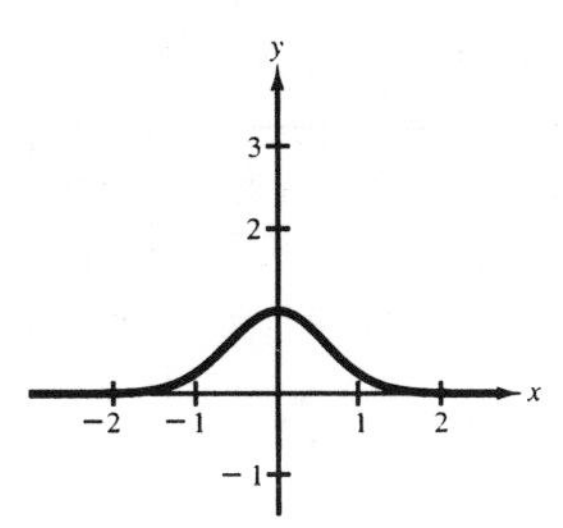

19. Complete the table in the textbook to determine the amount of money P that should be invested at a rate $r = 8\%$, compounded continuously, to produce a final balance of \$200,000 in t years.

Solution

$$200{,}000 = Pe^{0.08t}$$

$$P = \frac{200{,}000}{e^{0.08t}}$$

t	1	10	20	30	40	50
P	\$184,623.27	\$89,865.79	\$40,379.30	\$18,143.59	\$8,152.44	\$3,663.13

21. *Trust Fund* Suppose the day your child was born you deposited \$50,000 in a trust fund that paid 8.75% interest, compounded continuously. Between what birthdays would your child become a millionaire?

Solution

$$1{,}000{,}000 = 50{,}000e^{0.0875t}$$

$$\ln 20 = 0.0875t$$

$$t \approx 34.24$$

The child will become a millionaire between his/her 34th and 35th birthdays.

27. Evaluate $\ln e^{-3}$ without using a calculator.

Solution

$\ln e^{-3} = -3$

31. Use the fact that $f(x) = 10^x$ and $g(x) = \log_{10} x$ are inverses of each other to sketch their graphs on the same coordinate plane.

Solution

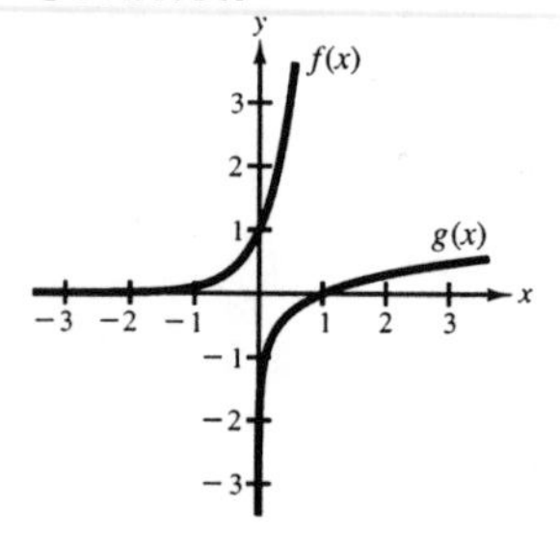

x	$f(x)$	x	$g(x)$
-1	$\frac{1}{10}$	$\frac{1}{10}$	-1
0	1	1	0
1	10	10	1

33. Find the domain, vertical asymptote, and x-intercept of $f(x) = \ln(x + 2)$ and sketch its graph.

Solution

Domain: $(-2, \infty)$

Vertical asymptote: $x = -2$

x-intercept $(-1, 0)$

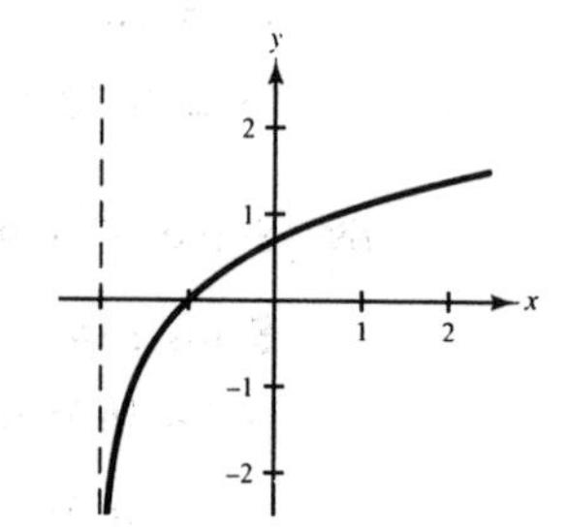

39. Evaluate $\log_4 9$ using the change-of-base formula. Do the problem twice, once with common logarithms and once with natural logarithms. Round your answers to three decimal places.

Solution

$$\log_4 9 = \frac{\log_{10} 9}{\log_{10} 4} = \frac{\ln 9}{\ln 4} \approx 1.585$$

43. Use the properties of logarithms to write $\log_{10}\left(\frac{x}{y}\right)$ as a sum, difference, and/or multiple of logarithms.

Solution

$$\log_{10}\left(\frac{x}{y}\right) = \log_{10} x - \log_{10} y$$

45. Use the properties of logarithms to write $\ln(x\sqrt{x-3})$ as a sum, difference, and/or multiple of logarithms.

Solution

$$\ln(x\sqrt{x-3}) = \ln x + \ln\sqrt{x-3} = \ln x + \tfrac{1}{2}\ln(x-3)$$

49. Write $\ln x + \ln 5$ as the logarithm of a single quantity.

Solution

$$\ln x + \ln 5 = \ln 5x$$

53. Write $\ln x - \ln(x-3) - \ln(x+1)$ as the logarithm of a single quantity.

Solution

$$\ln x - \ln(x-3) - \ln(x+1) = \ln x - [\ln(x-3) + \ln(x+1)]$$

$$= \ln x - \ln(x-3)(x+1) = \ln\left[\frac{x}{(x-3)(x+1)}\right] = \ln\left(\frac{x}{x^2-2x-3}\right)$$

57. Approximate $\log_b \sqrt{3}$ using the properties of logarithms given $\log_b 2 \approx 0.3562$, $\log_b 3 \approx 0.5646$, and $\log_b 5 \approx 0.8271$.

Solution

$$\log_b \sqrt{3} = \tfrac{1}{2}\log_b 3 = \tfrac{1}{2}(0.5646) = 0.2823$$

59. Find the exact value of $\log_3 27$.

Solution

$$\log_3 27 = \log_3 3^3 = 3$$

63. Solve $e^x = 12$.

Solution

$$e^x = 12$$

$$x = \ln 12 \approx 2.485$$

65. Solve $3e^{-5x} = 132$.

Solution

$$3e^{-5x} = 132$$

$$e^{-5x} = 44$$

$$-5x = \ln 44$$

$$x = \frac{\ln 44}{-5}$$

$$\approx -0.757$$

69. Solve $-2 + \ln 5x = 0$.

Solution

$$-2 + \ln 5x = 0$$

$$\ln 5x = 2$$

$$5x = e^2$$

$$x = \frac{e^2}{5}$$

$$x \approx 1.478$$

73. *Demand Function* The demand function for a certain product is given by

$$p = 600 - 0.3(e^{0.005x}).$$

Find the demand x for a price of (a) $p = \$500$ and (b) $p = \$400$.

Solution

(a)

$$500 = 600 - 0.3e^{0.005x}$$

$$e^{0.005x} = \frac{100}{0.3}$$

$$0.005x = \ln\left(\frac{100}{0.3}\right)$$

$$x = \frac{\ln\left(\frac{100}{0.3}\right)}{0.005}$$

$$x \approx 1162 \text{ units}$$

(b)

$$400 = 600 - 0.3e^{0.005x}$$

$$e^{0.005x} = \frac{200}{0.3}$$

$$0.005x = \ln\left(\frac{200}{0.3}\right)$$

$$x = \frac{\ln(200/0.3)}{0.005}$$

$$\approx 1300 \text{ units}$$

77. *Population* The population P of a city is given by $P = 270{,}000e^{0.019t}$ where t is the time in years with $t = 0$ corresponding to 1990. (a) Use a graphing utility to graph this equation. (b) According to this model, in what year will the city have a population of 300,000?

Solution

(a)

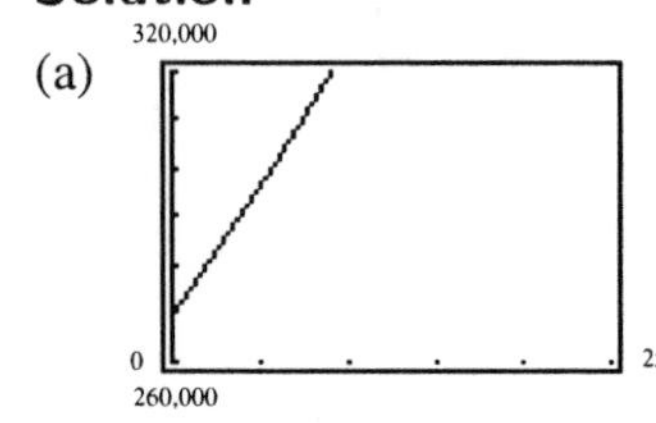

(b)

$$P = 270{,}000e^{0.019t}, \quad t \geq 0$$

$$300{,}000 = 270{,}000e^{0.019t}$$

$$\frac{10}{9} = e^{0.019t}$$

$$\frac{\ln(10/9)}{0.019} = t$$

$$t \approx 5.5453$$

$$1990 + 5.5453 = 1995.5453$$

The population will reach 300,000 during the middle of 1995.

81. *Learning Curve* The management at a factory has found that the maximum number of units a worker can produce in a day is 50. The learning curve for the number of units N produced per day after a new employee has worked t days is given by $N = 50(1 - e^{kt})$. After 20 days on the job, a particular worker produced 31 units in one day.

(a) Find the learning curve for this worker.

(b) How many days should pass before this worker is producing 45 units per day?

Solution

(a)

$$31 = 50\left(1 - e^{20k}\right)$$

$$0.38 = e^{20k}$$

$$\frac{\ln 0.38}{20} = k$$

$$N = 50\left(1 - e^{[\ln 0.38/20]t}\right)$$

$$\approx 50\left(1 - e^{-0.04838t}\right)$$

(b)

$$45 = 50\left(1 - e^{[\ln 0.38/20]t}\right)$$

$$\ln 0.1 = [\ln 0.38/20]t$$

$$t = \frac{20 \ln 0.1}{\ln 0.38}$$

$$\approx 48 \text{ days}$$

87. *Bacteria Growth* The number of bacteria N in a culture is given by the model $N = 200e^{Kt}$ where t is the time in hours, with $t = 0$ corresponding to the time when $N = 200$. When $t = 5$, there are 350 bacteria. How long does it take for the population to triple in size?

Solution

$$N = 200e^{Kt}$$

$$350 = 200e^{K(5)}$$

$$1.75 = e^{5K}$$

$$\ln 1.75 = 5K$$

$$\frac{\ln 1.75}{5} = K$$

$$K \approx 0.111923$$

$$3(200) = 200e^{0.111923t}$$

$$3 = e^{0.111923t}$$

$$\ln 3 = 0.111923t$$

$$\frac{\ln 3}{0.111923} = t$$

$$t \approx 9.81 \text{ hours}$$

Test for Chapter 5

1. Sketch the graph of $y = 2^x$.

Solution

x	-2	-1	0	1	2
y	$\frac{1}{4}$	$\frac{1}{2}$	1	2	4

Horizontal asymptote: x-axis

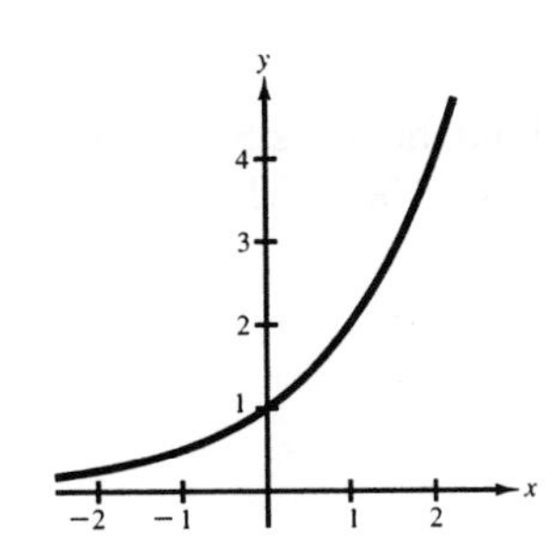

2. Sketch the graph of $y = 3^{-x}$.

Solution

x	-2	-1	0	1	2
y	9	3	1	$\frac{1}{3}$	$\frac{1}{9}$

Horizontal asymptote: x-axis

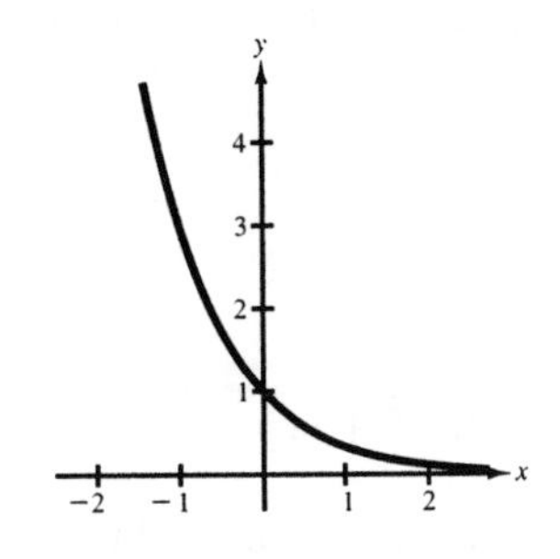

3. Sketch the graph of $y = \ln x$.

Solution

x	$\frac{1}{4}$	$\frac{1}{2}$	1	2	3
y	-1.39	-0.69	0	0.69	1.1

Vertical asymptote: x-axis

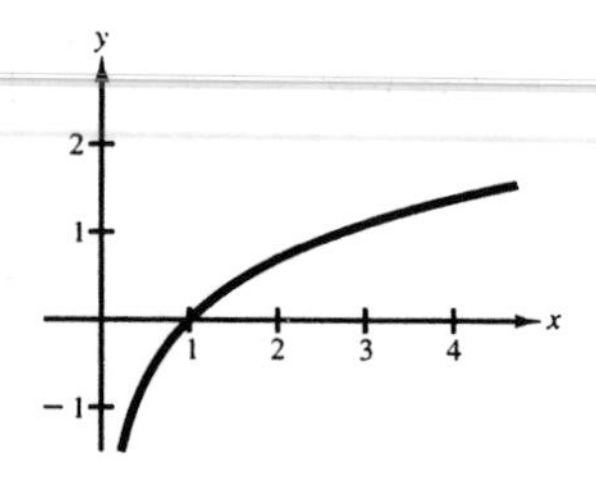

4. Sketch the graph of $y = \log_3(x - 1)$.

Solution

$y = \log_3(x - 1) \Rightarrow 3^y = x - 1 \Rightarrow 3^y + 1 = x$

x	$\frac{10}{9}$	$\frac{4}{3}$	2	4	10
y	-2	-1	0	1	2

Vertical asymptote: $x = 1$

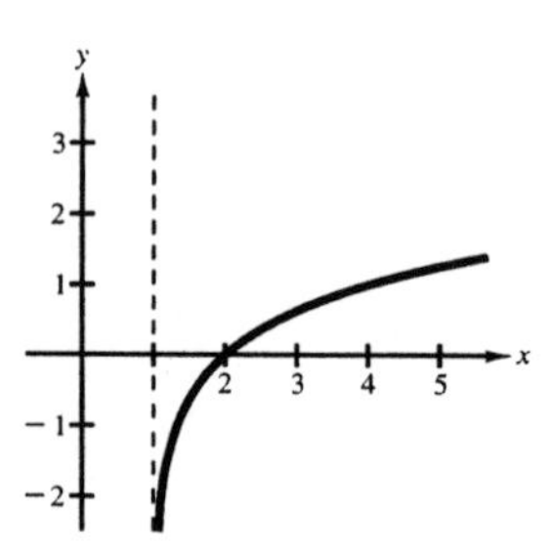

5. You deposit \$20,000 into a fund that pays 8.5% interest, compounded continuously. When will the balance be greater than \$100,000?

Solution

$$20{,}000e^{0.085t} > 100{,}000$$
$$e^{0.085t} > 5$$
$$0.085t > \ln 5$$
$$t > \frac{\ln 5}{0.085}$$
$$t > 18.9346 \text{ years}$$

The balance will be greater than \$100,000 after approximately 18.93 years.

6. Expand $\ln \frac{x^2y^3}{z}$.

Solution

$$\ln \frac{x^2y^3}{z} = \ln(x^2y^3) - \ln z = \ln x^2 + \ln y^3 - \ln z$$
$$= 2\ln x + 3\ln y - \ln z$$

7. Expand $\log_{10} 3xyz^2$.

Solution

$$\log_{10}(3xyz^2) = \log_{10} 3 + \log_{10} x + \log_{10} y + \log_{10} z^2$$
$$= \log_{10} 3 + \log_{10} x + \log_{10} y + 2\log_{10} z$$

8. Expand. $\ln(x\sqrt[3]{x-2})$.

Solution

$\ln(x\sqrt[3]{x-2}) = \ln x + \ln\sqrt[3]{x-2} = \ln x + \frac{1}{3}\ln(x-2)$

9. Condense $\ln y + 2\ln z - 3\ln x$.

Solution

$\ln y + 2\ln z - 3\ln x = \ln y + \ln z^2 - \ln x^3 = \ln\left(\frac{yz^2}{x^3}\right)$

10. Condense $\frac{2}{3}(\log_{10} x + \log_{10} y)$.

Solution

$\frac{2}{3}(\log_{10} x + \log_{10} y) = \frac{2}{3}\log_{10}(xy) = \log_{10}(xy)^{2/3} = \log_{10}\sqrt[3]{x^2y^2}$

11. Solve $e^{4x} = 21$.

Solution

$$e^{4x} = 21$$
$$4x = \ln 21$$
$$x = \frac{\ln 21}{4}$$
$$\approx 0.761$$

12. Solve $e^{2x} - 8e^x + 12 = 0$.

Solution

$$e^{2x} - 8e^x + 12 = 0$$
$$(e^x - 2)(e^x - 6) = 0$$
$$e^x = 2 \text{ or } e^x = 6$$
$$x = \ln 2 \quad \text{or} \quad x = \ln 6$$
$$\approx 0.693 \qquad \approx 1.792$$

13. Solve $-3 + \ln 4x = 0$.

Solution

$$-3 + \ln 4x = 0$$
$$\ln 4x = 3$$
$$4x = e^3$$
$$x = e^3/4$$
$$\approx 5.021$$

14. Solve $\ln\sqrt{x+2} = 3$.

Solution

$$\ln\sqrt{x+2} = 3$$
$$\sqrt{x+2} = e^3$$
$$x + 2 = e^6$$
$$x = e^6 - 2$$
$$\approx 401.429$$

In Exercises 15–17, use the following information.

Students in a psychology class were given an exam and then retested monthly with an equivalent exam. The average score for the class is given by the human memory model

$$f(t) = 87 - 15\log_{10}(t+1), \quad 0 \le t \le 4$$

where t is the time in months.

15. What was the average score on the original exam?

Solution

$f(0) = 87 - 15\log_{10}(0+1) = 87$

16. What was the average score after one month?

Solution

$f(1) = 87 - 15\log_{10}(1+1) \approx 82.5$

17. What was the average score after four months?

Solution

$f(4) = 87 - 15\log_{10}(4+1) \approx 76.5$

18. The population P of a city is given by $P = 70{,}000e^{0.023t}$ where $t = 0$ represents 1990. When will the city have a population of 100,000? Explain.

Solution

$$70{,}000e^{0.023t} = 100{,}000$$

$$e^{0.023t} = \frac{10}{7}$$

$$0.023t = \ln\left(\frac{10}{7}\right)$$

$$t = \frac{\ln(10/7)}{0.023} \approx 15.5076$$

The population will be 100,000 when $t = 15.5076$ which corresponds to the middle of the year 2005.

19. The number of bacteria N in a culture is given by $N = 100e^{kt}$ where t is the time in hours with $t = 0$ corresponding to the time when $N = 100$. When $t = 8$, $N = 500$. How long does it take the population to double?

Solution

$$N = 100e^{kt}$$

$$500 = 100e^{8k}$$

$$5 = e^{8k}$$

$$\ln 5 = 8k$$

$$k = \frac{\ln 5}{8}$$

$$N = 100e^{[(\ln 5)/8]t}$$

$$200 = 100e^{[(\ln 5)/8]t}$$

$$2 = e^{[(\ln 5)/8]t}$$

$$\ln 2 = \left(\frac{\ln 5}{8}\right)t$$

$$\frac{8\ln 2}{\ln 5} = t$$

$$t \approx 3.4 \text{ hours}$$

20. Carbon 14 has a half-life of 5,730 years. You have an initial quantity of 10 grams. How many grams will remain after 10,000 years?

Solution

$$A = 10e^{kt}$$

$$5 = 10e^{5730k}$$

$$\ln\left(\frac{1}{2}\right) = 5730k$$

$$k = \frac{\ln(1/2)}{5730}$$

$$A = 10e^{[\ln(1/2)/5730]t}$$

$$A(10{,}000) = 10e^{[\ln(1/2)/5730](10{,}000)} \approx 2.98 \text{ grams}$$

Practice Test for Chapter 5

1. Solve for x : $x^{3/5} = 8$.

2. Solve for x : $3^{x-1} = \frac{1}{81}$.

3. Graph $f(x) = 2^{-x}$.

4. Graph $g(x) = e^x + 1$.

5. If $5,000 is invested at 9% interest, find the amount after three years if the interest is compounded

(a) monthly (b) quarterly (c) continuously.

6. Write the equation in logarithmic form: $7^{-2} = \frac{1}{49}$.

7. Solve for x: $x - 4 = \log_2 \frac{1}{64}$.

8. Given $\log_b 2 = 0.3562$ and $\log_b 5 = 0.8271$, evaluate $\log_b \sqrt[4]{8/25}$.

9. Write $5 \ln x - \frac{1}{2} \ln y + 6 \ln z$ as a single logarithm.

10. Using your calculator and the change-of-base formula, evaluate $\log_9 28$.

11. Use your calculator to solve for N: $\log_{10} N = 0.6646$.

12. Graph $y = \log_4 x$.

13. Determine the domain of $f(x) = \log_3(x^2 - 9)$.

14. Graph $y = \ln(x - 2)$.

15. True or False: $\dfrac{\ln x}{\ln y} = \ln(x - y)$.

16. Solve for x: $5^x = 41$.

17. Solve for x: $x - x^2 = \log_5 \frac{1}{25}$.

18. Solve for x: $\log_2 x + \log_2(x - 3) = 2$.

19. Solve for x: $\dfrac{e^x + e^{-x}}{3} = 4$.

20. $6,000 is deposited into a fund at an annual percentage rate of 13%. Find the time required for the investment to double if the interest is compounded continuously.

Cumulative Test for Chapters 3–5

1. Given $f(x) = 2x^2 - 3$ and $g(x) = 3x + 5$, find $(f + g)(x)$.

Solution

$(f + g)(x) = f(x) + g(x) = (2x^2 - 3) + (3x + 5) = 2x^2 + 3x + 2$

2. Given $f(x) = 2x^2 - 3$ and $g(x) = 3x + 5$, find $(f - g)(x)$.

Solution

$(f - g)(x) = f(x) - g(x) = (2x^2 - 3) - (3x + 5) = 2x^2 - 3x - 8$

3. Given $f(x) = 2x^2 - 3$ and $g(x) = 3x + 5$, find $(fg)(x)$.

Solution

$(fg)(x) = f(x) \cdot g(x) = (2x^2 - 3)(3x + 5) = 6x^3 + 10x^2 - 9x - 15$

4. Given $f(x) = 2x^2 - 3$ and $g(x) = 3x + 5$, find $(f/g)(x)$.

Solution

$$\left(\frac{f}{g}\right)(x) = \frac{f(x)}{g(x)} = \frac{2x^2 - 3}{3x + 5},\ x \neq -\frac{5}{3}$$

5. Given $f(x) = 2x^2 - 3$ and $g(x) = 3x + 5$, find $(f \circ g)(x)$.

Solution

$$\begin{aligned}(f \circ g)(x) &= f(g(x)) = f(3x + 5) = 2(3x + 5)^2 - 3\\ &= 2(9x^2 + 30x + 25) - 3 = 18x^2 + 60x + 47\end{aligned}$$

6. Given $f(x) = 2x^2 - 3$ and $g(x) = 3x + 5$, find $(g \circ f)(x)$.

Solution

$(g \circ f)(x) = g(f(x)) = g(2x^2 - 3) = 3(2x^2 - 3) + 5 = 6x^2 - 4$

7. Sketch the graph of $f(x) = (x - 2)^2 + 3$.

Solution

Parabola with vertex $(2, 3)$

x	1	0	3	4
y	4	7	4	7

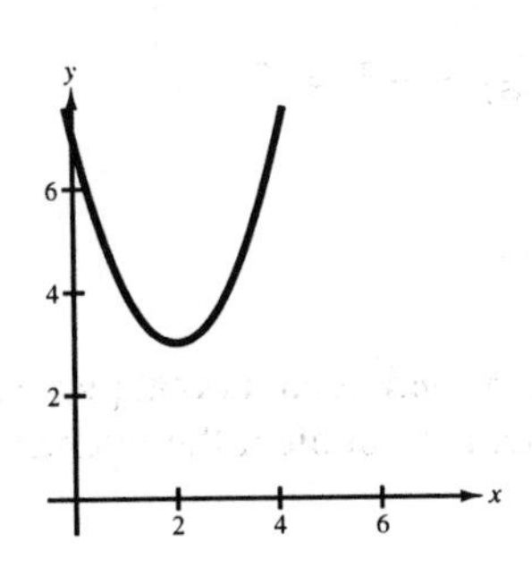

8. Sketch the graph of $g(x) = \dfrac{2}{x - 3}$.

Solution

Vertical asymptote: $x = 3$

Horizontal asymptote: x-axis

x	0	1	2	4	5
y	$-\frac{2}{3}$	-1	-2	2	1

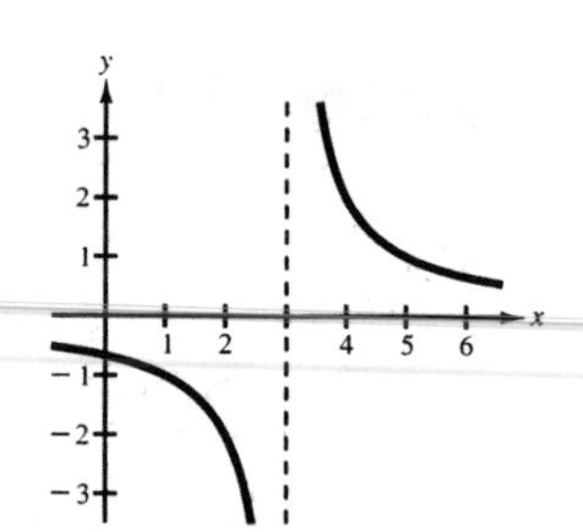

9. Sketch the graph of $h(x) = 2^{-x}$.

Solution

Horizontal asymptote: x-axis

x	-2	-1	0	1	2
y	4	2	1	$\frac{1}{2}$	$\frac{1}{4}$

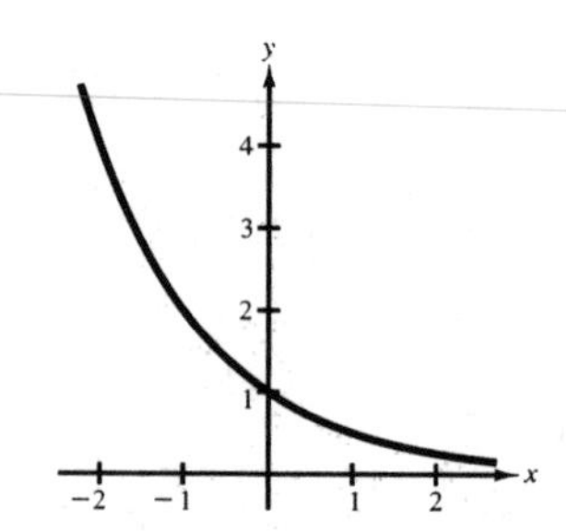

10. Sketch the graph of $f(x) = \log_4(x - 1)$.

Solution

$f(x) = \log_4(x - 1) \Rightarrow 4^y = x - 1 \Rightarrow 4^y + 1 = x$

10. –CONTINUED–

Vertical asymptote: $x = 1$

x	$\frac{17}{16}$	$\frac{5}{4}$	2	5	3
y	-2	-1	0	1	$\frac{1}{2}$

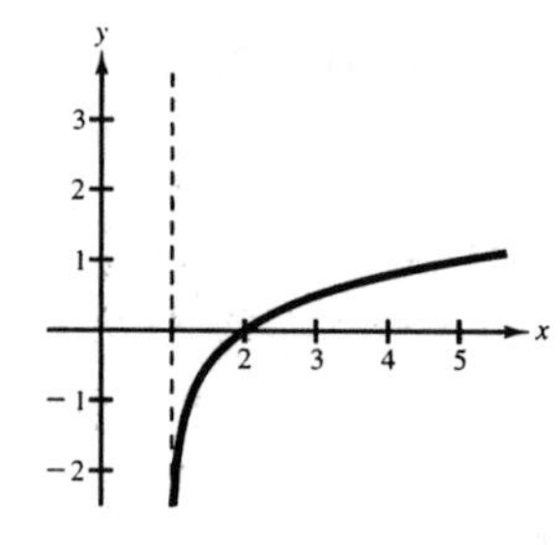

11. Sketch the graph of $g(x) = \begin{cases} x + 5, & x < 0 \\ 5, & x = 0 \\ x^2 + 5, & x > 0 \end{cases}$

Solution

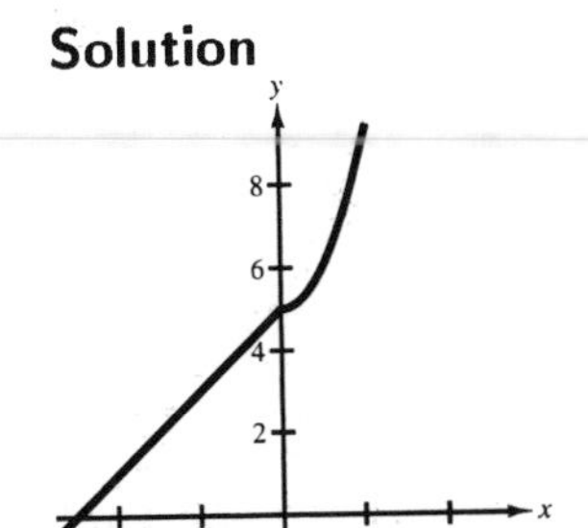

12. The profit for a company is given by $P = 500 + 20x - 0.0025x^2$ where x is the number of units produced. What production level will yield a maximum profit?

Solution

$$P = 500 + 20x - 0.0025x^2 = -0.0025(x^2 - 8000x) + 500$$

$$= -0.0025(x^2 - 8000x + 16{,}000{,}000 - 16{,}000{,}000) + 500$$

$$= -0.0025(x^2 - 4000)^2 + 40{,}500$$

The profit will be maximum ($40,500) when $x = 4000$ units.

13. Perform the indicated operations on $(10 + 2i)(3 - 4i)$ and write the result in standard form.

Solution

$$(10 + 2i)(3 - 4i) = 30 - 40i + 6i - 8i^2 = 38 - 34i$$

14. Perform the indicated operations on $(5 + 4i)^2$ and write the result in standard form.

Solution

$$(5 + 4i)^2 = 25 + 40i + 16i^2 = 9 + 40i$$

15. Perform the indicated operations on $\dfrac{2+i}{3-i}$ and write the result in standard form.

Solution

$$\frac{2+i}{3-i} = \frac{2+i}{3-i} \cdot \frac{3+i}{3+i} = \frac{6 + 2i + 3i + i^2}{9 - i^2} = \frac{5 + 5i}{10} = \frac{1}{2} + \frac{1}{2}i$$

16. Use the Quadratic Formula to solve $3x^2 - 5x + 7 = 0$.

Solution

$$x = \frac{-(-5) \pm \sqrt{(-5)^2 - 4(3)(7)}}{2(3)} = \frac{5 \pm \sqrt{-59}}{6} = \frac{5 \pm \sqrt{59}i}{6} = \frac{5}{6} \pm \frac{\sqrt{59}}{6}i$$

17. Find all the zeros of $f(x) = x^4 + 10x^2 + 9$ given that $3i$ is a zero. Explain your reasoning.

Solution

Since $3i$ is a zero, so is $-3i$.

$3i$	1	0	10	0	9
		$3i$	-9	$3i$	-9
$-3i$	1	$3i$	1	$3i$	0
		$-3i$	0	$-3i$	
	1	0	1	0	

$f(x) = (x - 3i)(x + 3i)(x^2 + 1) = (x - 3i)(x + 3i)(x - i)(x + i)$

Zeros: $\pm 3i, \pm i$

18. Solve $e^{2x} - 11e^x + 24 = 0$.

Solution

$e^{2x} - 11e^x + 24 = 0$

$(e^x - 3)(e^x - 8) = 0$

$e^x = 3$ or $e^x = 8$

$x = \ln 3$ or $x = \ln 8$

19. Solve $\frac{1}{2}\ln(x - 3) = 4$.

Solution

$\frac{1}{2}\ln(x - 3) = 4$

$\ln(x - 3) = 8$

$x - 3 = e^8$

$x = 3 + e^8$

20. The population P of a city is given by $P = 80{,}000e^{0.035t}$ where t is the time in years with $t = 3$ corresponding to 1993. According to this model, in what year will the city have a population of 120,000?

Solution

$80{,}000e^{0.035t} = 120{,}000$

$e^{0.035t} = 1.5$

$0.035t = \ln 1.5$

$$t = \frac{\ln 1.5}{0.035} r \approx 11.5847$$

The population will reach 120,000 when $t \approx 11.5847$, which corresponds to the middle of the year 2001. $(1993 + 8.5847 = 2001.5847)$

CHAPTER SIX
Systems of Equations and Inequalities

6.1 Systems of Equations

- You should be able to solve systems of equations by the method of substitution.
  (a) Solve one of the equations for one of the variables.
  (b) Substitute this expression into the other equation so that you now have an equation in one variable.
  (c) Solve this equation.
  (d) Back-substitute into the first equation to find the value of the other variable.
  (e) Check your answer in both original equations.
- You should be able to solve systems of equations by graphically finding points of intersection.

SOLUTIONS TO SELECTED EXERCISES

5. Solve the following system by the method of substitution. (See graph in textbook to confirm your solution.)

$x - y = -3$ Equation 1

$x^2 - y = -1$ Equation 2

Solution

Solve for y in Equation 1: $y = x + 3$

Substitute for y in Equation 2: $x^2 - (x + 3) = -1$

Solve for x: $x^2 - x - 3 = -1 \quad \Rightarrow \quad (x - 2)(x + 1) = 0 \quad \Rightarrow \quad x = 2, \ -1$

Back-substitute $x = 2$: $y = x + 3 = 2 + 3 = 5$

Back-substitute $x = -1$: $y = x + 3 = -1 + 3 = 2$

Answer: $(-1, \ 2), \ (2, \ 5)$

9. Solve the following system by the method of substitution. (See graph in textbook to confirm your solution.)

$$x^2 - y = 0 \qquad \text{Equation 1}$$

$$x^2 - 4x + y = 0 \qquad \text{Equation 2}$$

Solution

Solve for y in Equation 1: $y = x^2$

Substitute for y in Equation 2: $x^2 - 4x + x^2 = 0$

Solve for x: $2x^2 - 4x = 0 \quad \Rightarrow \quad 2x(x-2) = 0 \quad \Rightarrow \quad x = 0,\ 2$

Back-substitute $x = 0$: $y = x^2 = 0$

Back-substitute $x = 2$: $y = x^2 = 4$

Answer: $(0, 0)$, $(2, 4)$

13. Solve the following system by the method of substitution.

$$x - y = 0 \qquad \text{Equation 1}$$

$$5x - 3y = 10 \qquad \text{Equation 2}$$

Solution

Solve for y in Equation 1: $y = x$

Substitute for y in Equation 2: $5x - 3x = 10$

Solve for x: $2x = 10 \quad \Rightarrow \quad x = 5$

Back-substitute in Equation 1: $y = x = 5$

Answer: $(5, 5)$

17. Solve the following system by the method of substitution.

$$30x - 40y - 33 = 0 \qquad \text{Equation 1}$$

$$10x + 20y - 21 = 0 \qquad \text{Equation 2}$$

Solution

Solve for x in Equation 2: $x = -2y + 2.1$

Substitute for x in Equation 1: $30(-2y + 2.1) - 40y - 33 = 0$

Solve for y: $-100y + 30 = 0 \quad \Rightarrow \quad y = 0.3$

Back-substitute $y = 0.3$: $x = -2y + 2.1 = -2(0.3) + 2.1 = 1.5$

Answer: $(1.5,\ 0.3)$

19. Solve the following system by the method of substitution.

$$\tfrac{1}{5}x + \tfrac{1}{2}y = 8 \qquad \text{Equation 1}$$

$$x + y = 20 \qquad \text{Equation 2}$$

Solution

Solve for x in Equation 2: $x = 20 - y$

Substitute for x in Equation 1: $\frac{1}{5}(20 - y) + \frac{1}{2}y = 8$

Solve for y: $4 + \frac{3}{10}y = 8 \quad \Rightarrow \quad y = \frac{40}{3}$

Back-substitute $y = \frac{40}{3}$: $x = 20 - y = 20 - \frac{40}{3} = \frac{20}{3}$

Answer: $\left(\frac{20}{3},\ \frac{40}{3}\right)$

23. Solve the following system by the method of substitution.

$y = 2x$ Equation 1

$y = x^2 + 1$ Equation 2

Solution

Substitute for y in Equation 2: $2x = x^2 + 1$

Solve for x: $x^2 - 2x + 1 = (x - 1)^2 = 0 \quad \Rightarrow \quad x = 1$

Back-substitute $x = 1$ in Equation 1: $y = 2x = 2$

Answer: $(1, 2)$

25. Solve the following system by the method of substitution.

$3x - 7y + 6 = 0$ Equation 1

$x^2 - y^2 = 4$ Equation 2

Solution

Solve for y in Equation 1: $y = \dfrac{3x + 6}{7}$

Substitute for y in Equation 2: $x^2 - \left(\dfrac{3x + 6}{7}\right)^2 = 4$

Solve for x: $x^2 - \left(\dfrac{9x^2 + 36x + 36}{49}\right) = 4$

$$49x^2 - (9x^2 + 36x + 36) = 196$$

$$40x^2 - 36x - 232 = 0$$

$$10x^2 - 9x - 58 = 0 \quad \Rightarrow \quad x = \frac{9 \pm \sqrt{81 + 40(58)}}{20} \quad \Rightarrow \quad x = \frac{29}{10}, \ -2$$

Back-substitute $x = \dfrac{29}{10}$: $y = \dfrac{3x + 6}{7} = \dfrac{3(29/10) + 6}{7} = \dfrac{21}{10}$

Back-substitute $x = -2$: $y = \dfrac{3x + 6}{7} = \dfrac{3(-2) + 6}{7} = 0$

Answer: $\left(\dfrac{29}{10}, \dfrac{21}{10}\right)$, $(-2, 0)$

29. Solve the following system by the method of substitution.

$y = x^4 - 2x^2 + 1$ Equation 1

$y = 1 - x^2$ Equation 2

Solution

Substitute for y in Equation 1: $1 - x^2 = x^4 - 2x^2 + 1$

Solve for x: $x^4 - x^2 = 0 \quad \Rightarrow \quad x^2(x^2 - 1) = 0 \quad \Rightarrow \quad x = 0, \ \pm 1$

Back-substitute $x = 0$: $1 - x^2 = 1 - 0^2 = 1$

Back-substitute $x = 1$: $1 - x^2 = 1 - 1^2 = 0$

Back-substitute $x = -1$: $1 - x^2 = 1 - (-1)^2 = 0$

Answer: $(0, 1)$, $(\pm 1, 0)$

33. Use a graphing utility to find all points of intersection of the graphs. Confirm algebraically.

$$y = x^2 + 3x - 1$$

$$y = -x^2 - 2x + 2$$

Solution

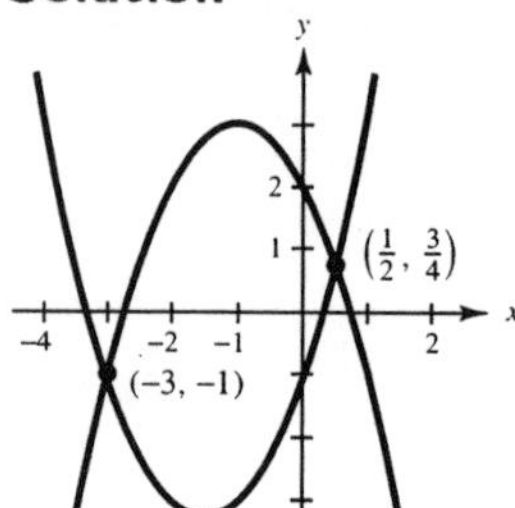

Answer: $(-3, -1), \left(\frac{1}{2}, \frac{3}{4}\right)$

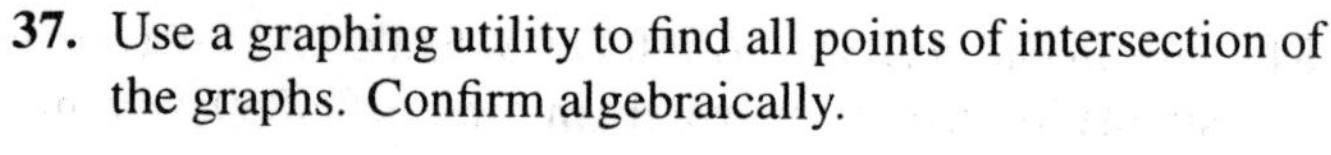

37. Use a graphing utility to find all points of intersection of the graphs. Confirm algebraically.

$$y = e^x$$

$$x - y + 1 = 0$$

Solution

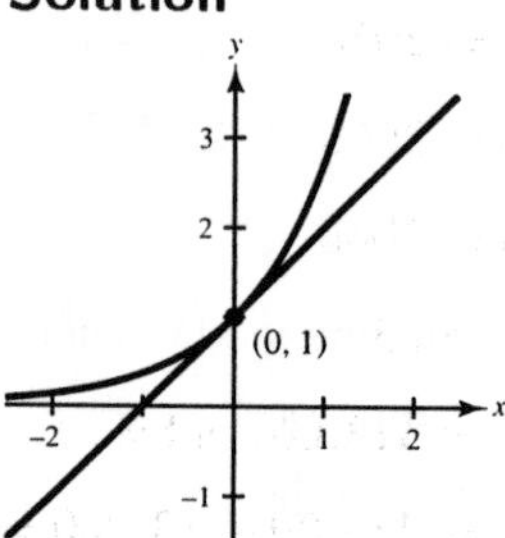

Answer: (0, 1)

41. *Break-Even Analysis* Find the sale necessary to break even ($R = C$) for the cost $C = 8650x + 250{,}000$ of x units, and the revenue $R = 9950x$ obtained by selling x units. (Round your answer to the nearest whole unit.)

Solution

$$C = 8650x + 250{,}000, \quad R = 9950x$$

$$R = C$$

$$9950x = 8650x + 250{,}000$$

$$1300x = 250{,}000$$

$$x \approx 193 \text{ units}$$

$$R \approx \$1{,}920{,}350$$

47. *Comparing Populations* From 1987 to 1993, the Northeastern part of the United States was growing at a rate that was slower than the Western part. (*Source: U.S. Bureau of Census.*) Two models that represent the population of the two regions are

$$P = 50{,}845.9 + 158.3t - 1.25t^2 \qquad \text{Northeast}$$

$$P = 52{,}922.1 + 1062.3t \qquad \text{West}$$

where P is the population in thousands and $t = 0$ represents 1990. Use a graphing utility to determine when the population of the West overtook the population of the Northeast.

Solution

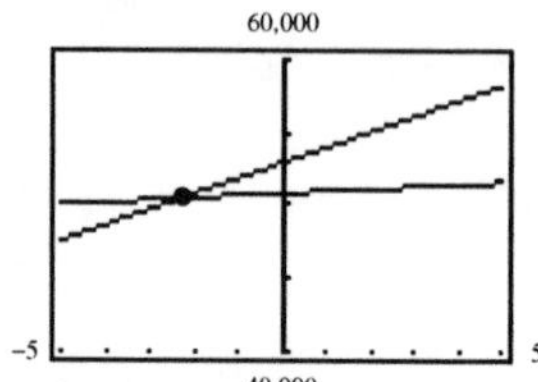

The curves intersect when $t \approx -2.3$ which corresponds to late 1987.

51. *Investment Portfolio* A total of \$25,000 is invested in two funds paying 8% and 8.5% simple interest. If the yearly interest is \$2,060, how much of the \$25,000 is invested at each rate?

Solution

$$x + y = 25{,}000 \quad \Rightarrow \quad y = 25{,}000 - x$$

$$0.08x + 0.085y = 2060$$

$$0.08x + 0.085(25{,}000 - x) = 2060$$

$$-0.005x + 2125 = 2060$$

$$0.005x = 65$$

$$x = \$13{,}000 \text{ at } 8\%$$

$$y = 25{,}000 - 13{,}000 = \$12{,}000 \text{ at } 8.5\%$$

55. *Color or Monochrome?* The sales of monochrome and color computer monitors from 1981 to 1988 can be approximated by the models

$$y = -0.927 + 1.524t + 0.1176t^2 \qquad \text{Monochrome}$$

$$y = -0.010 - 0.2749t + 0.2610t^2 \qquad \text{Color}$$

where y represents the number of monitors (in millions) and $t = 1$ represents 1981. According to these models, when will the sales of color monitors equal the sale of monochrome monitors? (*Source: Future Computing/Datapro, Inc.*)

Solution

$$-0.010 - 0.2749t + 0.2610t^2 = -0.927 + 1.524t + 0.1176t^2, \quad t \geq 1$$

$$0.1434t^2 - 1.7989t + 0.917 = 0$$

$$t = \frac{1.7989 \pm \sqrt{2.71005001}}{0.2868}$$

Using $t \approx 12$, the sales of color monitors will equal the sales of monocrome monitors in 1992.

6.2 Systems of Linear Equations in Two Variables

- You should be able to solve a linear system by the method of elimination.
  - (a) Obtain coefficients for x (or y) that differ only in sign by multiplying all terms of one or both equations by constants.
  - (b) Add the equations to eliminate one of the variables and then solve for the other variable.
  - (c) Back-substitute this value into either of the original equations to solve for the other variable.
  - (d) Check your answer.
- You should know that for a system of two linear equations, one of the following is true.
  - (a) There are infinitely many solutions; the lines are identical.
  - (b) There is no solution; the lines are parallel.
  - (c) There is one solution; the lines intersect at one point.

SOLUTIONS TO SELECTED EXERCISES

3. Solve the linear system by elimination. Label each line in the graph with the appropriate equation. (See graph in the textbook.)

$x - y = 0$ Equation 1

$3x - 2y = -1$ Equation 2

Solution

Multiply Equation 1 by (-2): $-2x + 2y = 0$

Add this to Equation 2 to eliminate y: $x = -1$

Substitute $x = -1$ in Equation 1: $-1 - y = 0 \quad \Rightarrow \quad y = -1$

Answer: $(-1, -1)$

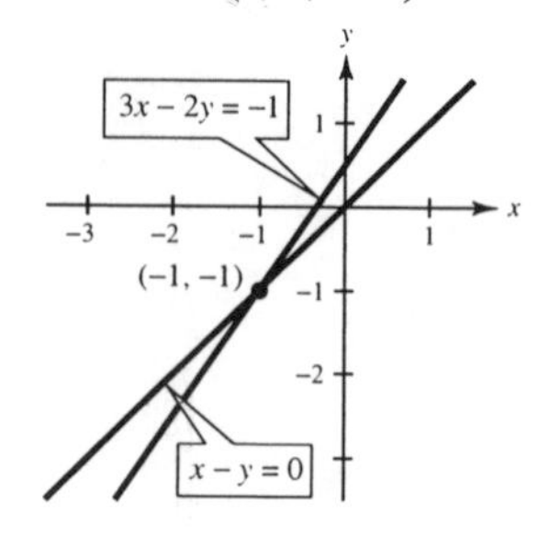

7. Solve the linear system by elimination. Label each line in the graph with the appropriate equation. (See graph in the textbook.)

$$3x - 2y = 6 \quad \text{Equation 1}$$
$$-6x + 4y = -12 \quad \text{Equation 2}$$

Solution

Multiply Equation 1 by 2 and add to Equation 2: $0 = 0$

The equations are dependent. The solution set consists of all points (x, y) lying on the line

$$3x - 2y = 6$$
$$y = \tfrac{3}{2}x - 3$$

Let $x = 2a$, then $y = \tfrac{3}{2}(2a) - 3 = 3a - 3$.

Answer: $(2a,\ 3a - 3)$ for any real number a.

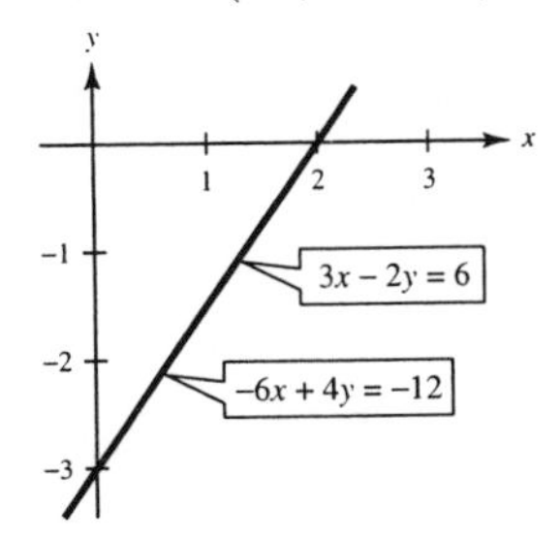

13. Solve the system by elimination.

$$2x + 3y = 18 \quad \text{Equation 1}$$
$$5x - y = 11 \quad \text{Equation 2}$$

Solution

Multiply Equation 2 by 3: $15x - 3y = 33$

Add this to Equation 1 to eliminate y: $17x = 51 \quad \Rightarrow \quad x = 3$

Substitute $x = 3$ in Equation 1: $6 + 3y = 18 \quad \Rightarrow \quad y = 4$

Answer: $(3, 4)$

17. Solve the system by elimination.

$$2u + v = 120 \quad \text{Equation 1}$$
$$u + 2v = 120 \quad \text{Equation 2}$$

Solution

Multiply Equation 2 by (-2): $-2u - 4v = -240$

Add this to Equation 1 to eliminate u: $-3v = -120$

$$v = 40$$

Substitute $v = 40$ in Equation 2: $u + 80 = 120 \quad \Rightarrow \quad u = 40$

Answer: $(40, 40)$

21. Solve the system by elimination.

$$\frac{x}{4} + \frac{y}{6} = 1 \qquad \text{Equation 1}$$

$$x - y = 3 \qquad \text{Equation 2}$$

Solution

Multiply Equation 1 by 6: $\frac{3}{2}x + y = 6$

Add this to Equation 2 to eliminate y: $\frac{5}{2}x = 9 \quad \Rightarrow \quad x = \frac{18}{5}$

Substitute $x = \frac{18}{5}$ in Equation 2: $\frac{18}{5} - y = 3$

$$y = \frac{3}{5}$$

Answer: $\left(\frac{18}{5}, \frac{3}{5}\right)$

25. Solve the system by elimination.

$$2.5x - 3y = 1.5 \qquad \text{Equation 1}$$

$$10x - 12y = 6 \qquad \text{Equation 2}$$

Solution

Multiply Equation 1 by (-4): $-10x + 12y = -6$

Add this to Equation 2 to eliminate x: $0 = 0$ (Dependent)

The solution set consists of all points (x, y) lying on the line

$10x - 12y = 6$

$$y = \tfrac{5}{6}x - \tfrac{1}{2}$$

Let $x = a$, then $y = \frac{5}{6}a - \frac{1}{2}$.

Answer: $\left(a, \frac{5}{6}a - \frac{1}{2}\right)$, where a is any real number

29. Solve the system by elimination.

$$4b + 3m = 3 \qquad \text{Equation 1}$$

$$3b + 11m = 13 \qquad \text{Equation 2}$$

Solution

Multiply Equation 1 by 3 and Equation 2 by (-4):

$$12b + 9m = 9$$

$$-12b - 44m = -52$$

Add to eliminate b: $-35m = -43$

$$m = \tfrac{43}{35}$$

Substitute $m = \frac{43}{35}$ in Equation 1: $4b + 3\left(\frac{43}{35}\right) = 3 \quad \Rightarrow \quad b = -\frac{6}{35}$

Answer: $\left(-\frac{6}{35}, \frac{43}{35}\right)$

33. *Airplane Speed* An airplane flying into a headwind travels the 1,800-mile flying distance between two cities in 3 hours and 36 minutes. On the return flight, the distance is traveled in 3 hours. Find the ground speed of the plane and the speed of the wind, assuming that both remain constant.

Solution

Let $x =$ the ground speed and $y =$ the wind speed.

$3.6(x - y) = 1800 \Rightarrow x - y = 500$ Equation 1

$3(x + y) = 1800 \Rightarrow x + y = 600$ Equation 2

$2x = 1100 \Rightarrow x = 550$

$550 - y = 500$

$y = 50$

Answer: $x = 550$ mph, $y = 50$ mph

35. *Acid Mixture* Ten gallons of a 30% acid solution are obtained by mixing a 20% solution with a 50% solution. How much of each must be used?

Solution

Let $x =$ the number of gallons at 20% and $y =$ the number of gallons at 50%.

$x + y = 10 \Rightarrow -2x - 2y = -20$ Equation 1

$0.2x + 0.5y = 0.3(10) \Rightarrow 2x + 5y = 30$ Equation 2

$3y = 10 \Rightarrow y = 3\frac{1}{3}$

By back substitution $x = 6\frac{2}{3}$.

Answer: $x = 6\frac{2}{3}$ gallons at 20%., $y = 3\frac{1}{3}$ gallons at 50%.

39. *Ticket Sales* You are the manager of a theater. On Saturday morning you are going over the ticket sales for Friday evening. Five hundred tickets were sold. The tickets for adults and children sold for \$7.50 and \$4.00, respectively, and the receipts for the performance were \$3,312.50. However, your assistant manager did not record how many of each type of ticket were sold. From the information you have, can you determine how many of each type were sold?

Solution

Let $x =$ number of adult tickets sold and $y =$ number of child tickets sold.

$x + y = 500 \Rightarrow -4x - 4y = -2000$ Equation 1

$7.5x + 4y = \$3,312.50 \Rightarrow 7.5x + 4y = 3312.50$ Equation 2

$3.5x = 1312.50$

$x = 375$

By back-substitution $y = 125$.

Answer: $x = 375$ adult tickets, $y = 125$ child tickets

43. *Supply and Demand* Find the point of equilibrium for demand, $p = 140 - 0.00002x$ and supply, $p = 80 + 0.00001x$.

Solution

Demand = Supply

$$\begin{aligned} 140 - 0.00002x &= 80 + 0.00001x \\ 60 &= 0.00003x \\ x &= 2{,}000{,}000 \text{ units} \\ p &= \$100.00 \end{aligned}$$

47. *Fitting a Line to Data* Find the **least squares reqression line** $y = ax + b$ for the points $(x_1, y_1), (x_2, y_2), \ldots (x_n, y_n)$. To find the line, solve the following system for a and b. (If you are unfamiliar with summation notation, look at the discussion in Section 8.1 in the textbook.) (See graph in textbook.)

$$nb + \left(\sum_{i=1}^{n} x_i\right)a = \sum_{i=1}^{n} y_i$$

$$\left(\sum_{i=1}^{n} x_i\right)b + \left(\sum_{i=1}^{n} x_i^2\right)a = \sum_{i=1}^{n} x_i y_i$$

$$5b + 10a = 20.2$$
$$10b + 30a = 50.1$$

Solution

$$\begin{aligned} 5b + 10a = 20.2 \Rightarrow -10b - 20a &= -40.4 \\ 10b + 30a &= 50.1 \\ \hline 10a &= 9.7 \\ a &= 0.97 \end{aligned}$$

By back-substitution, $b = 2.1$.

Answer: $a = 0.97, \quad b = 2.1$

Least squares regression line: $y = 0.97x + 2.1$

53. *Restaurant Sales* The total sales (in billions of dollars) for full-service restaurants in the United States from 1990 to 1994 are shown in the table in the textbook. (*Source: National Restaurant Association*)

(a) Use the technique demonstrated in Exercises 47–52 to find the line that best fits the data.

(b) Graph the line on a graphing utility and use the TRACE feature to predict the total restaurant sales in 1996.

53. –CONTINUED–

Solution

(a) (0, 76.1), (1, 78.4), (2, 80.3), (3, 83.1), (4, 85.5)

$$n = 5,\ \sum_{i=1}^{5} x_i = 10,\ \sum_{i=1}^{5} y_i = 403.4$$

$$\sum_{i=1}^{5} x_i^2 = 30,\ \sum_{i=1}^{5} x_i y_i = 830.3$$

Solve the system $\begin{aligned} 5b + 10a &= 403.4 \\ 10b + 30a &= 830.3 \end{aligned}$

Answer: $b = 75.98,\ a = 2.35$

Least squares regression line: $y = 2.35x + 75.98$

(b)

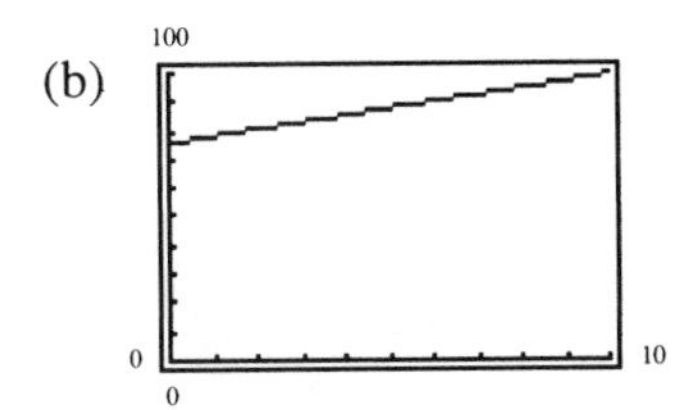

When $x = 6$, $y \approx \$90.1$ billion

6.3 Linear Systems in Three or More Variables

- You should know the operations that lead to equivalent systems of equations:
  - (a) Interchange any two equations.
  - (b) Multiply all terms of an equation by a nonzero constant.
  - (c) Replace an equation by the sum of itself and a constant multiple of any other equation in the system.
- You should be able to use the method of elimination.

SOLUTIONS TO SELECTED EXERCISES

3. Solve the system of linear equations.

$$\begin{aligned} 4x + y - 3z &= 11 \\ 2x - 3y + 2z &= 9 \\ x + y + z &= -3 \end{aligned}$$

Solution

$$\begin{aligned} x + y + z &= -3 \quad \text{Interchange the equations.} \\ 4x + y - 3z &= 11 \\ 2x - 3y + 2z &= 9 \end{aligned}$$

$$\begin{aligned} x + y + z &= -3 \\ -3y - 7z &= 23 \quad -4 \text{ Equation 1} + \text{Equation 2} \\ -5y &= 15 \quad -2 \text{ Equation 1} + \text{Equation 3} \end{aligned}$$

$$\begin{aligned} -5y = 15 &\Rightarrow y = -3 \quad \text{Solve Equation 3.} \\ -3(-3) - 7z = 23 &\Rightarrow z = -2 \quad \text{Back-substitute.} \\ x - 3 - 2 = -3 &\Rightarrow x = 2 \end{aligned}$$

Answer: $(2, -3, -2)$

7. Solve the system of linear equations.

$$\begin{aligned} 3x - 2y + 4z &= 1 \\ x + y - 2z &= 3 \\ 2x - 3y + 6z &= 8 \end{aligned}$$

Solution

$$\begin{aligned} x + y - 2z &= 3 \quad \text{Interchange the equations.} \\ 3x - 2y + 4z &= 1 \\ 2x - 3y + 6z &= 8 \end{aligned}$$

$$\begin{aligned} x + y - 2z &= 3 \\ -5y + 10z &= -8 \quad -3 \text{ Equation 1} + \text{Equation 2} \\ -5y + 10z &= 2 \quad -2 \text{ Equation 1} + \text{Equation 3} \end{aligned}$$

$$\begin{aligned} x + y - 2z &= 3 \\ -5y + 10z &= -8 \\ 0 &= 10 \quad \rightarrow\leftarrow \ -1 \text{ Equation 2} + \text{Equation 3} \end{aligned}$$

No solution, inconsistent

11. Solve the system of linear equations.

$$\begin{aligned} x + 2y - 7z &= -4 \\ 2x + y + z &= 13 \\ 3x + 9y - 36z &= -33 \end{aligned}$$

Solution

$$\begin{aligned} x + 2y - 7z &= -4 \\ 2x + y + z &= 13 \\ 3x + 9y - 36z &= -33 \end{aligned}$$

$$\begin{aligned} x + 2y - 7z &= -4 \\ -3y + 15z &= 21 \quad && -2 \text{ Equation 1} + \text{Equation 2} \\ 3y - 15z &= -21 \quad && -3 \text{ Equation 1} + \text{Equation 3} \end{aligned}$$

$$\begin{aligned} x + 2y - 7z &= -4 \\ -3y + 15z &= 21 \\ 0 &= 0 \quad && \text{Equation 2} + \text{Equation 3} \end{aligned}$$

$$\begin{aligned} x + 2y - 7z &= -4 \\ y - 5z &= -7 \quad && -\tfrac{1}{3} \text{ Equation 2} \end{aligned}$$

$$\begin{aligned} x + 3z &= 10 \quad && -2 \text{ Equation 2} + \text{Equation 1} \\ y - 5z &= -7 \end{aligned}$$

$z = a$

$y = 5a - 7$ Back-substitute into Equation 2.

$x = -3a + 10$ Back-substitute into Equation 1.

Answer: $(-3a + 10,\ 5a - 7,\ a)$

15. Solve the system of linear equations.

$$\begin{aligned} x - 2y + 5z &= 2 \\ 3x + 2y - z &= -2 \end{aligned}$$

Solution

$$\begin{aligned} x - 2y + 5z &= 2 \\ 3x + 2y - z &= -2 \end{aligned}$$

$$\begin{aligned} x - 2y + 5z &= 2 \\ 8y - 16z &= -8 \quad -3 \text{ Equation 1} + \text{Equation 2} \end{aligned}$$

$$\begin{aligned} x - 2y + 5z &= 2 \\ y - 2z &= -1 \quad \tfrac{1}{8} \text{ Equation 2} \end{aligned}$$

$$\begin{aligned} x + z &= 0 \quad 2 \text{ Equation 2} + \text{Equation 1} \\ y - 2z &= -1 \end{aligned}$$

$z = a$

$y = 2a - 1$ Back-substitute into Equation 2.

$x = -a$ Back-substitute into Equation 1.

Answer: $(-a,\ 2a - 1,\ a)$

19. Solve the system of linear equations.

$$\begin{aligned} x \qquad\qquad + 3w &= 4 \\ 2y - z - w &= 0 \\ 3y \qquad - 2w &= 1 \\ 2x - y + 4z \qquad &= 5 \end{aligned}$$

Solution

$$\begin{aligned} x \qquad\qquad + 3w &= 4 \\ 2y - z - w &= 0 \\ 3y \qquad - 2w &= 1 \\ 2x - y + 4z \qquad &= 5 \end{aligned}$$

$$\begin{aligned} x \qquad\qquad + 3w &= 4 \\ 2y - z - w &= 0 \\ 3y \qquad - 2w &= 1 \\ -y + 4z - 6w &= -3 \end{aligned}$$ -2 Equation 1 $+$ Equation 4

$$\begin{aligned} x \qquad\qquad + 3w &= 4 \\ y - 4z + 6w &= 3 \\ 2y - z - w &= 0 \\ 3y \qquad - 2w &= 1 \end{aligned}$$ Multiply Equation 4 by -1 and interchange the equations.

$$\begin{aligned} x \qquad\qquad + 3w &= 4 \\ y - 4z + 6w &= 3 \\ 7z - 13w &= -6 \\ 12z - 20w &= -8 \end{aligned}$$ -2 Equation 2 $+$ Equation 3; -3 Equation 2 $+$ Equation 4

$$\begin{aligned} x \qquad\qquad + 3w &= 4 \\ y - 4z + 6w &= 3 \\ z - 3w &= -2 \\ 12z - 20w &= -8 \end{aligned}$$ $-\frac{1}{2}$ Equation 4 $+$ Equation 3

$$\begin{aligned} x \qquad\qquad + 3w &= 4 \\ y - 4z + 6w &= 3 \\ z - 3w &= -2 \\ 16w &= 16 \end{aligned}$$ -12 Equation 3 $+$ Equation 4

$16w = 16 \Rightarrow w = 1$

$z - 3(1) = -2 \Rightarrow z = 1$ Back-substitute into Equation 3.

$y - 4(1) + 6(1) = 3 \Rightarrow y = 1$ Back-substitute into Equation 2.

$x + 3(1) = 4 \Rightarrow x = 1$ Back-substitute into Equation 1.

Answer: $(1, 1, 1, 1)$

23. Solve the system of linear equations.

$$4x + 3y + 17z = 0$$
$$5x + 4y + 22z = 0$$
$$4x + 2y + 19z = 0$$

Solution

$5x + 4y + 22z = 0$ Interchange the equations.

$4x + 3y + 17z = 0$

$4x + 2y + 19z = 0$

$x + y + 5z = 0$ -1 Equation 2 + Equation 1

$4x + 3y + 17z = 0$

$4x + 2y + 19z = 0$

$x + y + 5z = 0$

$-y - 3z = 0$ -4 Equation 1 + Equation 2

$-2y - z = 0$ -4 Equation 1 + Equation 3

$x + y + 5z = 0$

$y + 3z = 0$ -1 Equation 2

$5z = 0$ 2 Equation 2 + Equation 3

$5z = 0 \Rightarrow z = 0$

$y + 3(0) = 0 \Rightarrow y = 0$ Back-substitute into Equation 2.

$x + 0 + 5(0) = 0 \Rightarrow x = 0$ Back-substitute into Equation 1.

Answer: (0, 0, 0)

29. Find the equation of the parabola $y = ax^2 + bx + c$ that passes through (1, 0), (2, −1) and (3, 0). (See graph in textbook.)

Solution

$y = ax^2 + bx + c$

$x = 1,\ y = 0:\quad 0 = a + b + c$ Equation 1

$x = 3,\ y = 0:\quad 0 = 9a + 3b + c$ Equation 2

$x = 2,\ y = -1:\ -1 = 4a + 2b + c$ Equation 3

Answer: $a = 1,\ b = -4,\ c = 3$

The equation of the parabola is: $y = x^2 - 4x + 3$.

35. *Regular Polygons* The total number of sides and diagonals of a regular polygon with three, four, and five sides are three, six, and ten. Find a quadratic function $y = ax^2 + bx + c$ that fits this data. Then check to see if it gives the correct answer for a polygon with six sides. (See figure in textbook.)

Solution

$y = ax^2 + bx + c$

Passing through (3, 3), (4, 6), and (5, 10)

$$3 = 9a + 3b + c$$
$$6 = 16a + 4b + c$$
$$10 = 25a + 5b + c$$

Answer: $a = \frac{1}{2}, \quad b = -\frac{1}{2}, \quad c = 0$

The equation of the parabola is: $y = \frac{1}{2}x^2 - \frac{1}{2}x$.

When $x = 6, \; y = 15$ which is the correct answer for a polygon with six sides.

37. *Investment* An inheritance of \$16,000 was divided among three investments yielding a total of \$990 in interest per year. The interest rates for the three investments were 5%, 6%, and 7%. Find the amount placed in each investment if the 5% and 6% investments were \$3,000 and \$2,000 less than the 7% investment, respectively.

Solution

Let $x =$ amount at 5%, $y =$ amount at 6%, and $z =$ amount at 7%.

$$x + y + z = 16{,}000$$
$$0.05x + 0.06y + 0.07z = 990$$
$$x + 3000 = z$$
$$y + 2000 = z$$

$$(z - 3000) + (z - 2000) + z = 16{,}000$$
$$3z = 21{,}000$$
$$z = 7000$$

$x = 4000, \quad y = 5000$

Check: $0.05(4000) + 0.06(5000) + 0.07(7000) = 990$

Answer: $x = \$4{,}000$ at 5%, $y = \$5{,}000$ at 6%, $z = \$7{,}000$ at 7%

41. *Grades of Paper* A manufacturer sells a 50-pound package of paper that consists of three grades of computer paper. Grade C costs \$3.50 per pound, Grade B costs \$4.50 per pound, and Grade A costs \$6.00 per pound. Half of the 50-pound package consists of the two cheaper grades. The cost of the 50-pound package is \$252.50. How many pounds of each grade of paper should be used to form the 50-pound package?

Solution

Let $x =$ amount Grade A paper, $y =$ amount Grade B paper and,
$z =$ amount Grade C paper.

$$x + y + z = 50$$

$$y + z = 25$$

$$x = 25$$

$$6x + 4.5y + 3.5z = 252.50$$

$$4.5y + 3.5(25 - y) = 252.50 - 6(25)$$

$$y = 15 \Rightarrow z = 10$$

Answer: 25 pounds of Grade A paper, 15 pounds of Grade B paper, 10 pounds of Grade C paper.

45. *Investment Portfolio* You have a portfolio totaling \$500,000, that is to be invested among the following types of investment.

Let $C =$ amount in certificates of deposit.

Let $M =$ amount in municipal bonds.

Let $B =$ amount in blue-chip stocks.

Let $G =$ amount in growth or speculative stocks.

How much should be allocated to each type of investment when given the following information?
The certificates of deposit pay 10% annually and the municipal bonds pay 8% annually. Over a five-year period, the investor expects the blue-chip stocks to return 12% annually, and expects the growth stocks to return 13% annually. You want a combined annual return of 10% and also want to have only one-fourth of the portfolio in stocks.

Solution

$$C + M + B + G = 500{,}000$$

$$0.10C + 0.08M + 0.12B + 0.13G = 0.10(500{,}000)$$

$$B + G = \tfrac{1}{4}(500{,}000)$$

This system has infinitely many solutions.

Let $G = a$, then $B = 125{,}000 - a$, $M = 125{,}000 + \frac{1}{2}a$, and $C = 250{,}000 - \frac{1}{2}a$.

Answer: $\left(250{,}000 - \frac{1}{2}a,\ 125{,}000 + \frac{1}{2}a,\ 125,000 - a,\ a\right)$ where $0 \le a \le 125{,}000$

49. *Fitting a Parabola to Data* Find the **least squares regression parabola** $y = ax^2 + bx + c$ for the points $(x_1, y_1), (x_2, y_2), \ldots (x_n, y_n)$. To find the parabola, solve the following system of linear equations for a, b, and c. (See graph in the textbook.)

$$nc + \left(\sum_{i=1}^{n} x_i\right)b + \left(\sum_{i=1}^{n} x_i^2\right)a = \sum_{i=1}^{n} y_i$$

$$\left(\sum_{i=1}^{n} x_i\right)c + \left(\sum_{i=1}^{n} x_i^2\right)b + \left(\sum_{i=1}^{n} x_i^3\right)a = \sum_{i=1}^{n} x_i y_i$$

$$\left(\sum_{i=1}^{n} x_i^2\right)c + \left(\sum_{i=1}^{n} x_i^3\right)b + \left(\sum_{i=1}^{n} x_i^4\right)a = \sum_{i=1}^{n} x_i^2 y_i$$

$$\begin{aligned} 6c + 3b + 19a &= 23.9 \\ 3c + 19b + 27a &= -7.2 \\ 19c + 27b + 115a &= 48.8 \end{aligned}$$

Solution

Using the methods of this section, we have

$c = \frac{1779}{350}$, $b = -\frac{621}{700}$, $a = -\frac{29}{140}$

Least squares regression parabola: $y = -\frac{29}{140}x^2 - \frac{621}{700}x + \frac{1,779}{350} \approx -0.207x^2 - 0.887x + 5.083$

Mid-Chapter Quiz for Chapter 6

1. Write a system that has (3, 2) as a solution.

Solution

Answers will vary.

$x = 3, \; y = 2$

$$\begin{aligned} 3x + y &= 11 \\ x - 2y &= -1 \end{aligned}$$

2. Write a system that has (5, 3) as a solution.

Solution

Answers will vary.

$x = 5, \; y = 3$

$$\begin{aligned} x + 2y &= 11 \\ x - y &= 7 \end{aligned}$$

3. Use a graphing utility to solve the system.

$$y = 2\sqrt{x} + 1$$
$$y = x - 2$$

Solution

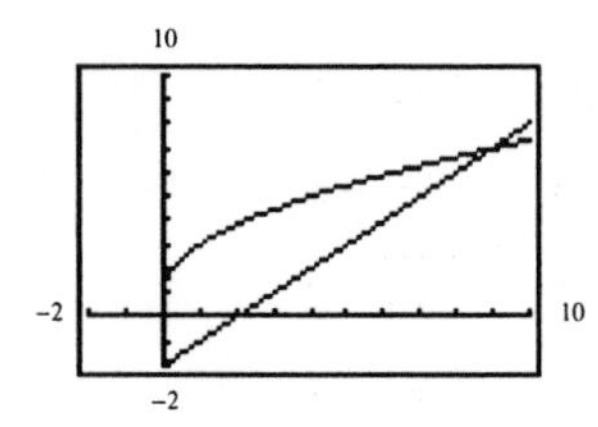

Answer: $(9, 7)$

4. Use a graphing utility to solve the system.

$$x^2 + y^2 = 9$$
$$y = 2x + 1$$

Solution

Enter

$y_1 = \sqrt{9 - x^2}$

$y_2 = -\sqrt{9 - x^2}$

$y_3 = 2x + 1$

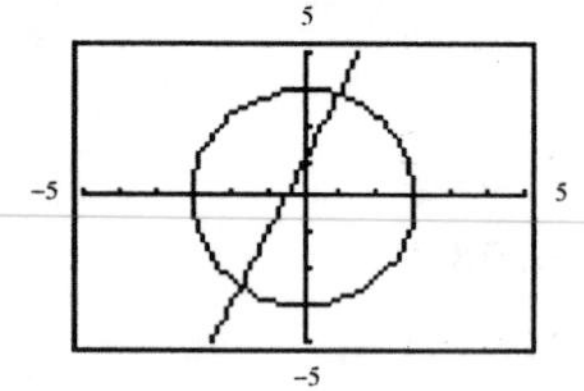

Answer: $(-1.727, -2.454)$, $(0.927, 2.854)$

5. Find the number of sales necessary to break even.

$C = 12.50x + 10{,}000, \quad R = 19.95x$

Solution

$$R = C$$
$$19.95x = 12.50x + 10{,}000$$
$$7.45x = 10{,}000$$
$$x \approx 1342.28$$
$$x \approx 1342 \text{ units}$$

6. Find the number of sales necessary to break even.

$C = 3.79x + 400{,}000, \quad R = 4.59x$

Solution

$$R = C$$
$$4.59x = 3.79x + 400{,}000$$
$$0.80x = 400{,}000$$
$$x = 500{,}000 \text{ units}$$

7. Solve the system by elimination. Verify the solution graphically.

$$2.5x - \quad y = 6$$
$$3x + 4y = 2$$

Solution

$$\begin{aligned} 2.5x - y = 6 \Rightarrow \quad 10x - 4y &= 24 \\ 3x + 4y &= 2 \\ \hline 13x \qquad &= 26 \\ x \qquad &= 2 \\ 3(2) + 4y &= 2 \\ 4y &= -4 \\ y &= -1 \end{aligned}$$

Answer: $(2, -1)$

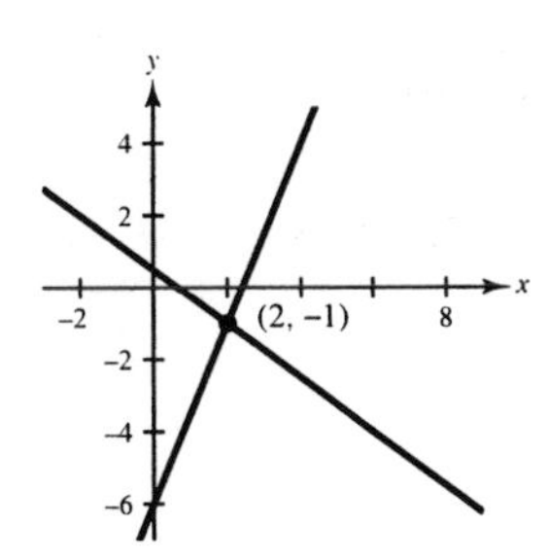

8. Solve the system by elimination. Verify the solution graphically.

$$\tfrac{1}{2}x + \tfrac{1}{3}y = \quad 1$$
$$x - 2y = -2$$

Solution

$$\begin{aligned} \tfrac{1}{2}x + \tfrac{1}{3}y = 1 \Rightarrow \quad 3x + \quad 2y &= \quad 6 \\ x - \quad 2y &= -2 \\ \hline 4x \qquad &= \quad 4 \\ x \qquad &= \quad 1 \\ 1 - \quad 2y &= -2 \\ -2y &= -3 \\ y = \tfrac{3}{2} &= 1\tfrac{1}{2} \end{aligned}$$

Answer: $\left(1, 1\tfrac{1}{2}\right)$

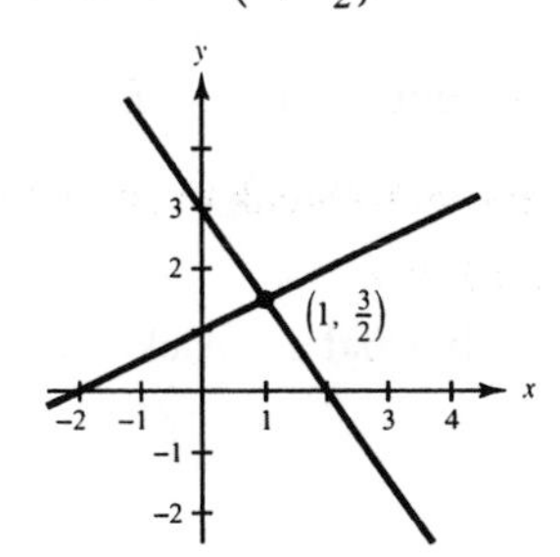

9. Find the point of equilibrium.

Demand: $p = 45 - 0.001x$
Supply: $p = 23 + 0.0002x$

Solution

$$\text{Supply} = \text{Demand}$$
$$23 + 0.0002x = 45 - 0.001x$$
$$0.0012x = 22$$
$$x \approx 18{,}333 \text{ units}$$

10. Find the point of equilibrium.

Demand: $p = 95 - 0.0002x$
Supply: $p = 80 + 0.00001x$

Solution

$$\text{Supply} = \text{Demand}$$
$$80 + 0.00001x = 95 - 0.0002x$$
$$0.00021x = 15$$
$$x \approx 71{,}429 \text{ units}$$

11. Solve the system of equations.

$$\begin{aligned} 2x + 3y - z &= -7 \\ x + 3z &= 10 \\ 2y + z &= -1 \end{aligned}$$

Solution

$$\begin{aligned} x + 3z &= 10 \\ 2x + 3y - z &= -7 \\ 2y + z &= -1 \end{aligned}$$ Interchange the equations.

$$\begin{aligned} x + 3z &= 10 \\ 3y - 7z &= -27 \\ 2y + z &= -1 \end{aligned}$$ -2 Equation 1 $+$ Equation 2

$$\begin{aligned} x + 3z &= 10 \\ y - 8z &= -26 \\ 17z &= 51 \end{aligned}$$ -1 Equation 3 $+$ Equation 2; -2 Equation 2 $+$ Equation 3

$z = 3$ Solve and use back-substitution.

$y - 8(3) = -26 \Rightarrow y = -2$

$x + 3(3) = 10 \Rightarrow x = 1$

Answer: $(1, -2, 3)$

12. Solve the system of equations.

$$\begin{aligned} x + y + 3z &= 11 \\ 2x - y - z &= -11 \\ 2y + 3z &= 17 \end{aligned}$$

Solution

$$\begin{aligned} x + y + 3z &= 11 \\ -3y - 7z &= -33 \quad -2 \text{ Equation 1} + \text{Equation 2} \\ 2y + 3z &= 17 \end{aligned}$$

$$\begin{aligned} x + y + 3z &= 11 \\ y - z &= 1 \quad 2 \text{ Equation 3} + \text{Equation 2} \\ 5z &= 15 \quad -2 \text{ Equation 2} + \text{Equation 3} \end{aligned}$$

$$z = 3 \quad \text{Solve and use back-substitution.}$$

$$y - 3 = 1 \Rightarrow y = 4$$

$$x + 4 + 3(3) = 11 \Rightarrow x = -2$$

Answer: $(-2, 4, 3)$

13. Solve the system of equations.

$$\begin{aligned} 2x + 3y \quad &= 12 \\ -x + 2y + z &= 0 \\ y - z &= 7 \end{aligned}$$

Solution

$$\begin{aligned} x + 5y + z &= 12 \quad \text{Equation 2} + \text{Equation 1} \\ 7y + 2z &= 12 \quad \text{Equation 1} + \text{Equation 2} \\ y - z &= 7 \end{aligned}$$

$$\begin{aligned} x + 5y + z &= 12 \\ y - z &= 7 \quad \text{Interchange the equations.} \\ 7y + 2z &= 12 \end{aligned}$$

$$\begin{aligned} x + 5y + z &= 12 \\ y - z &= 7 \\ 9z &= -37 \quad -7 \text{ Equation 2} + \text{Equation 3} \end{aligned}$$

$$z = -\tfrac{37}{9} = -4.1\overline{1} \quad \text{Solve and use back-substitution.}$$

$$y - \left(-\tfrac{37}{9}\right) = 7 \Rightarrow y = \tfrac{26}{9} = 2.8\overline{8}$$

$$x + 5\left(\tfrac{26}{9}\right) + \left(-\tfrac{37}{9}\right) = 12 \Rightarrow x = \tfrac{15}{9} = 1.6\overline{6}$$

Answer: $(1.6\overline{6}, 2.8\overline{8}, -4.1\overline{1})$

14. Write three ordered triples of the form $(a, a-2, 3a)$.

Solution

Answers will vary.

$(a, a-2, 3a)$

$a = 1 : (1, -1, 3)$

$a = 0 : (0, -2, 0)$

$a = 3 : (3, 1, 9)$

15. Write three ordered triples of the form $(2a, a+5, a)$.

Solution

Answers will vary.

$(2a, a+5, a)$

$a = 0 : (0, 5, 0)$

$a = 2 : (4, 7, 2)$

$a = 1 : (2, 6, 1)$

16. When solving a square system of equations in three variables, you arrive at an equation $0 = -3$. What conclusion can you make?

Solution

Since $0 = -3$ is a false statement, you conclude that the system is inconsistent. It has no solution.

17. How many solutions does the system $2x - 3y + z = 4,\ y - z = 7$ have?

Solution

Since you only have two equations and three variables, there are infinitely many solutions.

18. *Acid Mixture* Ten gallons of a 35% acid solution is obtained by mixing a 25% solution with a 40% solution. How much of each must be used?

Solution

$$x + y = 10$$

$$0.25x + 0.40y = 0.35(10)$$

$$-25x - 25y = -250 \quad -25 \text{ Times Equation 1}$$

$$25x + 40y = 350 \quad 100 \text{ Times Equation 2}$$

$$15y = 100$$

$$y = \tfrac{100}{15} = 6\tfrac{2}{3}$$

$$x + 6\tfrac{2}{3} = 10 \Rightarrow x = 3\tfrac{1}{3}$$

Use $3\frac{1}{3}$ gallons of the 25% solution and $6\frac{2}{3}$ gallons of the 40% solution.

19. *True or False?* A system of equations may have infinitely many solutions.

Solution

True. See Exercise 17.

20. *True or False?* A square linear system must have exactly one solution.

Solution

False. It may have infinitely many or no solutions.

6.4 Systems of Inequalities

- You should be able to sketch the graph of an inequality in two variables:
  (a) Replace the inequality with an equal sign and graph the equation. Use a dashed line for $<$ or $>$, a solid line for $\leq$ or $\geq$.
  (b) Test a point in each region formed by the graph. If the point satisfies the inequality, shade the whole region.
- You should be able to solve systems of inequalitites:
  (a) Sketch each inequality.
  (b) Find the vertices.
  (c) Shade the region that is common to every graph in the system.

SOLUTIONS TO SELECTED EXERCISES

3. Match the inequality $2x + 3y \leq 6$ with its graph. [The graphs are labeled (a), (b), (c), (d), (e), and (f) in the textbook.]

Solution

$2x + 3y \leq 6$

The line $2x + 3y = 6$ has intercepts at $(3, 0)$ and $(0, 2)$. The region below the line satisfies the inequality. Matches (d).

7. Sketch the graph of $x \geq 2$.

Solution

Graph the vertical line $x = 2$. Points to the right of the line satisfy the inequality.

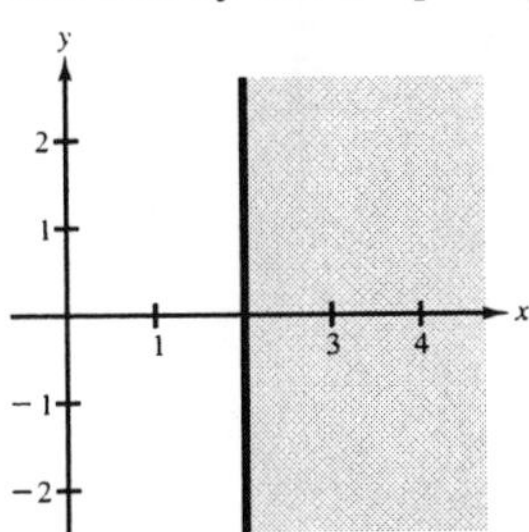

11. Sketch the graph of $y < 2 - x$.

Solution

Graph the line $y = 2 - x$ using dashes. Points to the left of the line satisfy the inequality.

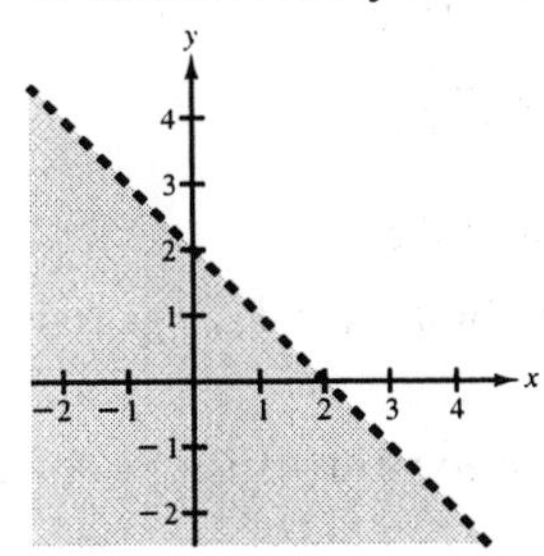

15. Sketch the graph of $y^2 + 2x > 0$.

Solution

Graph the parabola $y^2 = -2x$ using dashes. Points to the right of the parabola satisfy the inequality.

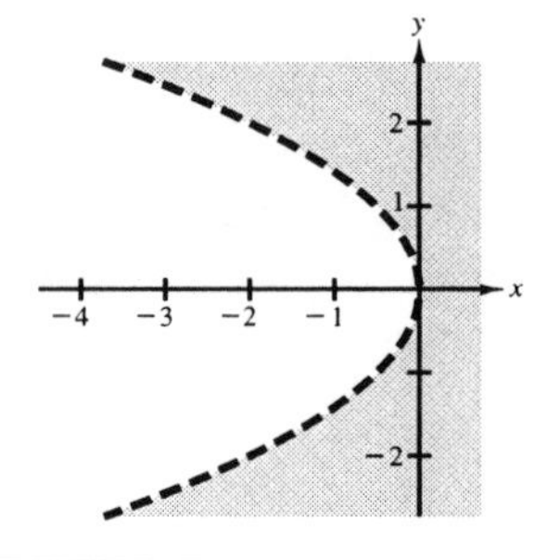

21. Sketch the graph of the solution set of the following system of inequalities.

$$x + y \le 1$$
$$-x + y \le 1$$
$$y \ge 0$$

Solution

Graph the lines $y = 1 - x$, $y = 1 + x$, and $y = 0$. The vertices are $(-1, 0)$, $(0, 1)$, and $(1, 0)$. The region inside the triangle satisfies all three inequalities.

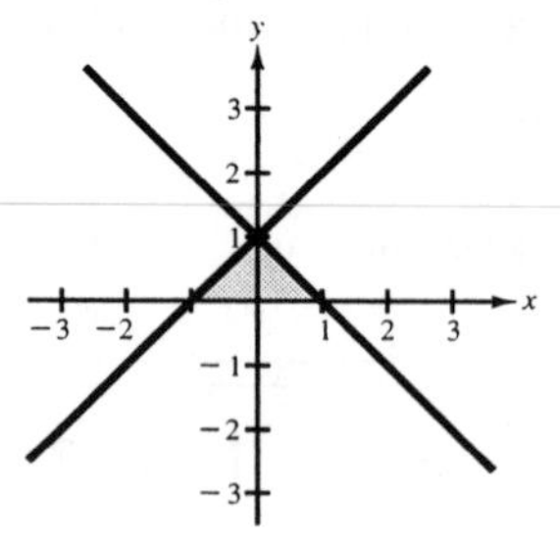

25. Sketch the graph of the solution set of the following system of inequalities.

$$-3x + 2y < 6$$
$$x + 4y > -2$$
$$2x + y < 3$$

Solution

Graph the lines $-3x + 2y = 6$, $x + 4y = -2$, and $2x + y = 3$ using dashes. The region inside the triangle satisfies all three inequalities.

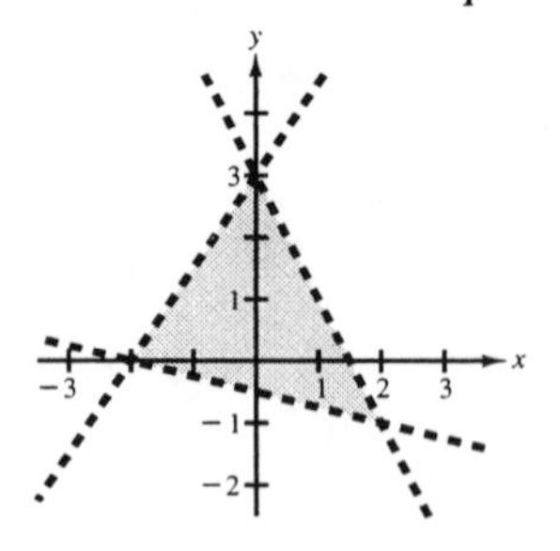

27. Sketch the graph of the solution set of the following system of inequalities.

$$2x + y > 2$$
$$6x + 3y < 2$$

Solution

Graph the lines $2x + y = 2$ and $6x + 3y = 2$ using dashes. The region between the lines satisfies both inequalities.

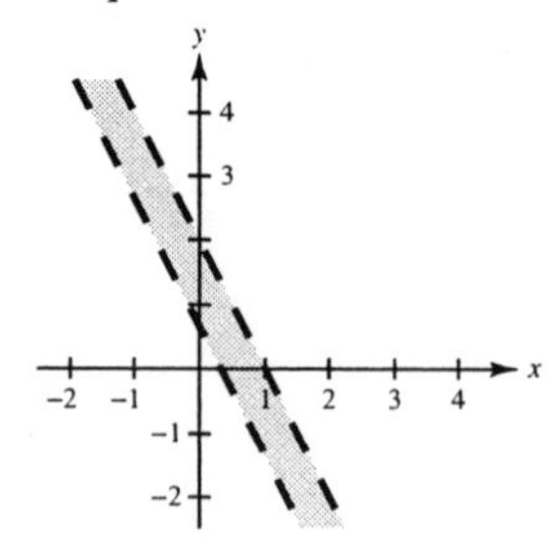

31. Sketch the graph of the solution set of the following system of inequalities.

$$x^2 + y^2 \leq 9$$

$$x^2 + y^2 \geq 1$$

Solution

Graph the circles $x^2 + y^2 = 9$ and $x^2 + y^2 = 1$. The region between the circles satisfies both inequalities.

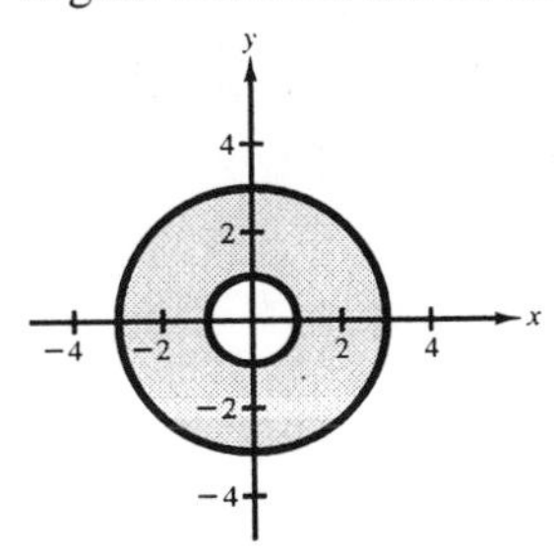

37. Sketch the graph of the solution set of the following system of inequalities.

$$y < x^3 - 2x + 1$$

$$y > -2x$$

$$x < 1$$

Solution

Graph $y = x^3 - 2x + 1$ and $y = -2x$ and $x = 1$ using dashes. The enclosed region satisfies all three inequalities.

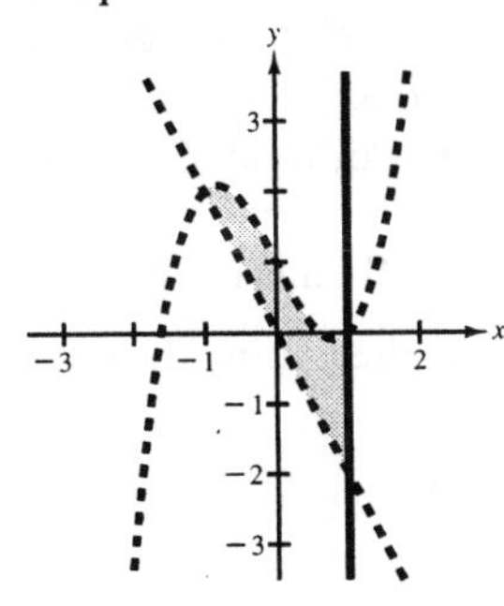

41. Derive a set of inequalities to describe a rectangular region with vertices at (2, 1), (5, 1), (5, 7), and (2, 7).

Solution

$2 \leq x \leq 5$

$1 \leq y \leq 7$

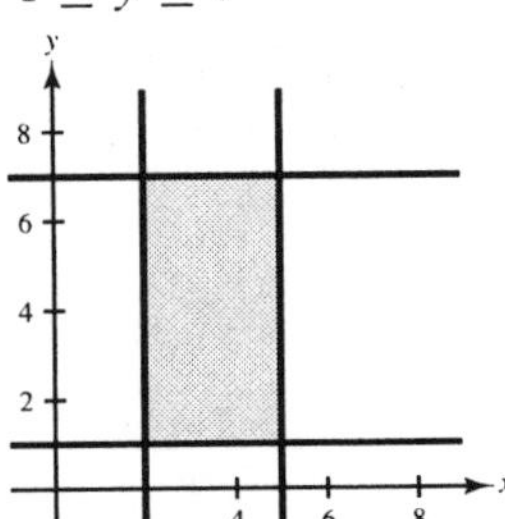

47. *Furniture Production* A furniture company can sell all the tables and chairs it produces. Each table requires 1 hour in the assembly center and $1\frac{1}{3}$ hours in the finishing center. Each chair requires $1\frac{1}{2}$ hours in the assembly center and $1\frac{1}{2}$ hours in the finishing center. The company's assembly center is available 12 hours per day, and its finishing center is available 15 hours per day. If x is the number of tables produced per day and y is the number of chairs, find a system of inequalities describing all possible production levels. Sketch the graph of the system.

Solution

47. –CONTINUED–

$x =$ number of tables

$y =$ number of chairs

Assembly center: $x + \frac{3}{2}y \leq 12$

Finishing center: $\frac{4}{3}x + \frac{3}{2}y \leq 15$

$x \geq 0$

$y \geq 0$

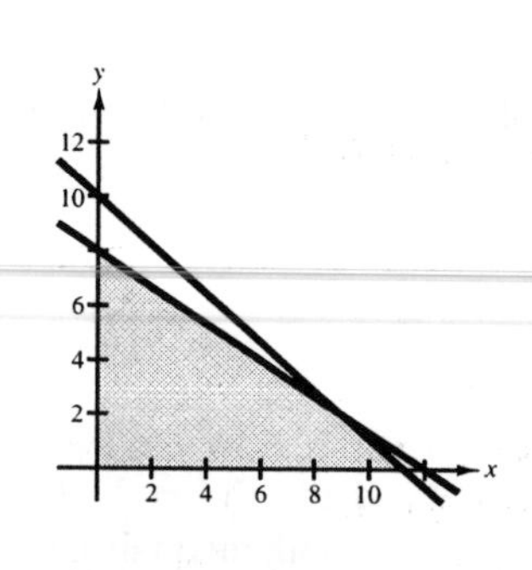

51. *Diet Supplement* A dietitian is asked to design a special diet supplement using two different foods. Each ounce of food X contains 20 units of calcium, 15 units of iron, and 10 units of vitamin B. Each ounce of food Y contains 10 units of calcium, 10 units of iron, and 20 units of vitamin B. The minimum daily requirements in the diet are 300 units of calcium, 150 units of iron, and 200 units of vitamin B.

(a) Find a system of inequalities describing the different amounts of food X and food Y that can be used in the diet.

(b) Sketch the graph of the system.

Solution

(a) $x =$ number of ounces of food X

$y =$ number of ounces of food Y

Calcium: $20x + 10y \geq 300$

Iron: $15x + 10y \geq 150$

Vitamin B: $10x + 20y \geq 200$

$x \geq 0$

$y \geq 0$

(b)

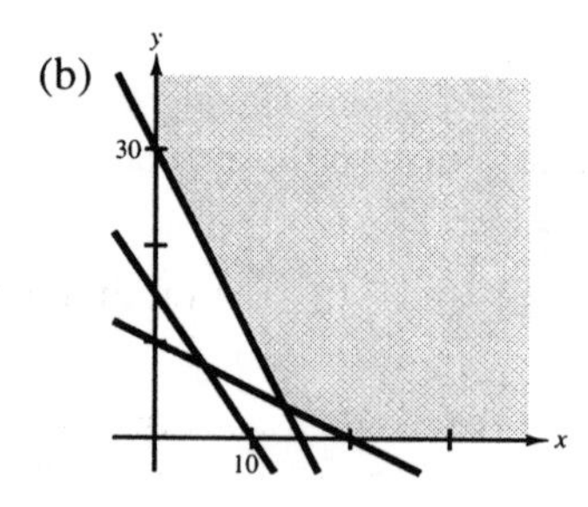

55. *Consumer and Producer Surplus* Find the consumer surplus and producer surplus when given the demand, $p = 140 - 0.00002x$ and supply $p = 80 + 0.00001x$.

Solution

Demand = Supply

$140 - 0.00002x = 80 + 0.00001x$

Solving, we have $x = 2{,}000{,}000$ units and $p = \$100$.

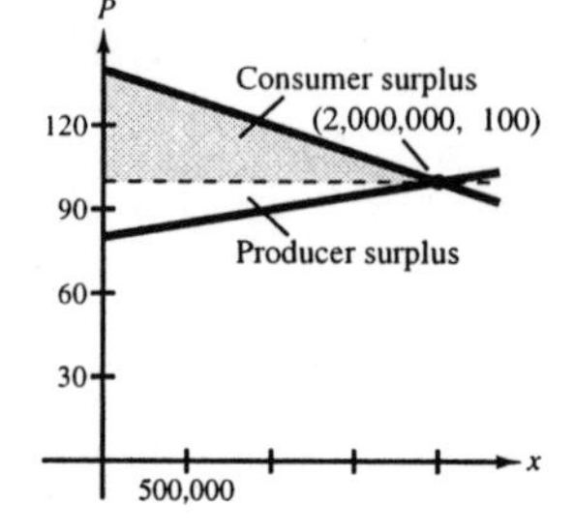

Use the formula for the area of a triangle.

$A = \frac{1}{2}(\text{base})(\text{height})$

$\text{Consumer Surplus} = \frac{1}{2}(40)(2{,}000{,}000)$

$= \$40{,}000{,}000$

$\text{Producer Surplus} = \frac{1}{2}(20)(2{,}000{,}000)$

$= \$20{,}000{,}000$

6.5 Linear Programming

- To solve a linear programming problem:
  1. Sketch the solution set for the system of constraints.
  2. Find the vertices of the region.
  3. Test the objective function at each of the vertices.

SOLUTIONS TO SELECTED EXERCISES

3. Find the minimum and maximum values of $z = 10x + 6y$, subject to the following constraints. (The graph of the region determined by the constraints is provided in the textbook.)

$$x \geq 0$$
$$y \geq 0$$
$$x + y \leq 6$$

Solution

At $(0, 6)$: $z = 10(0) + 6(6) = 36$
At $(0, 0)$: $z = 10(0) + 6(0) = 0$
At $(6, 0)$: $z = 10(6) + 6(0) = 60$
The maximum value is 60 at $(6, 0)$. The minimum value is 0 at $(0, 0)$.

9. Find the minimum and maximum values of $z = 10x + 7y$, subject to the following constraints. (The graph of the region determined by the constraints is provided in the textbook.)

$$0 \leq x \leq 60$$
$$0 \leq y \leq 45$$
$$5x + 6y \leq 420$$

Solution

At $(0, 45)$: $z = 10(0) + 7(45) = 315$
At $(30, 45)$: $z = 10(30) + 7(45) = 615$
At $(60, 20)$: $z = 10(60) + 7(20) = 740$
At $(60, 0)$: $z = 10(60) + 7(0) = 600$
At $(0, 0)$: $z = 10(0) + 7(0) = 0$
The maximum value is 740 at $(60, 20)$. The minimum value is 0 at $(0, 0)$.

13. Sketch the region determined by the constraints. Then find the minimum and maximum values of $z = 6x + 10y$, subject to these constraints.

$$x \geq 0$$
$$y \geq 0$$
$$2x + 5y \leq 10$$

Solution

At (0, 2): $z = 6(0) + 10(2) = 20$
At (5, 0): $z = 6(5) + 10(0) = 30$
At (0, 0): $z = 6(0) + 10(0) = 0$
The maximum value is 30 at (5, 0).
The minimum value is 0 at (0, 0).

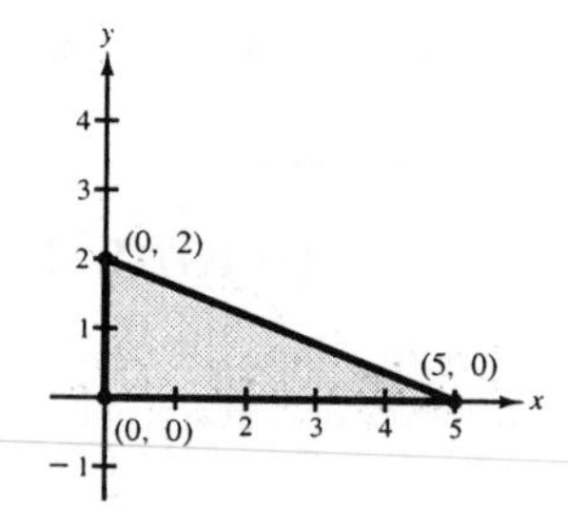

17. Sketch the region determined by the constraints. Then find the minimum and maximum values of $z = 4x + 5y$, subject to these constraints.

$$x \geq 0$$
$$y \geq 0$$
$$4x + 3y \geq 27$$
$$x + y \geq 8$$
$$3x + 5y \geq 30$$

Solution

At (10, 0): $z = 4(10) + 5(0) = 40$
At (5, 3): $z = 4(5) + 5(3) = 35$
At (3, 5): $z = 4(3) + 5(5) = 37$
At (0, 9): $z = 4(0) + 5(9) = 45$
C is unbounded. Therefore, there is no maximum. The minimum value is 35 at (5, 3).

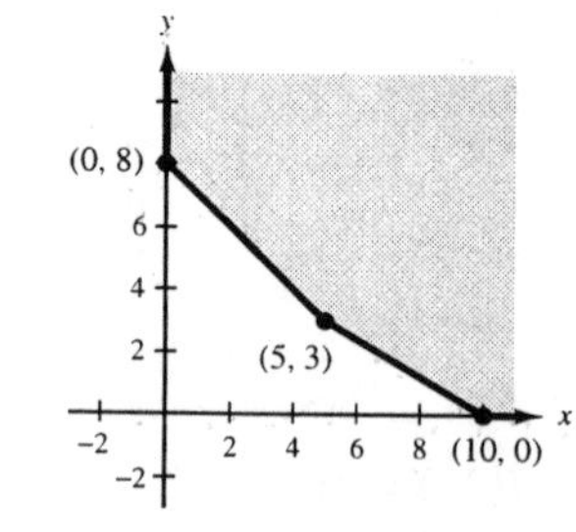

21. Sketch the region determined by the constraints. Then find the minimum and maximum values of $z = 4x + y$, subject to these constraints.

$$x \geq 0$$
$$y \geq 0$$
$$x + 2y \leq 40$$
$$x + y \geq 30$$
$$2x + 3y \geq 72$$

Solution

At $(36, 0)$: $z = 4(36) + 0 = 144$

At $(40, 0)$: $z = 4(40) + 0 = 160$

At $(24, 8)$: $z = 4(24) + 8 = 104$

The maximum value is 160 at $(40, 0)$. The minimum value is 104 at $(24, 8)$.

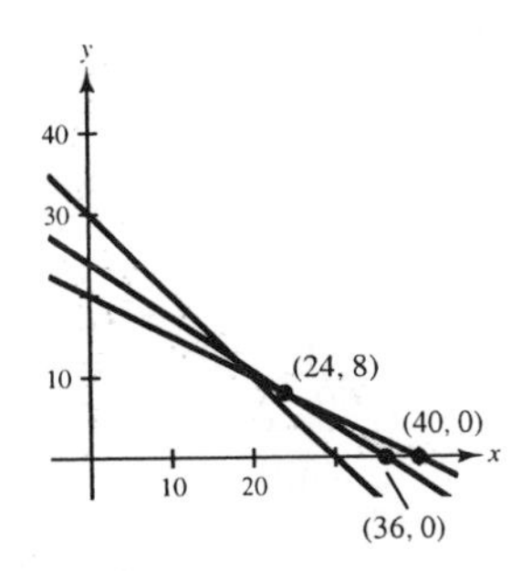

25. Maximize $z = 2x + y$ subject to the constraints $3x + y \leq 15$ and $4x + 3y \leq 30$, where $x \geq 0$ and $y \geq 0$.

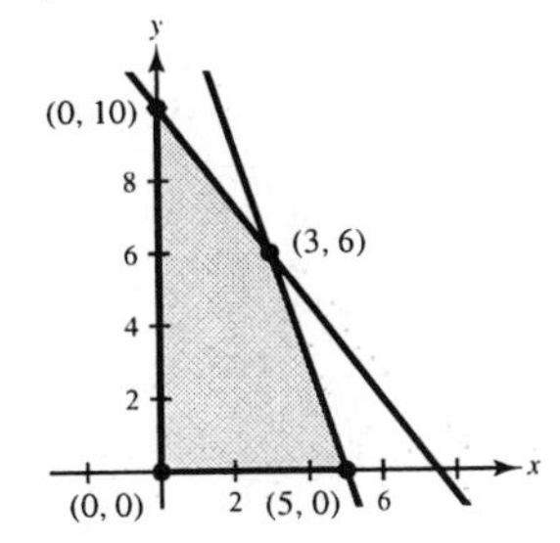

Solution

At $(0, 10)$: $z = 2(0) + (10) = 10$

At $(3, 6)$: $z = 2(3) + (6) = 12$

At $(5, 0)$: $z = 2(5) + (0) = 10$

At $(0, 0)$: $z = 2(0) + (0) = 0$

The maximum value is 12 at $(3, 6)$.

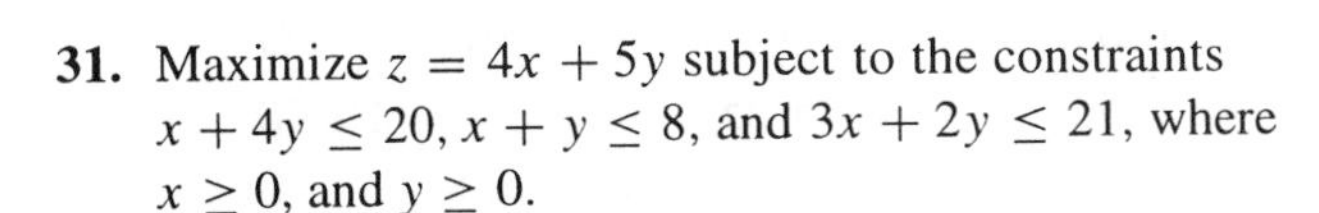

31. Maximize $z = 4x + 5y$ subject to the constraints $x + 4y \leq 20$, $x + y \leq 8$, and $3x + 2y \leq 21$, where $x \geq 0$, and $y \geq 0$.

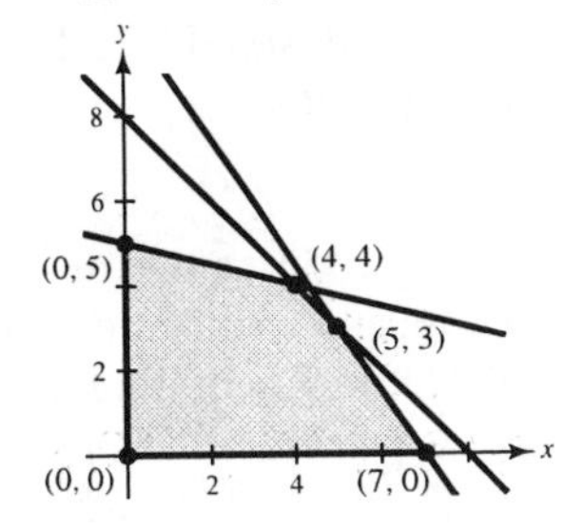

Solution

At $(0, 5)$: $z = 4(0) + 5(5) = 25$

At $(4, 4)$: $z = 4(4) + 5(4) = 36$

At $(5, 3)$: $z = 4(5) + 5(3) = 35$

At $(7, 0)$: $z = 4(7) + 5(0) = 28$

At $(0, 0)$: $z = 4(0) + 5(0) = 0$

The maximum value is 36 at $(4, 4)$.

33. *Maximum Profit* A store plans to sell two models of home computers. How many units of each model should be stocked to maximize profit, subject to the following constraints? The costs are \$250 and \$400, respectively. The merchant estimates that the total monthly demand will not exceed 250 units. The merchant does not want to invest more than \$70,000 in computer inventory. The \$250 model yields a profit of \$45 and the \$400 model yields a profit of \$50.

Solution

$x =$ number of \$250 models

$y =$ number of \$400 models

Constraints: $250x + 400y \leq 70{,}000$

$x + y \leq 250$

$x \geq 0$

$y \geq 0$

Objective function: $P = 45x + 50y$

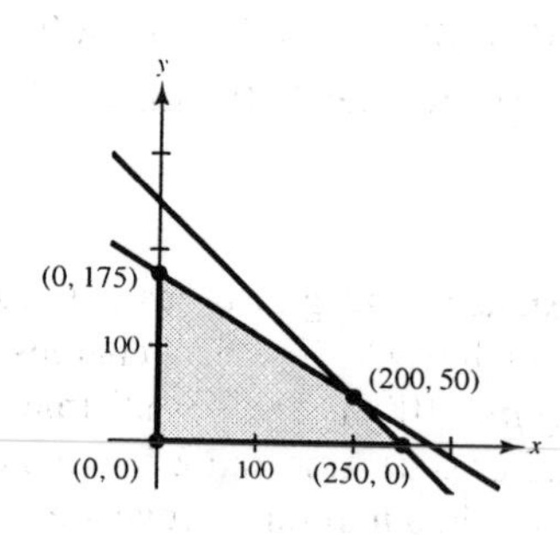

Vertices: (0, 175), (200, 50), (250, 0), (0, 0)
At (0, 175): $P = 45(0) + 50(175) = 8,750$
At (200, 50): $P = 45(200) + 50(50) = 11,500$
At (250, 0): $P = 45(250) + 50(0) = 11,250$
At (0, 0): $P = 45(0) + 50(0) = 0$
To maximize the profit, the merchant should stock 200 units of the model costing \$250 and 50 units of the model costing \$400.

37. *Maximum Profit* A manufacturer produces two model of bicycles. The time (in hours) required for assembling, painting, and packaging each model is shown in the table in the textbook. The total time available for assembling, painting, and packaging is 4,000 hours, 4,800 hours, and 1,500 hours, respectively. The profit per unit for each model is \$45 (model A) and \$50 (model B). How many of each type should be produced to obtain a maximum profit?

Solution

$x =$ number of Model A

$y =$ number of Model B

Constraints: $2x + 2.5y \leq 4000$

$4x + y \leq 4800$

$x + 0.75y \leq 1500$

$x \geq 0$

$y \geq 0$

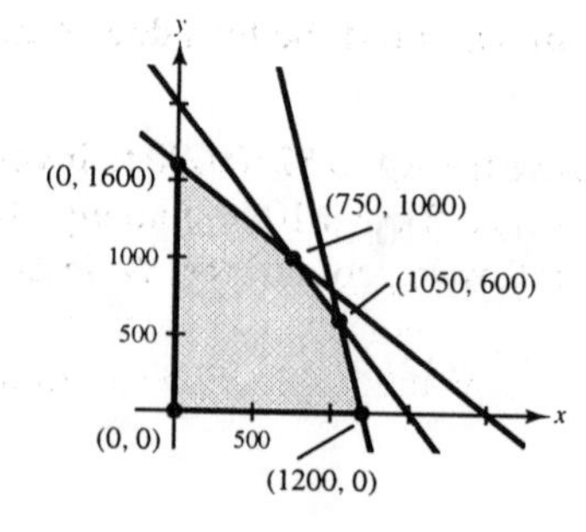

Objective function: $P = 45x + 50y$

Vertices: $(0,\ 0)$, $(0,\ 1600)$, $(750,\ 1000)$, $(1050,\ 600)$, $(1200,\ 0)$

At $(0, 0)$: $P = 45(0) + 50(0) = 0$

At $(0,\ 1600)$: $P = 45(0) + 50(1600) = 80{,}000$

At $(750,\ 1000)$: $P = 45(750) + 50(1000) = 83{,}750$

At $(1050,\ 600)$: $P = 45(1050) + 50(600) = 77{,}250$

At $(1200,\ 0)$: $P = 45(1200) + 50(0) = 54{,}000$

The maximum profit occurs when 750 units of Model A and 1,000 units of Model B are produced.

39. *Maximum Revenue* An accounting firm has 900 hours of staff time and 100 hours of review time available each week. The firm charges $2,000 for an audit and $300 for a tax return. Each audit requires 100 hours of staff time and 10 hours of review time, and each tax return requires 12.5 of staff time and 2.5 hours of review time. What number of audits and tax returns will bring in a maximum revenue?

Solution

Let $x =$ number of audits.

Let $y =$ number of tax returns.

Constraints: $100x + 12.5y \le 900$

$10x + 2.5y \le 100$

$x \ge 0$

$y \ge 0$

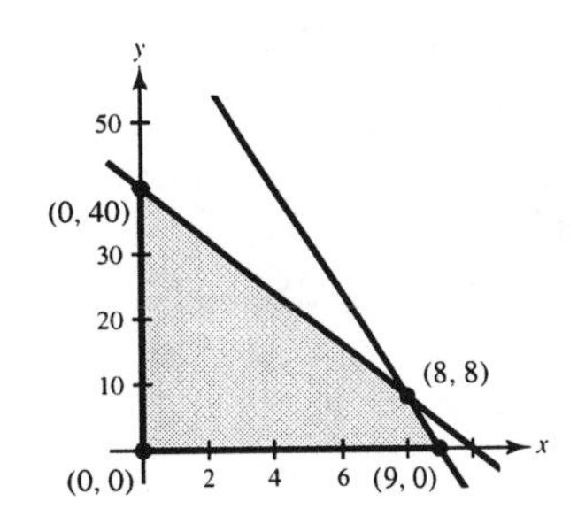

Objective function: $R = 2000x + 300y$

Vertices: $(0,\ 0)$, $(0,\ 40)$, $(8,\ 8)$, $(9,\ 0)$

At $(0, 0)$: $R = 2000(0) + 300(0) = 0$

At $(0, 40)$: $R = 2000(0) + 300(40) = 12{,}000$

At $(8, 8)$: $R = 2000(8) + 300(8) = 18{,}400$

At $(9, 0)$: $R = 2000(9) + 300(0) = 18{,}000$

The revenue will be maximum if the firm does 8 audits and 8 tax returns each week.

43. *Investments* An investor has up to $250,000 to invest in two types of investments. Type A pays 8% annually and type B pays 10% annually. To have a well-balanced portfolio, the investor imposes the following conditions. At least one-fourth of the total portfolio is to be allocated to type A investments and at least one-fourth of the portfolio is to be allocated to type B investments. How much should be allocated to each type of investment to obtain a maximum return?

Solution

43. –CONTINUED–

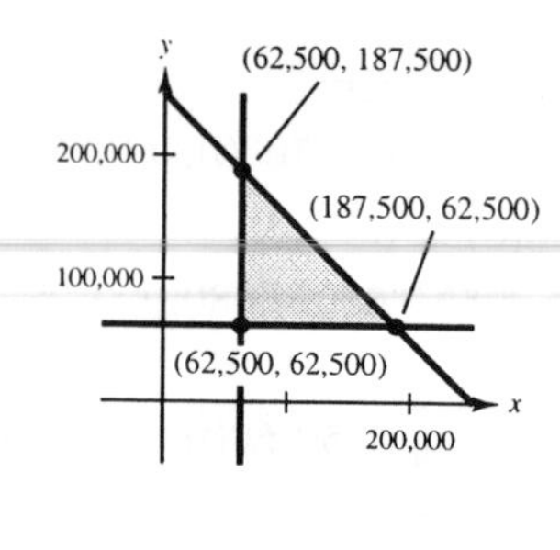

Let $x =$ amount in Type A investment.

Let $y =$ amount in Type B investment.

Constraints: $x \geq \frac{1}{4}(250{,}000)$

$y \geq \frac{1}{4}(250{,}000)$

$x + y \leq 250{,}000$

Objective function:

$$I = 0.08x + 0.10y$$

Vertices: (62,500, 62,500), (62,500, 187,500), (187,500, 62,500)

At (62,500, 62,500): $I = 0.80(62{,}500) + 0.10(62{,}500) = 11{,}250$

At (62,500, 187,500): $I = 0.08(62{,}500) + 0.10(187{,}500) = 23{,}750$

At (187,500, 62,500): $I = 0.08(187{,}500) + 0.10(62{,}500) = 21{,}250$

To obtain a maximum return, \$62,500 should be invested in Type A and \$187,500 in Type B.

47. The given linear programming problem has an unusual characteristic. Sketch a graph of the solution region for the problem and describe the unusual characteristic. The objective function is to be maximized.

Objective function: $z = -x + 2y$

Constraints: $x \geq 0$

$y \geq 0$

$x \leq 10$

$x + y \leq 7$

Solution

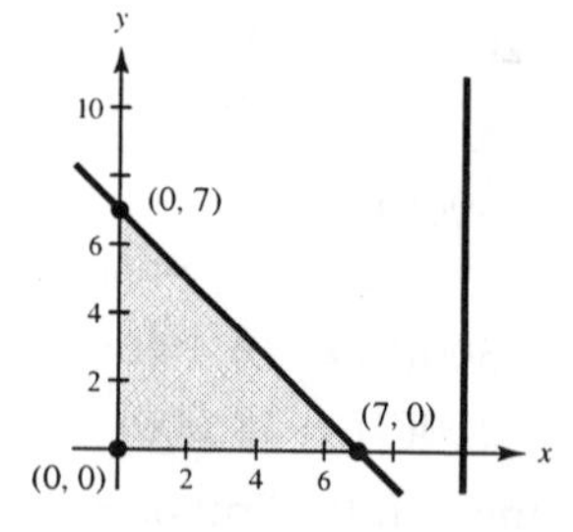

At (0, 0): $z = -0 + 2(0) = 0$

At (0, 7): $z = -0 + 2(7) = 14$

At (7, 0): $z = -7 + 2(0) = -7$

The constraint $x \leq 10$ is extraneous.

The maximum value of z occurs at (0, 7).

Review Exercises for Chapter 6

SOLUTIONS TO SELECTED EXERCISES

3. Solve the following system by the method of substitution.

$$\begin{aligned} \tfrac{1}{2}x + \tfrac{3}{5}y &= -2 \\ 2x + \phantom{\tfrac{3}{5}}y &= 6 \end{aligned}$$

Solution

$$\tfrac{1}{2}x + \tfrac{3}{5}y = -2 \Rightarrow 5x + 6y = -20$$

$$2x + y = 6 \Rightarrow y = 6 - 2x$$

$$5x + 6(6 - 2x) = -20$$

$$-7x = -56$$

$$x = 8$$

$$y = 6 - 2(8) = -10$$

Answer: $(8, -10)$

7. Use a graphing utility to find all points of intersection of the graphs of the equations.

$$y = x^2 - 3x + 11$$
$$y = -x^2 + 2x + 8$$

Solution

$(1, 9)$ and $(1.5, 8.75)$

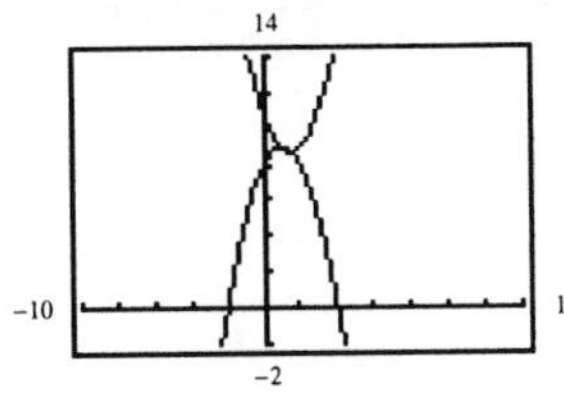

11. *Investment Portfolio* A total of \$34,000 is invested in two funds paying 7.0% and 7.5% simple interest. If the yearly interest is \$2,480, how much of the \$34,000 is invested at each rate?

Solution

$$x + y = 34{,}000 \Rightarrow y = 34{,}000 - x$$

$$0.07x + 0.075y = 2480$$

$$0.07x + 0.075(34{,}000 - x) = 2480$$

$$-0.005x = -70$$

$$x = 14{,}000$$

$$y = 20{,}000$$

Answer: \$14,000 at 7%, \$20,000 at 7.5%

15. Solve the system by elimination.

$$\begin{aligned} 1.25x - 2y &= 3.5 \\ 5x - 8y &= 14 \end{aligned}$$

Solution

$$\begin{aligned} 1.25x - 2y = 3.5 &\Rightarrow \quad 5x - 8y = 14 \\ 5x - 8y = 14 &\Rightarrow \ -5x + 8y = -14 \\ & \qquad\qquad 0 = 0 \end{aligned}$$

Infinite solutions. All points on the line $5x - 8y = 14$ are solutions.

17. Solve the system by elimination.

$$\begin{aligned} 1.5x + 2.5y &= 8.5 \\ 6x + 10y &= 24 \end{aligned}$$

Solution

$$\begin{aligned} 1.5x + 2.5y = 8.5 &\Rightarrow \quad 3x + 5y = 17 \\ 6x + 10y = 24 &\Rightarrow \ -3x - 5y = -12 \\ & \qquad\qquad 0 = 5 \ \rightarrow\leftarrow \end{aligned}$$

No solution

21. *Supply and Demand* Find the point of equilibrium for supply, $p = 22 + 0.00001x$, and demand, $p = 37 - 0.0002x$.

Solution

$$\text{Supply} = \text{Demand}$$

$$22 + 0.00001x = 37 - 0.0002x$$

$$0.00021x = 15$$

$$x \approx 71{,}429 \text{ units}$$

$$p \approx \$22.71$$

25. Solve the system.

$$\begin{aligned} x + y + z &= 10 \\ -2x + 3y + 4z &= 22 \end{aligned}$$

Solution

$$\begin{aligned} x + y + z = 10 &\Rightarrow 3x + 3y + 3z = 30 \\ -2x + 3y + 4z = 22 &\Rightarrow 2x - 3y - 4z = -22 \\ & \quad\ 5x \qquad\quad - z = 8 \Rightarrow x = \frac{8+z}{5} \end{aligned}$$

Let $z = a$, then $x = \dfrac{8+a}{5}$ and $\dfrac{8+a}{5} + y + a = 10 \Rightarrow y = \dfrac{42-6a}{5}$

Answer: $\left(\dfrac{8+a}{5}, \dfrac{42-6a}{5}, a\right) = \left(\dfrac{1}{5}a + \dfrac{8}{5}, -\dfrac{6}{5}a + \dfrac{42}{5}, a\right)$

29. Find the equation of the parabola $y = ax^2 + bx + c$ that passes through $(0, -6)$, $(1, -3)$, and $(2, 4)$.

Solution

$$\begin{array}{lllll}
\text{At } (0, -6): & -6 = c & & & \\
\text{At } (1, -3): & -3 = a + b + c & \Rightarrow a + b = 3 & \Rightarrow & -2a - 2b = -6 \\
\text{At } (2, 4): & 4 = 4a + 2b + c & \Rightarrow 4a + 2b = 10 & \Rightarrow & 4a + 2b = 10 \\
\hline
 & & & & 2a = 4 \\
 & & & & a = 2 \\
 & & & & b = 1
\end{array}$$

Answer: $y = 2x^2 + x - 6$

33. *Investment* Suppose you receive \$1,060 a year in interest from three investments. The interest rates for the three investments are 6%, 7%, and 8%. The 6% investment is $\frac{1}{5}$ the 7% investment, and the 8% investment is \$1,000 more than the 6% investment. What is the amount of each investment?

Solution

Let $x =$ amount at 6%.

Let $y =$ amount at 7%.

Let $z =$ amount at 8%.

$$0.06x + 0.07y + 0.08z = 1060$$

$$5x = y$$

$$z = x + 1000$$

$$0.06x + 0.07(5x) + 0.08(x + 1000) = 1060$$

$$0.49x = 980$$

$$x = 2000$$

$$y = 5x = 10{,}000$$

$$z = x + 1000 = 3000$$

Answer: $x =$ \$2,000 at 6%

$y =$ \$10,000 at 7%

$z =$ \$3,000 at 8%

37. *Fitting a Parabola to Data* Find the least squares regression parabola $y = ax^2 + bx + c$ for the points $(-2, 0.4)$, $(-1, 0.9)$, $(0, 1.9)$, $(1, 2.1)$, and $(2, 3.8)$. To find the parabola, solve the following system of linear equations for a, b, and c.

$$\begin{aligned} 5c + \quad + 10a &= 9.1 \\ 10b \quad &= 8.0 \\ 10c \quad + 34a &= 19.8 \end{aligned}$$

Solution

$$\begin{aligned} 5c \quad + 10a &= 9.1 \Rightarrow -10c - 20a = -18.2 \\ 10b \quad &= 8.0 \\ 10c \quad + 34a &= 19.8 \Rightarrow \quad 10c + 34a = 19.8 \end{aligned}$$

$$\begin{aligned} 14a &= 1.6 \\ a &= \tfrac{4}{35} \\ c &= \tfrac{557}{350} \end{aligned}$$

From Equation 2: $b = \frac{4}{5}$

Least squares regression parabola: $y = \frac{4}{35}x^2 + \frac{4}{5}x + \frac{557}{350} \approx 0.114x^2 + 0.8x + 1.59$

41. Sketch the graph of the solution set of the following system.

$$x + y > 4$$

$$3x + y < 10$$

Solution

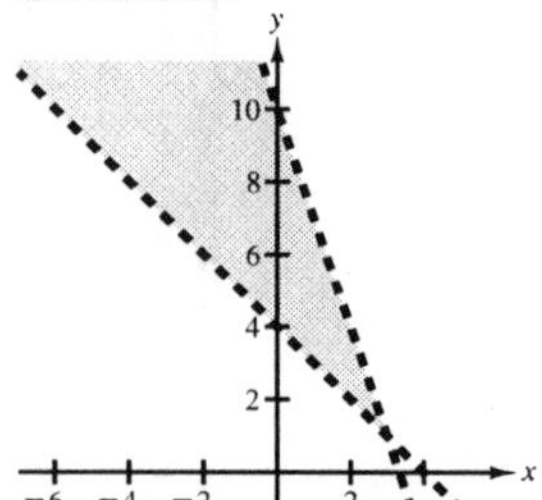

45. Derive a set of inequalities to describe the triangular region with vertices at $(1, 2)$, $(6, 7)$, and $(8, 1)$.

Solution

$$x - y \geq -1$$

$$3x + y \leq 25$$

$$x + 7y \geq 15$$

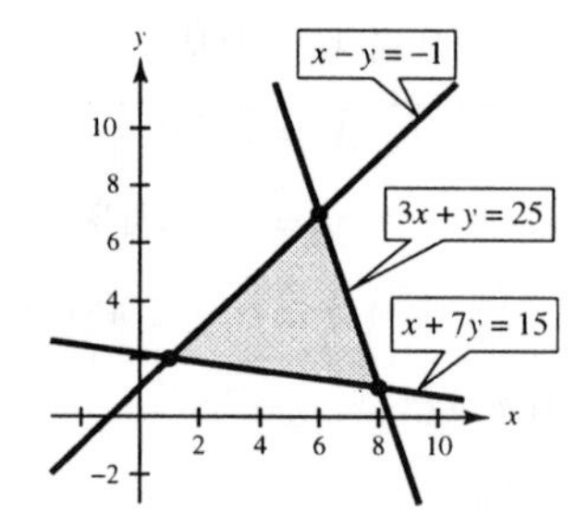

51. *Supply and Demand* Find the consumer surplus and producer surplus of demand, $p = 130 - 0.0002x$, and supply, $p = 30 + 0.0003x$.

Solution

Point of equilibrium: Demand = Supply

$$130 - 0.0002x = 30 + 0.0003x$$
$$100 = 0.0005x$$
$$x = 200{,}000 \text{ units}$$
$$p = \$90.00$$

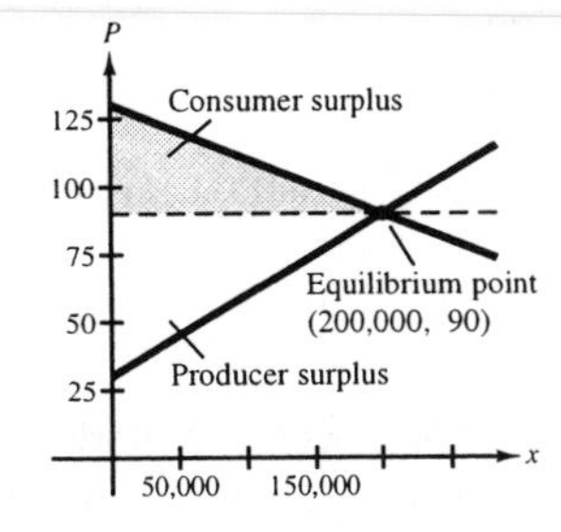

Area of a triangle:

$$\text{Area} = \tfrac{1}{2}(\text{base})(\text{height})$$

Consumer surplus:

$$\tfrac{1}{2}(40)(200{,}000) = \$4{,}000{,}000$$

Producer surplus:

$$\tfrac{1}{2}(60)(200{,}000) = \$6{,}000{,}000$$

55. Find the minimum and maximum values of $z = 50x + 60y$, subject to the following constraints. The graph of the region determined by the constraints is provided in the textbook.

$$x \geq 0$$
$$y \geq 0$$
$$3x + 4y \geq 1200$$
$$5x + 6y \leq 3000$$

Solution

At $(0, 300)$: $z = 50(0) + 60(300) = 18{,}000$
At $(0, 500)$: $z = 50(0) + 60(500) = 30{,}000$
At $(600, 0)$: $z = 50(600) + 60(0) = 30{,}000$
At $(400, 0)$: $z = 50(400) + 60(0) = 20{,}000$

The minimum value is 18,000 at (0, 300). The maximum value is 30,000 at every point on the line $5x + 6y = 3000$ between the points (0, 500) and (600, 0).

59. Sketch the region determined by the following constraints. Find the minimum and maximum values of $z = 4x + 5y$, subject to these constraints.

$$x \geq 0$$
$$y \geq 0$$
$$2x + 5y \leq 30$$
$$x + y \geq 3$$
$$2x + y \leq 14$$

Solution

59. –CONTINUED–

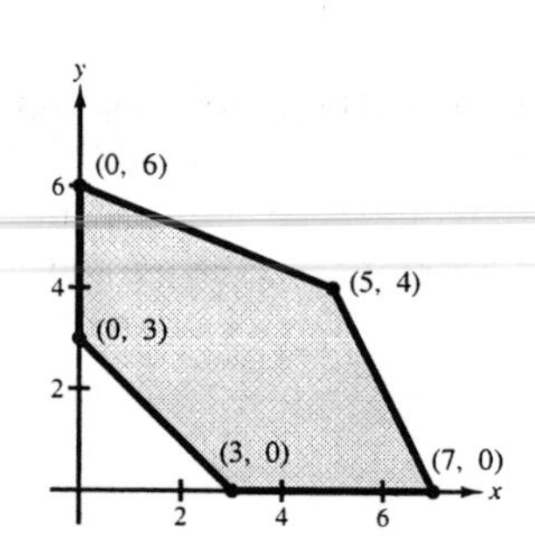

At (0, 3): $z = 4(0) + 5(3) = 15$
At (0, 6): $z = 4(0) + 5(6) = 30$
At (5, 4): $z = 4(5) + 5(4) = 40$
At (7, 0): $z = 4(7) + 5(0) = 28$
At (3, 0): $z = 4(3) + 5(0) = 12$

The minimum value is 12 at (3, 0). The maximum value is 40 at (5, 4).

63. *Maximum Profit* A manufacturer produces two models of a product. The time in hours required for assembling, finishing, and packaging each model is shown in the table in the textbook. The total time available for assembling, finishing, and packaging is 5,600 hours, 2,000 hours, and 910 hours, respectively. The profit per unit is \$80 (model A) and \$100 (model B). How many of each model should be produced to obtain a maximum profit?

Solution

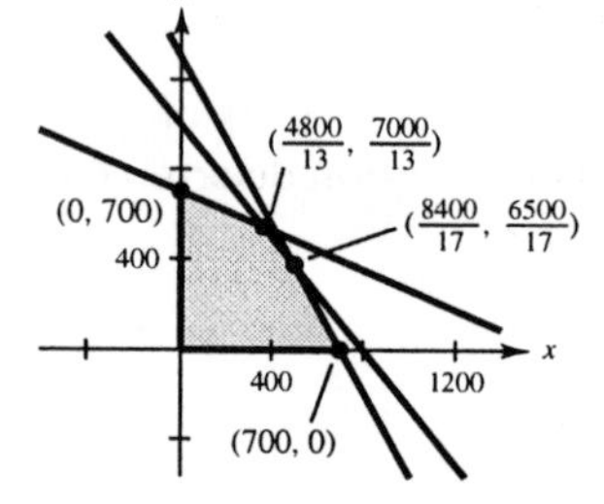

Let $x =$ number of Model A.

Let $y =$ number of Model B.

Constraints: $3.5x + 8y \le 5600$

$$2.5x + 2y \le 2000$$

$$1.3x + 0.7y \le 910$$

$$x \ge 0,\ y \ge 0$$

Objective function: $P = 80x + 100y$

At (0, 0): $P = 80(0) + 100(0) = 0$
At (0, 700): $P = 80(0) + 100(700) = 70{,}000$
At $\left(\frac{4800}{13}, \frac{7000}{13}\right)$: $P = 80\left(\frac{4800}{13}\right) + 100\left(\frac{7000}{13}\right) \approx 83{,}384.62$
At $\left(\frac{8400}{17}, \frac{6500}{17}\right)$: $P = 80\left(\frac{8{,}400}{17}\right) + 100\left(\frac{6500}{17}\right) \approx 77{,}764.71$
At (700, 0): $P = 80(700) + 100(0) = 56{,}000$

The profit is maximum when $x = \frac{4800}{13} \approx 369$ units and $y = \frac{7000}{13} \approx 538$ units are produced.

67. *Maximum Profit* An accounting firm has 800 hours of staff time and 90 hours of review time available each week. Each audit requires 100 hours of staff time and 10 hours of review time. Each tax return requires 10 hours of staff time and 2 hours of review time. The firm realizes a profit of \$800 for each audit and \$120 for each tax return. What combination of audits and tax returns will yield maximum profit?

Solution

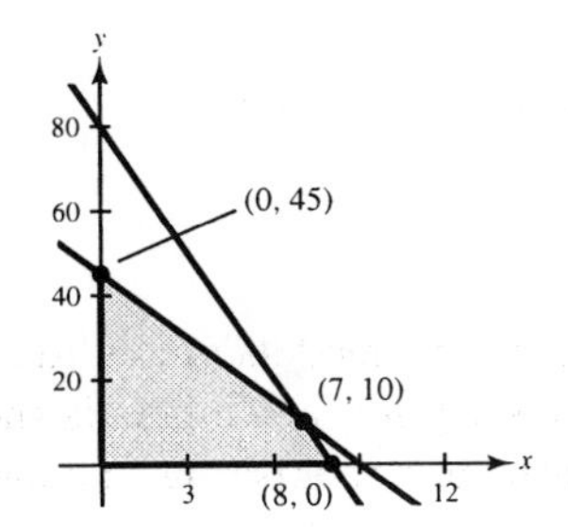

Let $x =$ number of audits.

Let $y =$ number of tax returns.

Constraints: $100x + 10y \le 800$

$$10x + 2y \le 90$$

$$x \ge 0,\ y \ge 0$$

Objective function: $P = 800x + 120y$

At $(0, 0)$: $P = 800(0) + 120(0) = 0$

At $(0, 45)$: $P = 800(0) + 120(45) = 5400$

At $(7, 10)$: $P = 800(7) + 120(10) = 6800$

At $(8, 0)$: $P = 800(8) + 120(0) = 6400$

The profit is maximum when the firm does 7 audits and 10 tax returns per week.

Test for Chapter 6

1. Solve the system by substitution.

$$\begin{aligned} 3x - 2y &= -2 \\ 4x + 3y &= 20 \end{aligned}$$

Solution

$$3x - 2y = -2 \Rightarrow y = \frac{3x + 2}{2}$$

$$4x + 3y = 20 \Rightarrow 4x + 3\left(\frac{3x + 2}{2}\right) = 20$$

$$4x + \frac{9}{2}x + 3 = 20$$

$$\frac{17}{2}x = 17$$

$$x = 2$$

$$y = \frac{3(2) + 2}{2} = 4$$

Answer: $(2, 4)$

2. Solve the system by substitution.

$$x + y = 3$$
$$x^2 + y = 9$$

Solution

$$x + y = 3 \Rightarrow y = 3 - x$$
$$x^2 + y = 9 \Rightarrow x^2 + (3 - x) = 9$$
$$x^2 - x - 6 = 0$$
$$(x - 3)(x + 2) = 0$$
$$x = 3 \text{ or } x = -2$$
$$y = 0 \qquad y = 5$$

Answer: (3, 0), (−2, 5)

3. Solve the system by substitution.

$$2x - y = 11$$
$$3x + 5y = 8$$

Solution

$$2x - y = 11 \Rightarrow y = 2x - 11$$
$$3x + 5y = 8 \Rightarrow 3x + 5(2x - 11) = 8$$
$$3x + 10x - 55 = 8$$
$$13x = 63$$
$$x = \tfrac{63}{13},\ y = -\tfrac{17}{13}$$

Answer: $\left(\frac{63}{13}, -\frac{17}{13}\right)$

4. Solve the system by graphing

$$5x - y = 6$$
$$2x^2 + y = 8$$

Solution

$$5x - y = 6 \Rightarrow y = 5x - 6$$
$$2x^2 + y = 8 \Rightarrow y = -2x^2 + 8$$

Answer: (−4.18, −26.89), (1.68, 2.38)

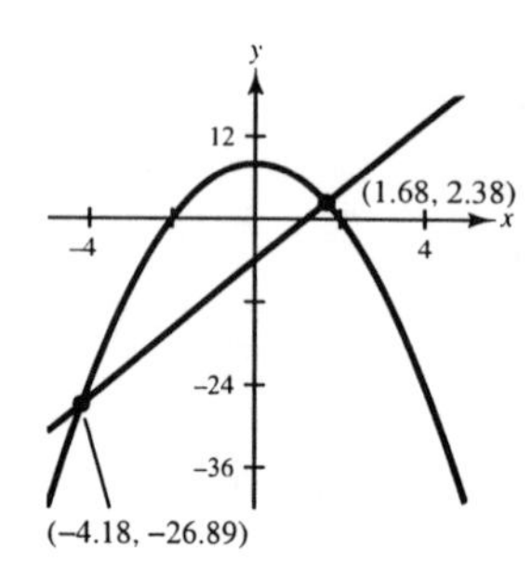

5. Solve the system by graphing.

$$1.5x - 2.25y = 8$$
$$2.5x + 2y = 5.75$$

Solution

$$1.5x - 2.25y = 8 \Rightarrow 6x - 9y = 32$$
$$2.5x + 2y = 5.75 \Rightarrow 10x + 8y = 23$$

Answer: (3.35, −1.32)

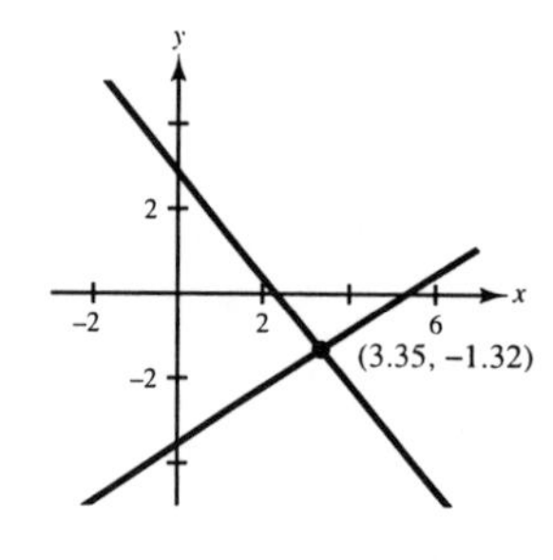

6. Solve the system by elimination.

$$\begin{aligned} 2x - 4y + z &= 11 \\ x + 2y + 3z &= 9 \\ 3y + 5z &= 12 \end{aligned}$$

Solution

$$\begin{aligned} x + 2y + 3z &= 9 \quad \text{Interchange the equations.} \\ 2x - 4y + z &= 11 \\ 3y + 5z &= 12 \end{aligned}$$

$$\begin{aligned} x + 2y + 3z &= 9 \\ -8y - 5z &= -7 \quad -2\text{ Eq. }1 + \text{Eq. }3 \\ 3y + 5z &= 12 \end{aligned}$$

$$\begin{aligned} x + 2y + 3z &= 9 \\ -8y - 5z &= -7 \\ -5y &= 5 \quad \text{Eq. }2 + \text{Eq. }3 \\ y &= -1 \quad \text{Solve and use back-substitution.} \end{aligned}$$

$-8(-1) - 5z = -7 \Rightarrow z = 3$

$x + 2(-1) + 3(3) = 9 \Rightarrow x = 2$

Answer: $(2, -1, 3)$

7. Solve the system by elimination.

$$\begin{aligned} 3x - 2y + z &= 16 \\ 5x - z &= 6 \\ 2x - y - z &= 3 \end{aligned}$$

Solution

$8x - 2y = 22$ Eq. 1 + Eq. 2

$5x - 3y = 19$ Eq. 1 + Eq. 3

$$\begin{aligned} 24x - 6y &= 66 \quad 3\text{ Eq. }1 \\ -10x + 6y &= -38 \quad -2\text{ Eq. }2 \\ \hline 14x &= 28 \\ x &= 2 \end{aligned}$$

$8(2) - 2y = 22 \Rightarrow y = -3$

$3(2) - 2(-3) + z = 16 \Rightarrow z = 4$

Answer: $(2, -3, 4)$

8. Solve the system by elimination.

$$\begin{aligned} -x + y - 2z &= 3 \\ x - 4y - 2z &= 1 \\ x + 2y + 6z &= 5 \end{aligned}$$

Solution

$$\begin{aligned} -x + y - 2z &= 3 \\ -3y - 4z &= 4 \quad \text{Eq. }1 + \text{Eq. }2 \\ 3y + 4z &= 8 \quad \text{Eq. }1 + \text{Eq. }3 \end{aligned}$$

$$\begin{aligned} -x + y - 2z &= 3 \\ -3y - 4z &= 4 \\ 0 &= 12 \quad \rightarrow\leftarrow \text{ Eq. }2 + \text{Eq. }3 \end{aligned}$$

Inconsistent, no solution

9. Find the point of equilibrium for a system that has a demand function $p = 45 - 0.0003x$ and a supply function of $p = 29 + 0.00002x$.

Solution

$$\begin{aligned} \text{Demand} &= \text{Supply} \\ 45 - 0.0003x &= 29 + 0.00002x \\ 16 &= 0.00032x \\ x &= 50,000 \text{ units} \\ p &= \$30.00 \end{aligned}$$

Point of equilibrium: (50,000, 30)

10. A system of linear equations reduces to $-23 = 0$. What can you conclude?

Solution

The system is inconsistent and has no solution.

11. A system of linear equations reduces to $0 = 0$. What can you conclude?

Solution

The system has infinite solutions.

12. Find the equation of the parabola $y = ax^2 + bx + c$ that passes through the points $(1, 10)$, $(-1, 4)$, and $(0, 5)$.

Solution

At $(1, 10)$: $10 = a + b + c$ Equation 1

At $(-1, 4)$: $4 = a - b + c$ Equation 2

At $(0, 5)$: $5 = c$ Equation 3

Equation 1: $10 = a + b + 5 \Rightarrow a + b = 5$

Equation 2: $4 = a - b + 5 \Rightarrow a - b = -1$

$$\begin{aligned} 2a &= 4 \\ a &= 2 \\ b &= 3 \end{aligned}$$

Answer: $y = 2x^2 + 3x + 5$

13. A total of \$35,000 is invested in two funds paying 8% and 8.5% simple interest. The interest is \$2,890. How much is invested in each fund?

Solution

Let $x =$ amount at 8% and $y =$ amount at 8.5%.

$x + y = 35,000 \Rightarrow y = 35,000 - x$

$$\begin{aligned} 0.08x + 0.085y &= 2890 \\ 0.08x + 0.085(35,000 - x) &= 2890 \\ 0.08x + 2975 - 0.085x &= 2890 \\ -0.005x &= -85 \\ x &= 17,000 \\ y &= 35,000 - 17,000 = 18,000 \end{aligned}$$

Answer: \$17,000 at 8%, \$18,000 at 8.5%

14. Sketch the inequality $x \geq 0$.

Solution

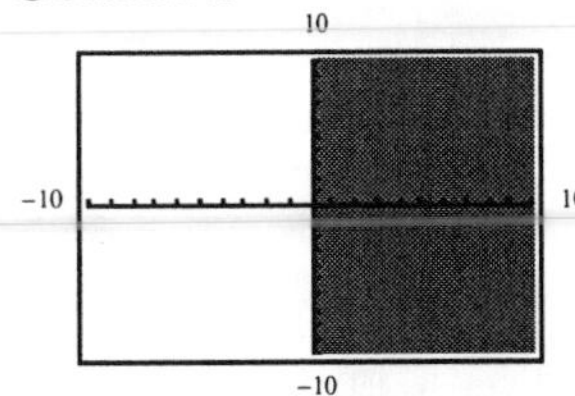

15. Sketch the inequality $y \geq 0$

Solution

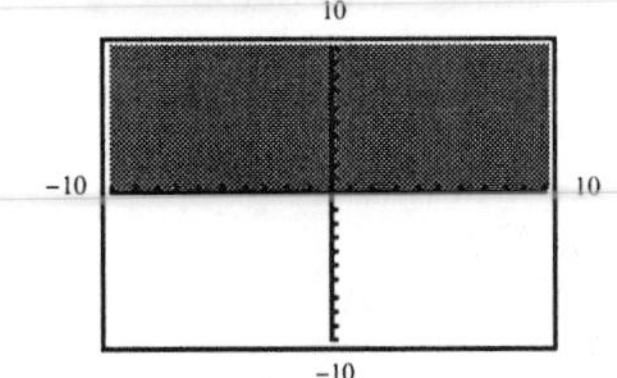

16. Sketch the inequality $x + 3y \leq 12$.

Solution

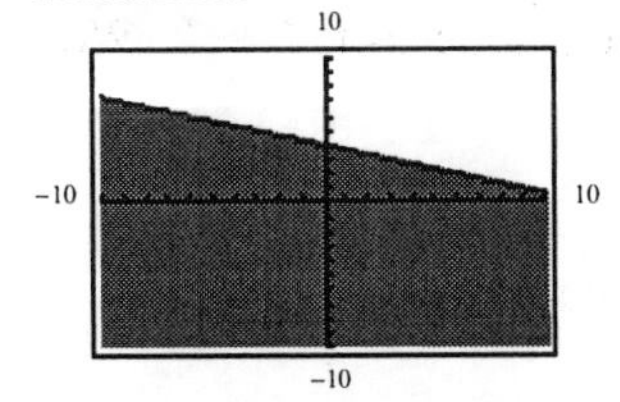

17. Sketch the inequality $3x + 2y \leq 15$.

Solution

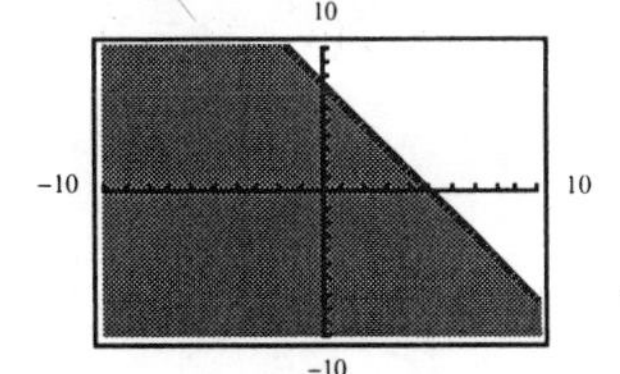

18. Sketch the solution of the system of inequalities given by the inequalities in Exercises 14–17.

Solution

$$\begin{aligned} x \quad\;\; &\geq 0 \\ y &\geq 0 \\ x + 3y &\leq 12 \\ 3x + 2y &\leq 15 \end{aligned}$$

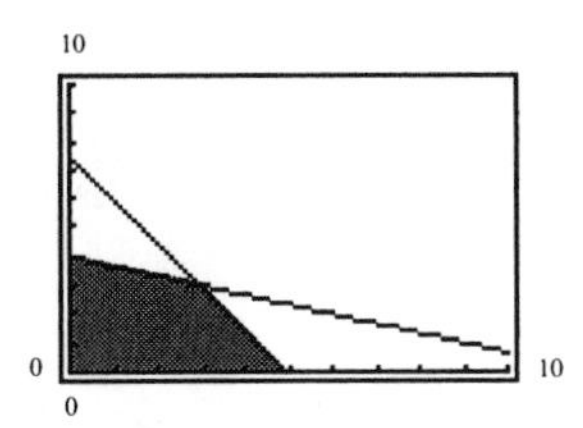

19. Find the minimum and maximum values of the objective function $z = 5x + 11y$, subject to the constraints given in Exercises 14–17.

Solution

At $(0, 0)$: $z = 5(0) + 11(0) = 0$
At $(0, 4)$: $z = 5(0) + 11(4) = 44$
At $(3, 3)$: $z = 5(3) + 11(3) = 48$
At $(5, 0)$: $z = 5(5) + 11(0) = 25$
Minimum of 0 at $(0, 0)$
Maximum of 48 at $(3, 3)$

20. A manufacturer produces two models of stair climbers. The time (in hours) required for assembling, painting, and packaging each model is shown in the textbook. The total time available for assembling, painting, and packaging is 5,600 hours, 2000 hours, and 900 hours, respectively. The profit per units is $100 (model A) and $150 (model B). How many of each model should be produced?

Solution

Objective function: $z = 100x + 150y$

At $(0, 0)$: $z = 0$

At $(0, 700)$: $z = 105{,}000$

At $(692, 0)$: $z = 69{,}200$

At $(297, 570)$: $z = 115{,}200$

Profit is maximized when $x = 297$ units of Model A and $y = 570$ units of Model B.

Constraints: $3.5x + 8y \le 5600$

$2.5x + 2y \le 2000$

$1.3x + 0.9y \le 900$

$x \ge 0$

$y \ge 0$

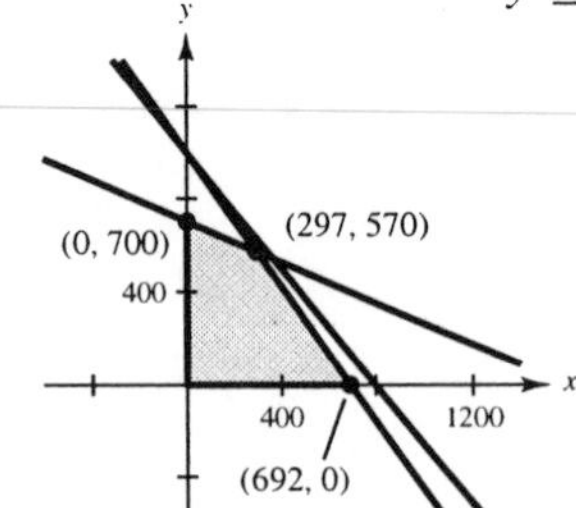

Practice Test for Chapter 6

For Exercises 1–3, solve the given system by the method of substitution.

1. $x + y = 1$

$3x - y = 15$

2. $x - 3y = -3$

$x^2 + 6y = 5$

3. $x + y + z = 6$

$2x - y + 3z = 0$

$5x + 2y - z = -3$

4. Find two numbers whose sum is 110 and product is 2,800.

5. Find the dimensions of a rectangle if its perimeter is 170 feet and its area is 2,800 square feet.

For Exercises 6–8, solve the linear system by elimination.

6. $2x + 15y = 4$

$x - 3y = 23$

7. $x + y = 2$

$38x - 19y = 7$

8. $0.4x + 0.5y = 0.112$

$0.3x - 0.7y = 0.131$

9. Herbert invests $17,000 in two funds that pay 11% and 13% simple interest, respectively. If he receives $2,080 in yearly interest, how much is invested in each fund?

10. Find the least squares regression line for the points (4, 3), (1, 1), (−1, −2), and (−2, −1).

For Exercises 11–13, solve the system of equations.

11.
$$\begin{aligned} x + &= -2 \\ 2x - y + z &= 11 \\ 4y - 3z &= -20 \end{aligned}$$

12.
$$\begin{aligned} 4x - y + 5z &= 4 \\ 2x + y - z &= 0 \\ 2x + 4y + 8z &= 0 \end{aligned}$$

13.
$$\begin{aligned} 3x + 2y - z &= 5 \\ 6x - y + 5z &= 2 \end{aligned}$$

14. Find the equation of the parabola $y = ax^2 + bx + c$ passing through the points (0, −1), (1, 4), and (2, 13).

15. Find the position equation $s = \frac{1}{2}at^2 + v_0t + s_0$ given that $s = 12$ feet after 1 second, $s = 5$ feet after 2 seconds, and $s = 4$ feet after 3 seconds.

16. Graph $x^2 + y^2 \geq 9$.

17. Graph the solution of the system.

$$\begin{aligned} x + y &\leq 6 \\ x &\geq 2 \\ y &\geq 0 \end{aligned}$$

18. Derive a set of inequalities to describe the triangle with vertices (0, 0), (0, 7), and (2, 3).

19. Find the maximum value of the objective function $C = 30x + 26y$ subject to the following constraints.

$$\begin{aligned} x &\geq 0 \\ y &\geq 0 \\ 2x + 3y &\leq 21 \\ 5x + 3y &\leq 30 \end{aligned}$$

20. Graph the system of ineqalities.

$$\begin{aligned} x^2 + y^2 &\leq 4 \\ (x - 2)^2 + y^2 &\geq 4 \end{aligned}$$

CHAPTER SEVEN
Matrices and Determinants

7.1 Matrices and Systems of Linear Equations

- You should be able to use elementary row operations to produce a row-echelon form of an augmented matrix.
  (a) Interchange two rows.
  (b) Multiply a row by a nonzero constant.
  (c) Add a multiple of a row to another row.
- You should be able to transform a matrix into reduced row-echelon form. This is called Gauss-Jordan elimination.

SOLUTIONS TO SELECTED EXERCISES

5. Determine the order of the matrix.

$$\begin{bmatrix} 33 & 45 \\ -9 & 20 \\ 12 & 15 \\ 16 & -2 \end{bmatrix}$$

Solution

Since the matrix has 4 rows and 2 columns, its order is 4×2.

9. Determine whether the following matrix is in row-echelon form.

$$\begin{bmatrix} 2 & 0 & 4 & 0 \\ 0 & -1 & 3 & 6 \\ 0 & 0 & 1 & 5 \end{bmatrix}$$

Solution

This matrix is not in row-echelon form since the first nonzero entry in both row 1 and row 2 is not 1.

13. Write the matrix in row-echelon form. Remember that the row-echelon form for a given matrix is not unique.

$$\begin{bmatrix} 1 & 1 & 0 & 5 \\ -2 & -1 & 2 & -10 \\ 3 & 6 & 7 & 14 \end{bmatrix}$$

Solution

$$\begin{bmatrix} 1 & 1 & 0 & 5 \\ -2 & -1 & 2 & -10 \\ 3 & 6 & 7 & 14 \end{bmatrix}$$

$$\begin{matrix} \\ 2R_1 + R_2 \rightarrow \\ -3R_1 + R_3 \rightarrow \end{matrix} \begin{bmatrix} 1 & 1 & 0 & 5 \\ 0 & 1 & 2 & 0 \\ 0 & 3 & 7 & -1 \end{bmatrix}$$

$$\begin{matrix} \\ \\ -3R_2 + R_3 \rightarrow \end{matrix} \begin{bmatrix} 1 & 1 & 0 & 5 \\ 0 & 1 & 2 & 0 \\ 0 & 0 & 1 & -1 \end{bmatrix}$$

17. Write the matrix in *reduced* row-echelon form.

$$\begin{bmatrix} 3 & 3 & 3 \\ -1 & 0 & -4 \\ 2 & 4 & -2 \end{bmatrix}$$

Solution

$$\begin{bmatrix} 3 & 3 & 3 \\ -1 & 0 & -4 \\ 2 & 4 & -2 \end{bmatrix}$$

$$\begin{matrix} \frac{1}{3}R_1 \rightarrow \\ \\ \\ \end{matrix} \begin{bmatrix} 1 & 1 & 1 \\ -1 & 0 & -4 \\ 2 & 4 & -2 \end{bmatrix}$$

$$\begin{matrix} \\ R_1 + R_2 \rightarrow \\ -2R_1 + R_3 \rightarrow \end{matrix} \begin{bmatrix} 1 & 1 & 1 \\ 0 & 1 & -3 \\ 0 & 2 & -4 \end{bmatrix}$$

$$\begin{matrix} -R_2 + R_1 \rightarrow \\ \\ -2R_2 + R_3 \rightarrow \end{matrix} \begin{bmatrix} 1 & 0 & 4 \\ 0 & 1 & -3 \\ 0 & 0 & 2 \end{bmatrix}$$

$$\begin{matrix} \\ \\ \frac{1}{2}R_3 \rightarrow \end{matrix} \begin{bmatrix} 1 & 0 & 4 \\ 0 & 1 & -3 \\ 0 & 0 & 1 \end{bmatrix}$$

$$\begin{matrix} -4R_3 + R_1 \rightarrow \\ 3R_3 + R_2 \rightarrow \\ \\ \end{matrix} \begin{bmatrix} 1 & 0 & 0 \\ 0 & 1 & 0 \\ 0 & 0 & 1 \end{bmatrix}$$

21. Write the system of linear equations represented by the following augmented matrix. (Use variables x, y, and z.)

$$\left[\begin{array}{cc:c} 4 & 3 & 8 \\ 1 & -2 & 3 \end{array}\right]$$

Solution

$$4x + 3y = 8$$

$$x - 2y = 3$$

25. Write the system of linear equations represented by the following reduced augmented matrix. Then use back-substitution to find the solution. (Use variables x, y, and z.)

$$\begin{bmatrix} 1 & -2 & \vdots & 4 \\ 0 & 1 & \vdots & -3 \end{bmatrix}$$

Solution

$$x - 2y = 4$$

$$y = -3$$

$$x - 2(-3) = 4$$

$$x = -2$$

Answer: $(-2, \ -3)$

29. An augmented matrix that represents a system of linear equations (in variables x, y, and z) has been reduced using Gauss-Jordan elimination. Write the solution represented by the following augmented matrix.

$$\begin{bmatrix} 1 & 0 & \vdots & 7 \\ 0 & 1 & \vdots & -5 \end{bmatrix}$$

Solution

$$\begin{bmatrix} 1 & 0 & \vdots & 7 \\ 0 & 1 & \vdots & -5 \end{bmatrix}$$

$x = 7$

$y = -5$

Answer: $(7, \ -5)$

33. Write the augmented matrix for the following system of linear equations.

$$4x - 5y = -2$$

$$-x + 8y = 10$$

Solution

$$\begin{bmatrix} 4 & -5 & \vdots & -2 \\ -1 & 8 & \vdots & 10 \end{bmatrix}$$

37. Solve the system of equations. Use Gaussian elimination with back-substitution or Gauss-Jordan elimination.

$$x + 2y = 7$$
$$2x + y = 8$$

Solution

$$\begin{bmatrix} 1 & 2 & \vdots & 7 \\ 2 & 1 & \vdots & 8 \end{bmatrix}$$

$$-2R_1 + R_2 \rightarrow \begin{bmatrix} 1 & 2 & \vdots & 7 \\ 0 & -3 & \vdots & -6 \end{bmatrix}$$

$$-\tfrac{1}{3}R_2 \rightarrow \begin{bmatrix} 1 & 2 & \vdots & 7 \\ 0 & 1 & \vdots & 2 \end{bmatrix}$$

$y = 2$

$x + 2(2) = 7 \quad \Rightarrow \quad x = 3$

Answer: $(3, 2)$

41. Solve the system of equations. Use Gaussian elimination with back-substitution or Gauss-Jordan elimination.

$$8x - 4y = 7$$
$$5x + 2y = 1$$

Solution

$$\begin{bmatrix} 8 & -4 & \vdots & 7 \\ 5 & 2 & \vdots & 1 \end{bmatrix}$$

$$\begin{matrix} 3R_1 \rightarrow \\ 5R_2 \rightarrow \end{matrix} \begin{bmatrix} 24 & -12 & \vdots & 21 \\ 25 & 10 & \vdots & 5 \end{bmatrix}$$

$$-R_2 + R_1 \rightarrow \begin{bmatrix} -1 & -22 & \vdots & 16 \\ 25 & 10 & \vdots & 5 \end{bmatrix}$$

$$25R_1 + R_2 \rightarrow \begin{bmatrix} -1 & -22 & \vdots & 16 \\ 0 & -540 & \vdots & 405 \end{bmatrix}$$

$$\begin{matrix} -R_1 \rightarrow \\ -\tfrac{1}{540}R_2 \rightarrow \end{matrix} \begin{bmatrix} 1 & 22 & \vdots & -16 \\ 0 & 1 & \vdots & -\tfrac{3}{4} \end{bmatrix}$$

$y = -\tfrac{3}{4}$

$x + 22\left(-\tfrac{3}{4}\right) = -16 \quad \Rightarrow \quad x = \tfrac{1}{2}$

Answer: $\left(\tfrac{1}{2}, -\tfrac{3}{4}\right)$

45. Solve the system of equations. Use Gaussian elimination with back-substitution or Gauss-Jordan elimination.

$$\begin{aligned} 2x \qquad + 3z &= 3 \\ 4x - 3y + 7z &= 5 \\ 8x - 9y + 15z &= 9 \end{aligned}$$

Solution

$$\left[\begin{array}{ccc:c} 2 & 0 & 3 & 3 \\ 4 & -3 & 7 & 5 \\ 8 & -9 & 15 & 9 \end{array}\right]$$

$$\begin{array}{r} \\ -2R_1 + R_2 \rightarrow \\ -4R_1 + R_3 \rightarrow \end{array} \left[\begin{array}{ccc:c} 2 & 0 & 3 & 3 \\ 0 & -3 & 1 & -1 \\ 0 & -9 & 3 & -3 \end{array}\right]$$

$$\begin{array}{r} \frac{1}{2}R_1 \rightarrow \\ -\frac{1}{3}R_2 \rightarrow \\ 9R_2 + R_3 \rightarrow \end{array} \left[\begin{array}{ccc:c} 1 & 0 & \frac{3}{2} & \frac{3}{2} \\ 0 & 1 & -\frac{1}{3} & \frac{1}{3} \\ 0 & 0 & 0 & 0 \end{array}\right]$$

Infinite solutions

Let $z = a$, then $y - \frac{1}{3}a = \frac{1}{3} \Rightarrow y = \frac{1}{3}a + \frac{1}{3}$ and $x + \frac{3}{2}a = \frac{3}{2} \Rightarrow x = -\frac{3}{2}a + \frac{3}{2}$

Answer: $\left(-\frac{3}{2}a + \frac{3}{2}, \frac{1}{3}a + \frac{1}{3}, a\right)$

49. Solve the system of equations. Use Gaussian elimination with back-substitution or Gauss-Jordan elimination.

$$\begin{aligned} x + 2y + z &= 8 \\ 3x + 7y + 6z &= 26 \end{aligned}$$

Solution

$$\left[\begin{array}{ccc:c} 1 & 2 & 1 & 8 \\ 3 & 7 & 6 & 26 \end{array}\right]$$

$$\begin{array}{r} \\ -3R_1 + R_2 \rightarrow \end{array} \left[\begin{array}{ccc:c} 1 & 2 & 1 & 8 \\ 0 & 1 & 3 & 2 \end{array}\right]$$

$$\begin{array}{r} -2R_2 + R_1 \rightarrow \\ \end{array} \left[\begin{array}{ccc:c} 1 & 0 & -5 & 4 \\ 0 & 1 & 3 & 2 \end{array}\right]$$

$z = a$

$y = -3a + 2$

$x = 5a + 4$

Answer: $(5a + 4, -3a + 2, a)$

55. Solve the system of equations. Use Gaussian elimination with back-substitution or Gauss-Jordan elimination.

$$\begin{aligned} x + 2y &= 0 \\ -x - y &= 0 \end{aligned}$$

Solution

$$\left[\begin{array}{cc:c} 1 & 2 & 0 \\ -1 & -1 & 0 \end{array}\right]$$

$$R_1 + R_2 \rightarrow \left[\begin{array}{cc:c} 1 & 2 & 0 \\ 0 & 1 & 0 \end{array}\right]$$

$$-2R_2 + R_1 \rightarrow \left[\begin{array}{cc:c} 1 & 0 & 0 \\ 0 & 1 & 0 \end{array}\right]$$

$x = 0$

$y = 0$

Answer: (0, 0)

59. *Borrowing Money* A small corporation borrowed \$500,000 to expand its product line. Some of the money was borrowed at 9%, some at 10%, and some at 12%. How much was borrowed at each rate if the annual interest was \$52,000 and the amount borrowed at 10% was two and one-half times the amount borrowed at 9%.

Solution

x = amount at 9%

y = amount at 10%

z = amount at 12%

$$\begin{aligned} x + y + z &= 500{,}000 \\ 0.09x + 0.10y + 0.12z &= 52{,}000 \\ 2.5x - y &= 0 \end{aligned}$$

$$\left[\begin{array}{ccc:c} 1 & 1 & 1 & 500{,}000 \\ 0.09 & 0.10 & 0.12 & 52{,}000 \\ 2.5 & -1 & 0 & 0 \end{array}\right]$$

$$\begin{array}{r} \\ -0.09R_1 + R_2 \rightarrow \\ -2.5R_1 + R_3 \rightarrow \end{array} \left[\begin{array}{ccc:c} 1 & 1 & 1 & 500{,}000 \\ 0 & 0.01 & 0.03 & 7{,}000 \\ 0 & -3.5 & -2.5 & -1{,}250{,}000 \end{array}\right]$$

$$\begin{array}{r} \\ 100R_2 \rightarrow \\ 3.5R_2 + R_3 \rightarrow \end{array} \left[\begin{array}{ccc:c} 1 & 1 & 1 & 500{,}000 \\ 0 & 1 & 3 & 700{,}000 \\ 0 & 0 & 8 & 1{,}200{,}000 \end{array}\right]$$

$8z = 1{,}200{,}000 \Rightarrow z = 150{,}000$

$y + 3(150{,}000) = 700{,}000 \Rightarrow y = 250{,}000$

$x + 250{,}000 + 150{,}000 = 500{,}000 \Rightarrow x = 100{,}000$

Answer: \$100,000 at 9%

\$250,000 at 10%

\$150,000 at 12%

63. *Curve Fitting* Find the cubic $y = ax^3 + bx^2 + cx + d$ that passes through the points shown on the graph in the textbook.

Solution

$f(x) = ax^3 + bx^2 + cx + d$

At $(-2, 2): f(-2) = -8a + 4b - 2c + d = 2$

At $(-1, 17): f(-1) = -a + b - c + d = 17$

At $(0, 20): f(0) = d = 20$

At $(1, 23): f(1) = a + b + c + d = 23$

$$\begin{bmatrix} 1 & 1 & 1 & 1 & \vdots & 23 \\ -1 & 1 & -1 & 1 & \vdots & 17 \\ -8 & 4 & -2 & 1 & \vdots & 2 \\ 0 & 0 & 0 & 1 & \vdots & 20 \end{bmatrix}$$

$$\begin{matrix} \\ R_1 + R_2 \rightarrow \\ 8R_1 + R_3 \rightarrow \\ \\ \end{matrix} \begin{bmatrix} 1 & 1 & 1 & 1 & \vdots & 23 \\ 0 & 2 & 0 & 2 & \vdots & 40 \\ 0 & 12 & 6 & 9 & \vdots & 186 \\ 0 & 0 & 0 & 1 & \vdots & 20 \end{bmatrix}$$

$$\begin{matrix} \\ \frac{1}{2}R_2 \rightarrow \\ -12R_2 + R_3 \rightarrow \\ \\ \end{matrix} \begin{bmatrix} 1 & 1 & 1 & 1 & \vdots & 23 \\ 0 & 1 & 0 & 1 & \vdots & 20 \\ 0 & 0 & 6 & -3 & \vdots & -54 \\ 0 & 0 & 0 & 1 & \vdots & 20 \end{bmatrix}$$

$d = 20$

$6c - 3(20) = -54 \Rightarrow c = 1$

$b + 20 = 20 \Rightarrow b = 0$

$a + 0 + 1 + 20 = 23 \Rightarrow a = 2$

Answer: $f(x) = 2x^3 + x + 20$

7.2 Operations with Matrices

- $A = B$ if and only if they have the same order and $a_{ij} = b_{ij}$.
- You should be able to perform the operations of matrix addition, scalar multiplication, and matrix multiplication.
- Some properties of matrix addition, scalar multiplication, and matrix multiplication are:

  (a) $A + B = B + A$
  (b) $A + (B + C) = (A + B) + C$
  (c) $(cd)A = c(dA)$
  (d) $1A = A$
  (e) $c(A + B) = cA + cB$
  (f) $(c + d)A = cA + dA$
  (g) $A(BC) = (AB)C$
  (h) $A(B + C) = AB + AC$
  (i) $(A + B)C = AC + BC$
  (j) $c(AB) = (cA)B = A(cB)$
- You should remember that $AB \neq BA$ in general.

SOLUTIONS TO SELECTED EXERCISES

5. Find (a) $A + B$, (b) $A - B$, (c) $3A$, and (d) $3A - 2B$ for the following.

$$A = \begin{bmatrix} 1 & -1 \\ 2 & -1 \end{bmatrix}, \quad B = \begin{bmatrix} 2 & -1 \\ -1 & 8 \end{bmatrix}$$

Solution

(a) $A + B = \begin{bmatrix} 1 & -1 \\ 2 & -1 \end{bmatrix} + \begin{bmatrix} 2 & -1 \\ -1 & 8 \end{bmatrix} = \begin{bmatrix} 1+2 & -1-1 \\ 2-1 & -1+8 \end{bmatrix} = \begin{bmatrix} 3 & -2 \\ 1 & 7 \end{bmatrix}$

(b) $A - B = \begin{bmatrix} 1 & -1 \\ 2 & -1 \end{bmatrix} - \begin{bmatrix} 2 & -1 \\ -1 & 8 \end{bmatrix} = \begin{bmatrix} 1-2 & -1+1 \\ 2+1 & -1-8 \end{bmatrix} = \begin{bmatrix} -1 & 0 \\ 3 & -9 \end{bmatrix}$

(c) $3A = 3\begin{bmatrix} 1 & -1 \\ 2 & -1 \end{bmatrix} = \begin{bmatrix} 3(1) & 3(-1) \\ 3(2) & 3(-1) \end{bmatrix} = \begin{bmatrix} 3 & -3 \\ 6 & -3 \end{bmatrix}$

(d) $3A - 2B = \begin{bmatrix} 3 & -3 \\ 6 & -3 \end{bmatrix} - 2\begin{bmatrix} 2 & -1 \\ -1 & 8 \end{bmatrix} = \begin{bmatrix} 3 & -3 \\ 6 & -3 \end{bmatrix} + \begin{bmatrix} -4 & 2 \\ 2 & -16 \end{bmatrix} = \begin{bmatrix} -1 & -1 \\ 8 & -19 \end{bmatrix}$

11. Find (a) AB, (b) BA, and (c) A^2 (if possible). (*Note:* $A^2 = AA$.)

$$A = \begin{bmatrix} 1 & 2 \\ 4 & 2 \end{bmatrix}, \quad B = \begin{bmatrix} 2 & -1 \\ -1 & 8 \end{bmatrix}$$

Solution

(a) $$AB = \begin{bmatrix} 1 & 2 \\ 4 & 2 \end{bmatrix}\begin{bmatrix} 2 & -1 \\ -1 & 8 \end{bmatrix} = \begin{bmatrix} 2-2 & -1+16 \\ 8-2 & -4+16 \end{bmatrix} = \begin{bmatrix} 0 & 15 \\ 6 & 12 \end{bmatrix}$$

(b) $$BA = \begin{bmatrix} 2 & -1 \\ -1 & 8 \end{bmatrix}\begin{bmatrix} 1 & 2 \\ 4 & 2 \end{bmatrix} = \begin{bmatrix} 2-4 & 4-2 \\ -1+32 & -2+16 \end{bmatrix} = \begin{bmatrix} -2 & 2 \\ 31 & 14 \end{bmatrix}$$

(c) $$A^2 = \begin{bmatrix} 1 & 2 \\ 4 & 2 \end{bmatrix}\begin{bmatrix} 1 & 2 \\ 4 & 2 \end{bmatrix} = \begin{bmatrix} 1+8 & 2+4 \\ 4+8 & 8+4 \end{bmatrix} = \begin{bmatrix} 9 & 6 \\ 12 & 12 \end{bmatrix}$$

15. Find (a) AB, (b) BA, and (c) A^2 (if possible). (*Note:* $A^2 = AA$.)

$$A = \begin{bmatrix} 1 & -1 & 7 \\ 2 & -1 & 8 \\ 3 & 1 & -1 \end{bmatrix}, \quad B = \begin{bmatrix} 1 & 1 & 2 \\ 2 & 1 & 1 \\ 1 & -3 & 2 \end{bmatrix}$$

Solution

(a) $$AB = \begin{bmatrix} 1 & -1 & 7 \\ 2 & -1 & 8 \\ 3 & 1 & -1 \end{bmatrix}\begin{bmatrix} 1 & 1 & 2 \\ 2 & 1 & 1 \\ 1 & -3 & 2 \end{bmatrix}$$

$$= \begin{bmatrix} 1-2+7 & 1-1-21 & 2-1+14 \\ 2-2+8 & 2-1-24 & 4-1+16 \\ 3+2-1 & 3+1+3 & 6+1-2 \end{bmatrix} = \begin{bmatrix} 6 & -21 & 15 \\ 8 & -23 & 19 \\ 4 & 7 & 5 \end{bmatrix}$$

(b) $$BA = \begin{bmatrix} 1 & 1 & 2 \\ 2 & 1 & 1 \\ 1 & -3 & 2 \end{bmatrix}\begin{bmatrix} 1 & -1 & 7 \\ 2 & -1 & 8 \\ 3 & 1 & -1 \end{bmatrix}$$

$$= \begin{bmatrix} 1+2+6 & -1-1+2 & 7+8-2 \\ 2+2+3 & -2-1+1 & 14+8-1 \\ 1-6+6 & -1+3+2 & 7-24-2 \end{bmatrix} = \begin{bmatrix} 9 & 0 & 13 \\ 7 & -2 & 21 \\ 1 & 4 & -19 \end{bmatrix}$$

(c) $$A^2 = \begin{bmatrix} 1 & -1 & 7 \\ 2 & -1 & 8 \\ 3 & 1 & -1 \end{bmatrix}\begin{bmatrix} 1 & -1 & 7 \\ 2 & -1 & 8 \\ 3 & 1 & -1 \end{bmatrix}$$

$$= \begin{bmatrix} 1-2+21 & -1+1+7 & 7-8-7 \\ 2-2+24 & -2+1+8 & 14-8-8 \\ 3+2-3 & -3-1-1 & 21+8+1 \end{bmatrix} = \begin{bmatrix} 20 & 7 & -8 \\ 24 & 7 & -2 \\ 2 & -5 & 30 \end{bmatrix}$$

19. Find AB (if possible).

$$A = \begin{bmatrix} -1 & 3 \\ 4 & -5 \\ 0 & 2 \end{bmatrix}, \quad B = \begin{bmatrix} 1 & 2 \\ 0 & 7 \end{bmatrix}$$

Solution

A is 3×2, B is 2×2 $\Rightarrow$ AB is 3×2.

$$AB = \begin{bmatrix} -1 & 3 \\ 4 & -5 \\ 0 & 2 \end{bmatrix}\begin{bmatrix} 1 & 2 \\ 0 & 7 \end{bmatrix} = \begin{bmatrix} -1 & 19 \\ 4 & -27 \\ 0 & 14 \end{bmatrix}$$

23. Find AB (if possible).

$$A = \begin{bmatrix} 6 \\ -2 \\ 1 \\ 6 \end{bmatrix}, \quad B = [10 \quad 12]$$

Solution

A is 4×1, B is 1×2 $\Rightarrow$ AB is 4×2.

$$AB = \begin{bmatrix} 6 \\ -2 \\ 1 \\ 6 \end{bmatrix} [10 \quad 12] = \begin{bmatrix} 60 & 72 \\ -20 & -24 \\ 10 & 12 \\ 60 & 72 \end{bmatrix}$$

27. Given $2X + 3A = B$, solve for X given $A = \begin{bmatrix} -2 & -1 \\ 1 & 0 \\ 3 & -4 \end{bmatrix}$ and $B = \begin{bmatrix} 0 & 3 \\ 2 & 0 \\ -4 & -1 \end{bmatrix}$.

Solution

$$X = -\tfrac{3}{2}A + \tfrac{1}{2}B = -\tfrac{3}{2}\begin{bmatrix} -2 & -1 \\ 1 & 0 \\ 3 & -4 \end{bmatrix} + \tfrac{1}{2}\begin{bmatrix} 0 & 3 \\ 2 & 0 \\ -4 & -1 \end{bmatrix} = \begin{bmatrix} 3 & 3 \\ -\frac{1}{2} & 0 \\ -\frac{13}{2} & \frac{11}{2} \end{bmatrix}$$

33. Find matrices A, X, and B such that the system of linear equations can be written as the matrix equation $AX = B$. Solve the system of equations.

$$\begin{aligned} x - 2y + 3z &= 9 \\ -x + 3y - z &= -6 \\ 2x - 5y + 5z &= 17 \end{aligned}$$

Solution

$$A = \begin{bmatrix} 1 & -2 & 3 \\ -1 & 3 & -1 \\ 2 & -5 & 5 \end{bmatrix}, \quad B = \begin{bmatrix} 9 \\ -6 \\ 17 \end{bmatrix}, \quad X = \begin{bmatrix} x \\ y \\ z \end{bmatrix}$$

Solving by elimination, we have $x = 1$, $y = -1$, $z = 2$.

37. *Factory Production* A certain corporation has four factories, each of which manufactures two products. The number of units of product i produced at factory j in one day is represented by a_{ij} in the matrix

$$A = \begin{bmatrix} 100 & 90 & 70 & 30 \\ 40 & 20 & 60 & 60 \end{bmatrix}.$$

Find the production levels if production is increased by 10%. (*Hint:* Since an increase of 10% corresponds to 100% + 10%, multiply the given matrix by 1.10.)

Solution

$$1.10\begin{bmatrix} 100 & 90 & 70 & 30 \\ 40 & 20 & 60 & 60 \end{bmatrix} = \begin{bmatrix} 110 & 99 & 77 & 33 \\ 44 & 22 & 66 & 66 \end{bmatrix}$$

41. *Total Revenue* A manufacturer produces three different models of a given product, which are shipped to two different warehouses. The number of units of model i that are shipped to warehouse j is represented by a_{ij} in the matrix.

$$A = \begin{bmatrix} 5000 & 4000 \\ 6000 & 10{,}000 \\ 8000 & 5000 \end{bmatrix}$$

The price per unit is represented by the matrix

$$B = [\,\$20.50 \quad \$26.50 \quad \$29.50\,]$$

Find the product BA and state what each entry of the product represents.

Solution

$$BA = [\,\$20.50 \quad \$26.50 \quad \$29.50\,] \begin{bmatrix} 5000 & 4000 \\ 6000 & 10{,}000 \\ 8000 & 5000 \end{bmatrix}$$

$$= [\,\$497{,}500 \quad \$494{,}500\,]$$

Each entry represents the value of the inventory at each warehouse.

45. *Voting Preference* The matrix

$$P = \overset{\text{From}}{\overbrace{\begin{matrix} R & D & I \end{matrix}}} \\ \begin{bmatrix} 0.6 & 0.1 & 0.1 \\ 0.2 & 0.7 & 0.1 \\ 0.2 & 0.2 & 0.8 \end{bmatrix} \begin{matrix} R \\ D \\ I \end{matrix} \Bigg\} \text{To}$$

is called a stochastic matrix. Each entry p_{ij} $(i \neq j)$ represents the proportion of the voting population that changes from party i to party j, and p_{ii} represents the proportion that remains loyal to the party from one election to the next. Use a graphing utility to find P^2. (This matrix gives the transition probabilities from the first election to the third.)

Solution

$$P^2 = PP = \begin{bmatrix} 0.40 & 0.15 & 0.15 \\ 0.28 & 0.53 & 0.17 \\ 0.32 & 0.32 & 0.68 \end{bmatrix}$$

7.3 The Inverse of a Square Matrix

- You should be able to find the inverse, if it exists, of a square matrix.
  (a) Write the $n \times 2n$ matrix that consists of the given matrix A on the left and the $n \times n$ identity matrix I on the right to obtain $[A \;\vdots\; I]$. Note that we separate the matrices A and I by a dotted line. We call this process **adjoining** the matrices A and I.
  (b) If possible, row reduce A to I using elementary row operations on the *entire* matrix $[A \;\vdots\; I]$. The result will be the matrix $[I \;\vdots\; A^{-1}]$. If this is not possible, then A is not invertible.
  (c) Check your work by multiplying to see that $AA^{-1} = I = A^{-1}A$.
- You should be able to use inverse matrices to solve systems of equations.

SOLUTIONS TO SELECTED EXERCISES

3. Show that B is the inverse of A.

$$A = \begin{bmatrix} 1 & 2 \\ 3 & 4 \end{bmatrix}, \quad B = \begin{bmatrix} -2 & 1 \\ \frac{3}{2} & -\frac{1}{2} \end{bmatrix}$$

Solution

$$AB = \begin{bmatrix} 1 & 2 \\ 3 & 4 \end{bmatrix}\begin{bmatrix} -2 & 1 \\ \frac{3}{2} & -\frac{1}{2} \end{bmatrix} = \begin{bmatrix} -2+3 & 1-1 \\ -6+6 & 3-2 \end{bmatrix} = \begin{bmatrix} 1 & 0 \\ 0 & 1 \end{bmatrix}$$

$$BA = \begin{bmatrix} -2 & 1 \\ \frac{3}{2} & -\frac{1}{2} \end{bmatrix}\begin{bmatrix} 1 & 2 \\ 3 & 4 \end{bmatrix} = \begin{bmatrix} -2+3 & -4+4 \\ \frac{3}{2}-\frac{3}{2} & 3-2 \end{bmatrix} = \begin{bmatrix} 1 & 0 \\ 0 & 1 \end{bmatrix}$$

7. Show that B is the inverse of A.

$$A = \begin{bmatrix} -1 & 0 & 2 \\ 1 & -2 & 0 \\ 1 & 0 & 3 \end{bmatrix}, \quad B = \frac{1}{10}\begin{bmatrix} -6 & 0 & 4 \\ -3 & -5 & 2 \\ 2 & 0 & 2 \end{bmatrix}$$

Solution

$$AB = \frac{1}{10}\begin{bmatrix} -1 & 0 & 2 \\ 1 & -2 & 0 \\ 1 & 0 & 3 \end{bmatrix}\begin{bmatrix} -6 & 0 & 4 \\ -3 & -5 & 2 \\ 2 & 0 & 2 \end{bmatrix} = \frac{1}{10}\begin{bmatrix} 10 & 0 & 0 \\ 0 & 10 & 0 \\ 0 & 0 & 10 \end{bmatrix} = \begin{bmatrix} 1 & 0 & 0 \\ 0 & 1 & 0 \\ 0 & 0 & 1 \end{bmatrix}$$

$$BA = \frac{1}{10}\begin{bmatrix} -6 & 0 & 4 \\ -3 & -5 & 2 \\ 2 & 0 & 2 \end{bmatrix}\begin{bmatrix} -1 & 0 & 2 \\ 1 & -2 & 0 \\ 1 & 0 & 3 \end{bmatrix} = \frac{1}{10}\begin{bmatrix} 10 & 0 & 0 \\ 0 & 10 & 0 \\ 0 & 0 & 10 \end{bmatrix} = \begin{bmatrix} 1 & 0 & 0 \\ 0 & 1 & 0 \\ 0 & 0 & 1 \end{bmatrix}$$

11. Find the inverse of the matrix (if it exists).

$$\begin{bmatrix} 2 & 0 \\ 0 & 3 \end{bmatrix}$$

Solution

$$[A \; \vdots \; I] = \left[\begin{array}{cc:cc} 2 & 0 & 1 & 0 \\ 0 & 3 & 0 & 1 \end{array}\right]$$

$$\begin{array}{r} \frac{1}{2}R_1 \rightarrow \\ \frac{1}{3}R_2 \rightarrow \end{array} \left[\begin{array}{cc:cc} 1 & 0 & \frac{1}{2} & 0 \\ 0 & 1 & 0 & \frac{1}{3} \end{array}\right]$$

$$= [I \; \vdots \; A^{-1}]$$

$$A^{-1} = \begin{bmatrix} \frac{1}{2} & 0 \\ 0 & \frac{1}{3} \end{bmatrix} = \frac{1}{6}\begin{bmatrix} 3 & 0 \\ 0 & 2 \end{bmatrix}$$

15. Find the inverse of the matrix (if it exists).

$$\begin{bmatrix} -1 & 1 \\ -2 & 1 \end{bmatrix}$$

Solution

$$[A \; \vdots \; I] = \left[\begin{array}{cc:cc} -1 & 1 & 1 & 0 \\ -2 & 1 & 0 & 1 \end{array}\right]$$

$$-2R_1 + R_2 \rightarrow \left[\begin{array}{cc:cc} -1 & 1 & 1 & 0 \\ 0 & -1 & -2 & 1 \end{array}\right]$$

$$R_2 + R_1 \rightarrow \left[\begin{array}{cc:cc} -1 & 0 & -1 & 1 \\ 0 & -1 & -2 & 1 \end{array}\right]$$

$$\begin{array}{r} -R_1 \rightarrow \\ -R_2 \rightarrow \end{array} \left[\begin{array}{cc:cc} 1 & 0 & 1 & -1 \\ 0 & 1 & 2 & -1 \end{array}\right]$$

$$= [I \; \vdots \; A^{-1}]$$

$$A^{-1} = \begin{bmatrix} 1 & -1 \\ 2 & -1 \end{bmatrix}$$

19. Find the inverse of the matrix (if it exists).

$$\begin{bmatrix} 2 & 7 & 1 \\ -3 & -9 & 2 \end{bmatrix}$$

Solution

$$A = \begin{bmatrix} 2 & 7 & 1 \\ -3 & -9 & 2 \end{bmatrix}$$

A has no inverse because it is not square.

23. Find the inverse of the matrix (if it exists).

$$\begin{bmatrix} 1 & 2 & -1 \\ 3 & 7 & -10 \\ -5 & -7 & -15 \end{bmatrix}$$

Solution

$$[A \;\vdots\; I] = \left[\begin{array}{rrr|rrr} 1 & 2 & -1 & 1 & 0 & 0 \\ 3 & 7 & -10 & 0 & 1 & 0 \\ -5 & -7 & -15 & 0 & 0 & 1 \end{array}\right]$$

$$\begin{array}{r} \\ -3R_1 + R_2 \rightarrow \\ 5R_1 + R_3 \rightarrow \end{array} \left[\begin{array}{rrr|rrr} 1 & 2 & -1 & 1 & 0 & 0 \\ 0 & 1 & -7 & -3 & 1 & 0 \\ 0 & 3 & -20 & 5 & 0 & 1 \end{array}\right]$$

$$\begin{array}{r} -2R_2 + R_1 \rightarrow \\ \\ -3R_2 + R_3 \rightarrow \end{array} \left[\begin{array}{rrr|rrr} 1 & 0 & 13 & 7 & -2 & 0 \\ 0 & 1 & -7 & -3 & 1 & 0 \\ 0 & 0 & 1 & 14 & -3 & 1 \end{array}\right]$$

$$\begin{array}{r} -13R_3 + R_1 \rightarrow \\ 7R_3 + R_2 \rightarrow \\ \\ \end{array} \left[\begin{array}{rrr|rrr} 1 & 0 & 0 & -175 & 37 & -13 \\ 0 & 1 & 0 & 95 & -20 & 7 \\ 0 & 0 & 1 & 14 & -3 & 1 \end{array}\right]$$

$$= [I \;\vdots\; A^{-1}]$$

$$A^{-1} = \begin{bmatrix} -175 & 37 & -13 \\ 95 & -20 & 7 \\ 14 & -3 & 1 \end{bmatrix}$$

25. Find the inverse of the matrix (if it exists).

$$\begin{bmatrix} 1 & 1 & 2 \\ 3 & 1 & 0 \\ -2 & 0 & 3 \end{bmatrix}$$

Solution

$$[A \;\vdots\; I] = \left[\begin{array}{rrr|rrr} 1 & 1 & 2 & 1 & 0 & 0 \\ 3 & 1 & 0 & 0 & 1 & 0 \\ -2 & 0 & 3 & 0 & 0 & 1 \end{array}\right]$$

$$\begin{array}{r} \\ -3R_1 + R_2 \rightarrow \\ 2R_1 + R_3 \rightarrow \end{array} \left[\begin{array}{rrr|rrr} 1 & 1 & 2 & 1 & 0 & 0 \\ 0 & -2 & -6 & -3 & 1 & 0 \\ 0 & 2 & 7 & 2 & 0 & 1 \end{array}\right]$$

$$\begin{array}{r} \\ \\ R_2 + R_3 \rightarrow \end{array} \left[\begin{array}{rrr|rrr} 1 & 1 & 2 & 1 & 0 & 0 \\ 0 & -2 & -6 & -3 & 1 & 0 \\ 0 & 0 & 1 & -1 & 1 & 1 \end{array}\right]$$

$$\begin{array}{r} -2R_2 + R_1 \rightarrow \\ 6R_3 + R_2 \rightarrow \\ \\ \end{array} \left[\begin{array}{rrr|rrr} 1 & 1 & 0 & 3 & -2 & -2 \\ 0 & -2 & 0 & -9 & 7 & 6 \\ 0 & 0 & 1 & -1 & 1 & 1 \end{array}\right]$$

$$\begin{array}{r} -R_2 + R_1 \rightarrow \\ -\frac{1}{2}R_2 \rightarrow \\ \\ \end{array} \left[\begin{array}{rrr|rrr} 1 & 0 & 0 & -\frac{3}{2} & \frac{3}{2} & 1 \\ 0 & 1 & 0 & \frac{9}{2} & -\frac{7}{2} & -3 \\ 0 & 0 & 1 & -1 & 1 & 1 \end{array}\right]$$

$$= [I \;\vdots\; A^{-1}]$$

$$A^{-1} = \frac{1}{2}\begin{bmatrix} -3 & 3 & 2 \\ 9 & -7 & -6 \\ -2 & 2 & 2 \end{bmatrix}$$

29. Find the inverse of the matrix (if it exists).

$$\begin{bmatrix} 1 & 0 & 0 \\ 3 & 4 & 0 \\ 2 & 5 & 5 \end{bmatrix}$$

Solution

$$[A \ \vdots \ I] = \left[\begin{array}{ccc|ccc} 1 & 0 & 0 & 1 & 0 & 0 \\ 3 & 4 & 0 & 0 & 1 & 0 \\ 2 & 5 & 5 & 0 & 0 & 1 \end{array}\right]$$

$$\begin{array}{r} \\ -3R_1 + R_2 \rightarrow \\ -2R_1 + R_3 \rightarrow \end{array} \left[\begin{array}{ccc|ccc} 1 & 0 & 0 & 1 & 0 & 0 \\ 0 & 4 & 0 & -3 & 1 & 0 \\ 0 & 5 & 5 & -2 & 0 & 1 \end{array}\right]$$

$$\begin{array}{r} \\ \\ -\frac{5}{4}R_2 + R_3 \rightarrow \end{array} \left[\begin{array}{ccc|ccc} 1 & 0 & 0 & 1 & 0 & 0 \\ 0 & 4 & 0 & -3 & 1 & 0 \\ 0 & 0 & 5 & \frac{7}{4} & -\frac{5}{4} & 1 \end{array}\right]$$

$$\begin{array}{r} \\ \frac{1}{4}R_2 \rightarrow \\ \frac{1}{5}R_3 \rightarrow \end{array} \left[\begin{array}{ccc|ccc} 1 & 0 & 0 & 1 & 0 & 0 \\ 0 & 1 & 0 & -\frac{3}{4} & \frac{1}{4} & 0 \\ 0 & 0 & 1 & \frac{7}{20} & -\frac{1}{4} & \frac{1}{5} \end{array}\right]$$

$$= [I \ \vdots \ A^{-1}]$$

$$A^{-1} = \frac{1}{20}\begin{bmatrix} 20 & 0 & 0 \\ -15 & 5 & 0 \\ 7 & -5 & 4 \end{bmatrix} = \begin{bmatrix} 1 & 0 & 0 \\ -0.75 & 0.25 & 0 \\ 0.35 & -0.25 & 0.2 \end{bmatrix}$$

33. Find the inverse of the matrix (if it exists).

$$\begin{bmatrix} -8 & 0 & 0 & 0 \\ 0 & 1 & 0 & 0 \\ 0 & 0 & 4 & 0 \\ 0 & 0 & 0 & -5 \end{bmatrix}$$

Solution

$$[A \ \vdots \ I] = \left[\begin{array}{cccc|cccc} -8 & 0 & 0 & 0 & 1 & 0 & 0 & 0 \\ 0 & 1 & 0 & 0 & 0 & 1 & 0 & 0 \\ 0 & 0 & 4 & 0 & 0 & 0 & 1 & 0 \\ 0 & 0 & 0 & -5 & 0 & 0 & 0 & 1 \end{array}\right]$$

$$\begin{array}{r} -\frac{1}{8}R_1 \rightarrow \\ \\ \frac{1}{4}R_3 \rightarrow \\ -\frac{1}{5}R_4 \rightarrow \end{array} \left[\begin{array}{cccc|cccc} 1 & 0 & 0 & 0 & -\frac{1}{8} & 0 & 0 & 0 \\ 0 & 1 & 0 & 0 & 0 & 1 & 0 & 0 \\ 0 & 0 & 1 & 0 & 0 & 0 & \frac{1}{4} & 0 \\ 0 & 0 & 0 & 1 & 0 & 0 & 0 & -\frac{1}{5} \end{array}\right]$$

$$= [I \ \vdots \ A^{-1}]$$

$$A^{-1} = \frac{1}{40}\begin{bmatrix} -5 & 0 & 0 & 0 \\ 0 & 40 & 0 & 0 \\ 0 & 0 & 10 & 0 \\ 0 & 0 & 0 & -8 \end{bmatrix}$$

37. Use an inverse matrix to solve the system of linear equations. (Use the inverse matrix found in Exercise 15, Section 7.3.)

$$-x + y = 4$$
$$-2x + y = 0$$

Solution

$$\begin{bmatrix} x \\ y \end{bmatrix} = A^{-1}B = \begin{bmatrix} 1 & -1 \\ 2 & -1 \end{bmatrix}\begin{bmatrix} 4 \\ 0 \end{bmatrix} = \begin{bmatrix} 4 \\ 8 \end{bmatrix}$$

Answer: (4, 8)

41. Use an inverse matrix to solve the system of linear equations. (Use the inverse matrix found in Exercise 18, Section 7.3.)

$$2x + 3y = 5$$
$$x + 4y = 10$$

Solution

$$\begin{bmatrix} x \\ y \end{bmatrix} = A^{-1}B = \frac{1}{5}\begin{bmatrix} 4 & -3 \\ -1 & 2 \end{bmatrix}\begin{bmatrix} 5 \\ 10 \end{bmatrix} = \begin{bmatrix} -2 \\ 3 \end{bmatrix}$$

Answer: $(-2, 3)$

45. Use an inverse matrix to solve the system of linear equations. (Use the inverse matrix found in Exercise 26, Section 7.3)

$$3x + 2y + 2z = 0$$
$$2x + 2y + 2z = 5$$
$$-4x + 4y + 3z = 2$$

Solution

$$\begin{bmatrix} x \\ y \\ z \end{bmatrix} = A^{-1}B = \begin{bmatrix} 1 & -1 & 0 \\ 7 & -\frac{17}{2} & 1 \\ -8 & 10 & -1 \end{bmatrix}\begin{bmatrix} 0 \\ 5 \\ 2 \end{bmatrix} = \begin{bmatrix} -5 \\ -\frac{81}{2} \\ 48 \end{bmatrix}$$

Answer: $\left(-5, -\frac{81}{2}, 48\right)$

49. *Starting Salary for Engineers* From 1980 to 1993, the average starting salary for new 4-year graduates in engineering is shown in the figure in the textbook. The least squares regression line $y = a + bt$ for this data is found by solving the system

$$14a + 91b = 402.2$$

$$91a + 819b = 2853.7$$

where y is the average salary (in thousands of dollars) and $t = 0$ represents 1980. (*Source: Northwestern University Placement Center.*)

(a) Use a graphing utility to find an inverse matrix to solve this system and find the equation of the least squares regression line.

(b) Use the result of part (a) to approximate the average salary in 1996.

Solution

(a) $A = \begin{bmatrix} 14 & 91 \\ 91 & 819 \end{bmatrix}$

$$A^{-1} = \begin{bmatrix} \frac{9}{35} & -\frac{1}{35} \\ -\frac{1}{35} & \frac{2}{455} \end{bmatrix} \approx \begin{bmatrix} 0.2571 & -0.0286 \\ -0.0286 & 0.0044 \end{bmatrix}$$

$$A^{-1}B = \begin{bmatrix} \frac{9}{35} & -\frac{1}{35} \\ -\frac{1}{35} & \frac{2}{455} \end{bmatrix} \begin{bmatrix} 402.2 \\ 2853.7 \end{bmatrix} = \begin{bmatrix} \frac{7661}{350} \\ \frac{342}{325} \end{bmatrix} \approx \begin{bmatrix} 21.89 \\ 1.05 \end{bmatrix}$$

$y = 21.89 + 1.05t$

(b) For 1996, use $t = 16$.

$y = 21.89 + 1.05(16) = 38.69$ thousand dollars or $38,690.

53. *Bond Investment* You are investing in AAA-rated bonds, A-rated bonds, and B-rated bonds. The average yield is 8% on AAA-bonds, 6% on A-Bonds, and 7% on B-bonds. Twice as much is invested in B-bonds as in A-bonds. Moreover, the total annual return for all three types of bonds is \$2,800. The desired system of linear equations (where x, y, and z represent the amounts invested in AAA-, A-, and B-bonds) is as follows.

$$\begin{aligned} x + \quad y + \quad z &= \text{(Total investment)} \\ 0.08x + 0.06y + 0.07z &= 2800 \\ 2y - \quad z &= 0 \end{aligned}$$

Use the inverse of the coefficient matrix of this system to find the amount invested in each type of bond, when the total investment equals \$37,500.

Solution

$$[A \;\vdots\; I] = \left[\begin{array}{ccc:ccc} 1 & 1 & 1 & 1 & 0 & 0 \\ 0.08 & 0.06 & 0.07 & 0 & 1 & 0 \\ 0 & 2 & -1 & 0 & 0 & 1 \end{array}\right]$$

$$-0.8R_1 + R_2 \rightarrow \left[\begin{array}{ccc:ccc} 1 & 1 & 1 & 1 & 0 & 0 \\ 0 & -0.02 & -0.01 & -0.08 & 1 & 0 \\ 0 & 2 & -1 & 0 & 0 & 1 \end{array}\right]$$

$$\begin{array}{r} -R_2 + R_1 \rightarrow \\ -\frac{1}{0.02}R_2 \rightarrow \\ -2R_2 + R_3 \rightarrow \end{array} \left[\begin{array}{ccc:ccc} 1 & 0 & 0.5 & -3 & 50 & 0 \\ 0 & 1 & 0.5 & 4 & -50 & 0 \\ 0 & 0 & -2 & -8 & 100 & 1 \end{array}\right]$$

$$\begin{array}{r} -0.5R_3 + R_1 \rightarrow \\ -0.5R_3 + R_2 \rightarrow \\ -\frac{1}{2}R_3 \rightarrow \end{array} \left[\begin{array}{ccc:ccc} 1 & 0 & 0 & -5 & 75 & 0.25 \\ 0 & 1 & 0 & 2 & -25 & 0.25 \\ 0 & 0 & 1 & 4 & -50 & -0.50 \end{array}\right]$$

$$= [I \;\vdots\; A^{-1}]$$

$$A^{-1} = \begin{bmatrix} -5 & 75 & 0.25 \\ 2 & -25 & 0.25 \\ 4 & -50 & -0.50 \end{bmatrix}$$

$$\begin{bmatrix} x \\ y \\ z \end{bmatrix} = A^{-1}B = \begin{bmatrix} -5 & 75 & 0.25 \\ 2 & -25 & 0.25 \\ 4 & -50 & -0.50 \end{bmatrix} \begin{bmatrix} 37{,}500 \\ 2{,}800 \\ 0 \end{bmatrix} = \begin{bmatrix} 22{,}500 \\ 5{,}000 \\ 10{,}000 \end{bmatrix}$$

Answer: \$22,500 in AAA-bonds

\$5,000 in A-bonds

\$10,000 in B-bonds

57. *Acquisition of Raw Materials* Consider a company that produces computer chips, resistors, and transistors. Each computer chip requires 2 units of copper, 2 units of zinc, and 1 unit of glass. Each resistor requires 1 unit of copper, 3 units of zinc, and 2 units of glass. Each transistor requires 3 units of copper, 2 units of zinc, and 2 units of glass. The desired system of linear equations (where x, y, and z represent the number of computer chips, resistors, and transistors) is as follows.

$$2x + y + 3z = \text{(Units of copper)}$$

$$2x + 3y + 2z = \text{(Units of zinc)}$$

$$x + 2y + 2z = \text{(Units of glass)}$$

Use the inverse of the coefficient matrix of this system to find the number of computer chips, resistors, and transistors that the company can produce with 70 units of copper, 80 units of zinc, and 55 units of glass.

Solution

$$[A \;\vdots\; I] = \left[\begin{array}{ccc|ccc} 2 & 1 & 3 & 1 & 0 & 0 \\ 2 & 3 & 2 & 0 & 1 & 0 \\ 1 & 2 & 2 & 0 & 0 & 1 \end{array}\right]$$

$$\begin{array}{r} -R_3 + R_1 \to \\ -2R_1 + R_2 \to \\ -R_1 + R_3 \to \end{array} \left[\begin{array}{ccc|ccc} 1 & -1 & 1 & 1 & 0 & -1 \\ 0 & 2 & -1 & -1 & 1 & 0 \\ 0 & 3 & 1 & -1 & 0 & 2 \end{array}\right]$$

$$\begin{array}{r} R_2 + R_1 \to \\ R_3 - R_2 \to \\ -3R_2 + R_3 \to \end{array} \left[\begin{array}{ccc|ccc} 1 & 0 & 3 & 1 & -1 & 1 \\ 0 & 1 & 2 & 0 & -1 & 2 \\ 0 & 0 & -5 & -1 & 3 & -4 \end{array}\right]$$

$$\begin{array}{r} -3R_3 + R_1 \to \\ -2R_3 + R_2 \to \\ -\frac{1}{5}R_3 \to \end{array} \left[\begin{array}{ccc|ccc} 1 & 0 & 0 & \frac{2}{5} & \frac{4}{5} & -\frac{7}{5} \\ 0 & 1 & 0 & -\frac{2}{5} & \frac{1}{5} & \frac{2}{5} \\ 0 & 0 & 1 & \frac{1}{5} & -\frac{3}{5} & \frac{4}{5} \end{array}\right]$$

$$= [I \;\vdots\; A^{-1}]$$

$$A^{-1} = \frac{1}{5}\begin{bmatrix} 2 & 4 & -7 \\ -2 & 1 & 2 \\ 1 & -3 & 4 \end{bmatrix}$$

$$\begin{bmatrix} x \\ y \\ z \end{bmatrix} = A^{-1}B = \frac{1}{5}\begin{bmatrix} 2 & 4 & -7 \\ -2 & 1 & 2 \\ 1 & -3 & 4 \end{bmatrix}\begin{bmatrix} 70 \\ 80 \\ 55 \end{bmatrix} = \begin{bmatrix} 15 \\ 10 \\ 10 \end{bmatrix}$$

Answer : 15 computer chips

10 resistors

10 transistors

7.4 The Determinant of a Square Matrix

- You should be able to determine the determinant of a matrix of order 2 or of order 3 by using the products of the diagonals.
- You should be able to use expansion by cofactors to find the determinant of a matrix of order 3 or greater.
- The determinant of a triangular matrix equals the product of the entries on the main diagonal.

SOLUTIONS TO SELECTED EXERCISES

5. Find the determinant of the matrix.

$$\begin{bmatrix} 5 & 2 \\ -6 & 3 \end{bmatrix}$$

Solution

$$\begin{vmatrix} 5 & 2 \\ -6 & 3 \end{vmatrix} = 5(3) - 2(-6) = 15 + 12 = 27$$

9. Find the determinant of the matrix.

$$\begin{bmatrix} 2 & 6 \\ 0 & 3 \end{bmatrix}$$

Solution

$$\begin{vmatrix} 2 & 6 \\ 0 & 3 \end{vmatrix} = 2(3) - 6(0) = 6$$

13. Find the determinant of the matrix.

$$\begin{bmatrix} 0.3 & 0.2 & 0.2 \\ 0.2 & 0.2 & 0.2 \\ -0.4 & 0.4 & 0.3 \end{bmatrix}$$

Solution

$$\begin{vmatrix} 0.3 & 0.2 & 0.2 \\ 0.2 & 0.2 & 0.2 \\ -0.4 & 0.4 & 0.3 \end{vmatrix} = 0.3\begin{vmatrix} 0.2 & 0.2 \\ 0.4 & 0.3 \end{vmatrix} - 0.2\begin{vmatrix} 0.2 & 0.2 \\ -0.4 & 0.3 \end{vmatrix} + 0.2\begin{vmatrix} 0.2 & 0.2 \\ -0.4 & 0.4 \end{vmatrix}$$

$$= 0.3(-0.02) - 0.2(0.14) + 0.2(0.16) = -0.002$$

17. Find the determinant of the matrix.

$$\begin{bmatrix} 6 & 3 & -7 \\ 0 & 0 & 0 \\ 4 & -6 & 3 \end{bmatrix}$$

Solution

$$\begin{vmatrix} 6 & 3 & -7 \\ 0 & 0 & 0 \\ 4 & -6 & 3 \end{vmatrix} = 0\begin{vmatrix} 3 & -7 \\ -6 & 3 \end{vmatrix} - 0\begin{vmatrix} 6 & -7 \\ 4 & 3 \end{vmatrix} + 0\begin{vmatrix} 6 & 3 \\ 4 & -6 \end{vmatrix} = 0$$

21. Find the determinant of the matrix.

$$\begin{bmatrix} -1 & 0 & 0 & 0 \\ 2 & 3 & 0 & 0 \\ -4 & 5 & 3 & 0 \\ 1 & 0 & 2 & 2 \end{bmatrix}$$

Solution

$$\begin{vmatrix} -1 & 0 & 0 & 0 \\ 2 & 3 & 0 & 0 \\ -4 & 5 & 3 & 0 \\ 1 & 0 & 2 & 2 \end{vmatrix} = (-1)(3)(3)(2) = -18 \quad \text{Lower Triangular}$$

25. Find (a) all minors and (b) cofactors for the given matrix.

$$\begin{bmatrix} 3 & 4 \\ 2 & -5 \end{bmatrix}$$

Solution

(a) $M_{11} = -5$ (b) $C_{11} = M_{11} = -5$

$M_{12} = 2$ $C_{12} = -M_{12} = -2$

$M_{21} = 4$ $C_{21} = -M_{21} = -4$

$M_{22} = 3$ $C_{22} = M_{22} = 3$

29. Find the determinant of the matrix by the method of expansion by cofactors. Expand by using (a) Row 1 and (b) Column 2.

$$\begin{bmatrix} -3 & 2 & 1 \\ 4 & 5 & 6 \\ 2 & -3 & 1 \end{bmatrix}$$

Solution

(a) $$\begin{vmatrix} -3 & 2 & 1 \\ 4 & 5 & 6 \\ 2 & -3 & 1 \end{vmatrix} = -3\begin{vmatrix} 5 & 6 \\ -3 & 1 \end{vmatrix} - 2\begin{vmatrix} 4 & 6 \\ 2 & 1 \end{vmatrix} + \begin{vmatrix} 4 & 5 \\ 2 & -3 \end{vmatrix}$$

$$= -3(23) - 2(-8) - 22 = -75$$

(b) $$\begin{vmatrix} -3 & 2 & 1 \\ 4 & 5 & 6 \\ 2 & -3 & 1 \end{vmatrix} = -2\begin{vmatrix} 4 & 6 \\ 2 & 1 \end{vmatrix} + 5\begin{vmatrix} -3 & 1 \\ 2 & 1 \end{vmatrix} + 3\begin{vmatrix} -3 & 1 \\ 4 & 6 \end{vmatrix}$$

$$= -2(-8) + 5(-5) + 3(-22) = -75$$

33. Find the determinant of the matrix by the method of expansion by cofactors. Expand by using (a) Row 2 and (b) Column 2.

$$\begin{bmatrix} 6 & 0 & -3 & 5 \\ 4 & 13 & 6 & -8 \\ -1 & 0 & 7 & 4 \\ 8 & 6 & 0 & 2 \end{bmatrix}$$

Solution

(a) $\begin{vmatrix} 6 & 0 & -3 & 5 \\ 4 & 13 & 6 & -8 \\ -1 & 0 & 7 & 4 \\ 8 & 6 & 0 & 2 \end{vmatrix}$

$$= -4\begin{vmatrix} 0 & -3 & 5 \\ 0 & 7 & 4 \\ 6 & 0 & 2 \end{vmatrix} + 13\begin{vmatrix} 6 & -3 & 5 \\ -1 & 7 & 4 \\ 8 & 0 & 2 \end{vmatrix} - 6\begin{vmatrix} 6 & 0 & 5 \\ -1 & 0 & 4 \\ 8 & 6 & 2 \end{vmatrix} - 8\begin{vmatrix} 6 & 0 & -3 \\ -1 & 0 & 7 \\ 8 & 6 & 0 \end{vmatrix}$$

$$= -4(-282) + 13(-298) - 6(-174) - 8(-234) = 170$$

(b) $\begin{vmatrix} 6 & 0 & -3 & 5 \\ 4 & 13 & 6 & -8 \\ -1 & 0 & 7 & 4 \\ 8 & 6 & 0 & 2 \end{vmatrix}$

$$= 0\begin{vmatrix} 4 & 6 & -8 \\ -1 & 7 & 4 \\ 8 & 0 & 2 \end{vmatrix} + 13\begin{vmatrix} 6 & -3 & 5 \\ -1 & 7 & 4 \\ 8 & 0 & 2 \end{vmatrix} + 0\begin{vmatrix} 6 & -3 & 5 \\ 4 & 6 & -8 \\ 8 & 0 & 2 \end{vmatrix} + 6\begin{vmatrix} 6 & -3 & 5 \\ 4 & 6 & -8 \\ -1 & 7 & 4 \end{vmatrix}$$

$$= 0 + 13(-298) + 0 + 6(674) = 170$$

37. Find the determinant of the matrix. Use a graphing utility to confirm your result.

$$\begin{bmatrix} 2 & 4 & 6 \\ 0 & 3 & 1 \\ 0 & 0 & -5 \end{bmatrix}$$

Solution

$$\begin{vmatrix} 2 & 4 & 6 \\ 0 & 3 & 1 \\ 0 & 0 & -5 \end{vmatrix} = (2)(3)(-5) = -30 \text{ (Upper Triangular)}$$

41. Find the determinant of the matrix. Use a graphing utility to confirm your result.

$$\begin{bmatrix} 5 & 3 & 0 & 6 \\ 4 & 6 & 4 & 12 \\ 0 & 2 & -3 & 4 \\ 0 & 1 & -2 & 2 \end{bmatrix}$$

Solution

Expand by Column 1.

$$\begin{vmatrix} 5 & 3 & 0 & 6 \\ 4 & 6 & 4 & 12 \\ 0 & 2 & -3 & 4 \\ 0 & 1 & -2 & 2 \end{vmatrix} = 5\begin{vmatrix} 6 & 4 & 12 \\ 2 & -3 & 4 \\ 1 & -2 & 2 \end{vmatrix} - 4\begin{vmatrix} 3 & 0 & 6 \\ 2 & -3 & 4 \\ 1 & -2 & 2 \end{vmatrix} = 5(0) - 4(0) = 0$$

45. Find two 4×4 matrices whose determinant is -24. That is, find a 4×4 upper triangular matrix and then a 4×4 lower triangular matrix. Use a graphing utility to confirm your result.

Solution

There are many different answers. If the matrix is triangular, then the determinant is simply the product of the main diagonal entries. In this case, design triangular matrices whose diagonal entries multiply together to -24. Two examples are as follows.

$$\begin{bmatrix} 3 & 5 & 11 & 7 \\ 0 & -2 & -3 & 9 \\ 0 & 0 & 1 & -8 \\ 0 & 0 & 0 & 4 \end{bmatrix} \text{ Upper Triangular} \qquad \begin{bmatrix} 12 & 0 & 0 & 0 \\ 3 & -1 & 0 & 0 \\ 14 & -17 & -2 & 0 \\ -9 & 6 & 5 & -1 \end{bmatrix} \text{ Lower Triangular}$$

Mid-Chapter Quiz for Chapter 7

1. Write a matrix of order 3×4.

Solution

Any matrix with 3 rows and 4 columns will satisfy this requirement. Some examples are

$$\begin{bmatrix} 1 & 0 & 0 & 0 \\ 0 & 1 & 0 & 0 \\ 0 & 0 & 1 & 1 \end{bmatrix} \begin{bmatrix} 1 & 2 & 3 & 4 \\ 5 & 6 & 7 & 8 \\ 9 & 10 & 11 & 12 \end{bmatrix} \begin{bmatrix} -3 & 2 & 6 & 4 \\ 1 & 9 & 8 & -7 \\ 7 & -1 & 5 & -9 \end{bmatrix}.$$

2. Write a matrix of order 1×3.

Solution

Any matrix with 1 row and 3 columns will satisfy this requirement. Some examples are

$$\begin{bmatrix} 1 & 2 & 3 \end{bmatrix} \begin{bmatrix} a & b & c \end{bmatrix} \begin{bmatrix} 5 & -12 & 20 \end{bmatrix}$$

3. Write the system of linear equations represented by the augmented matrix. (Use variables x, y, and z.)

$$\left[\begin{array}{cc:c} 3 & 2 & -2 \\ 5 & -1 & 19 \end{array}\right]$$

Solution

$$\begin{aligned} 3x + 2y &= -2 \\ 5x - y &= 19 \end{aligned}$$

4. Write the system of linear equations represented by the augmented matrix. (Use variables x, y, and z.)

$$\left[\begin{array}{ccc:c} 1 & 0 & 3 & -5 \\ 1 & 2 & -1 & 3 \\ 3 & 0 & 4 & 0 \end{array}\right]$$

Solution

$$\begin{aligned} x + 3z &= -5 \\ x + 2y - z &= 3 \\ 3x + 4z &= 0 \end{aligned}$$

5. Use Gauss-Jordan elimination to solve the system in Exercise 3.

Solution

$$\begin{bmatrix} 3 & 2 & \vdots & -2 \\ 5 & -1 & \vdots & 19 \end{bmatrix} \quad \begin{array}{r} \frac{1}{3}R_1 \rightarrow \\ -5R_1 + R_2 \rightarrow \end{array} \begin{bmatrix} 1 & \frac{2}{3} & \vdots & -\frac{2}{3} \\ 0 & -\frac{13}{3} & \vdots & \frac{67}{3} \end{bmatrix}$$

$$\begin{array}{r} -\frac{2}{3}R_2 + R_1 \rightarrow \\ -\frac{3}{13}R_2 \rightarrow \end{array} \begin{bmatrix} 1 & 0 & \vdots & \frac{36}{13} \\ 0 & 1 & \vdots & -\frac{67}{13} \end{bmatrix}$$

Thus, $x = \frac{36}{13} \approx 2.769$ and $y = -\frac{67}{13} \approx -5.154$.

Answer: $(2.769, -5.154)$

6. Use Gauss-Jordan elimination to solve the system in Exercise 4.

Solution

$$\begin{bmatrix} 1 & 0 & 3 & \vdots & -5 \\ 1 & 2 & -1 & \vdots & 3 \\ 3 & 0 & 4 & \vdots & 0 \end{bmatrix} \quad \begin{array}{r} \\ -R_1 + R_2 \rightarrow \\ -3R_1 + R_3 \rightarrow \end{array} \begin{bmatrix} 1 & 0 & 3 & \vdots & -5 \\ 0 & 2 & -4 & \vdots & 8 \\ 0 & 0 & -5 & \vdots & 15 \end{bmatrix}$$

$$\begin{array}{r} \\ -\frac{1}{2}R_2 \rightarrow \\ -\frac{1}{5}R_3 \rightarrow \end{array} \begin{bmatrix} 1 & 0 & 3 & \vdots & -5 \\ 0 & 1 & -2 & \vdots & 4 \\ 0 & 0 & 1 & \vdots & -3 \end{bmatrix}$$

$$\begin{array}{r} -3R_3 + R_1 \rightarrow \\ 2R_3 + R_2 \rightarrow \\ \end{array} \begin{bmatrix} 1 & 0 & 0 & \vdots & 4 \\ 0 & 1 & 0 & \vdots & -2 \\ 0 & 0 & 1 & \vdots & -3 \end{bmatrix}$$

Thus, $x = 4$, $y = -2$, and $z = -3$.

Answer: $(4, -2, -3)$

In Exercises 7–12, use the matrices to find the indicated matrix.

$$A = \begin{bmatrix} 1 & -2 \\ 3 & 4 \end{bmatrix}, \; B = \begin{bmatrix} -1 & 2 & -3 \\ 2 & 0 & 5 \end{bmatrix}, \; C = \begin{bmatrix} 0 & -2 \\ 3 & 1 \end{bmatrix}$$

7. $2A + 3C$

Solution

$$2A + 3C = 2\begin{bmatrix} 1 & -2 \\ 3 & 4 \end{bmatrix} + 3\begin{bmatrix} 0 & -2 \\ 3 & 1 \end{bmatrix}$$

$$= \begin{bmatrix} 2 & -4 \\ 6 & 8 \end{bmatrix} + \begin{bmatrix} 0 & -6 \\ 9 & 3 \end{bmatrix}$$

$$= \begin{bmatrix} 2 & -10 \\ 15 & 11 \end{bmatrix}$$

8. AB

Solution

$$AB = \begin{bmatrix} 1 & -2 \\ 3 & 4 \end{bmatrix}\begin{bmatrix} -1 & 2 & -3 \\ 2 & 0 & 5 \end{bmatrix} = \begin{bmatrix} -5 & 2 & -13 \\ 5 & 6 & 11 \end{bmatrix}$$

9. $A - 2C$

Solution

$$A - 2C = \begin{bmatrix} 1 & -2 \\ 3 & 4 \end{bmatrix} - 2\begin{bmatrix} 0 & -2 \\ 3 & 1 \end{bmatrix}$$
$$= \begin{bmatrix} 1 & -2 \\ 3 & 4 \end{bmatrix} - \begin{bmatrix} 0 & -4 \\ 6 & 2 \end{bmatrix}$$
$$= \begin{bmatrix} 1 & 2 \\ -3 & 2 \end{bmatrix}$$

10. C^2

Solution

$$C^2 = \begin{bmatrix} 0 & -2 \\ 3 & 1 \end{bmatrix}\begin{bmatrix} 0 & -2 \\ 3 & 1 \end{bmatrix} = \begin{bmatrix} -6 & -2 \\ 3 & -5 \end{bmatrix}$$

11. A^{-1}

Solution

$$[A \vdots I] = \left[\begin{array}{cc|cc} 1 & -2 & 1 & 0 \\ 3 & 4 & 0 & 1 \end{array}\right] \quad -3R_1 + R_2 \to \left[\begin{array}{cc|cc} 1 & -2 & 1 & 0 \\ 0 & 10 & -3 & 1 \end{array}\right]$$

$$\begin{array}{r} 2R_2 + R_1 \to \\ \frac{1}{10}R_2 \to \end{array} \left[\begin{array}{cc|cc} 1 & 0 & \frac{4}{10} & \frac{2}{10} \\ 0 & 1 & -\frac{3}{10} & \frac{1}{10} \end{array}\right]$$

$$A^{-1} = \begin{bmatrix} 0.4 & 0.2 \\ -0.3 & 0.1 \end{bmatrix}$$

12. B^{-1}

Solution

Since B is not a square matrix, it does not have an inverse.

13. Solve for X using matrices A and C from Exercises 7–12.

$X = 2A - 3C$

Solution

$$x = 2\begin{bmatrix} 1 & -2 \\ 3 & 4 \end{bmatrix} - 3\begin{bmatrix} 0 & -2 \\ 3 & 1 \end{bmatrix}$$
$$= \begin{bmatrix} 2 & -4 \\ 6 & 8 \end{bmatrix} - \begin{bmatrix} 0 & -6 \\ 9 & 3 \end{bmatrix}$$
$$= \begin{bmatrix} 2 & 2 \\ -3 & 5 \end{bmatrix}$$

14. Solve for X using matrices A and C from Exercises 7–12.

$$2X = 3A - C$$

Solution

$$2X = 3\begin{bmatrix} 1 & -2 \\ 3 & 4 \end{bmatrix} - \begin{bmatrix} 0 & -2 \\ 3 & 1 \end{bmatrix}$$

$$= \begin{bmatrix} 3 & -6 \\ 9 & 12 \end{bmatrix} - \begin{bmatrix} 0 & -2 \\ 3 & 1 \end{bmatrix}$$

$$= \begin{bmatrix} 3 & -4 \\ 6 & 11 \end{bmatrix}$$

$$X = \frac{1}{2}\begin{bmatrix} 3 & -4 \\ 6 & 11 \end{bmatrix} = \begin{bmatrix} \frac{3}{2} & -2 \\ 3 & \frac{11}{2} \end{bmatrix}$$

15. Find matrices A, X, and B such that the system can be written as $AX = B$. Then solve for X.

$$\begin{aligned} x - 3y &= 10 \\ -2x + y &= -10 \end{aligned}$$

Solution

$$A = \begin{bmatrix} 1 & -3 \\ -2 & 1 \end{bmatrix}, X = \begin{bmatrix} x \\ y \end{bmatrix} \; B = \begin{bmatrix} 10 \\ -10 \end{bmatrix}$$

$$X = A^{-1}B = \begin{bmatrix} -0.2 & -0.6 \\ -0.4 & -0.2 \end{bmatrix}\begin{bmatrix} 10 \\ -10 \end{bmatrix} = \begin{bmatrix} 4 \\ -2 \end{bmatrix}$$

Answer: $x = 4$, $y = -2$

16. Find matrices A, X, and B such that the system can be written as $AX = B$. Then solve for X.

$$\begin{aligned} 2x - y + z &= 3 \\ 3x - z &= 15 \\ 4y + 3z &= -1 \end{aligned}$$

Solution

$$A = \begin{bmatrix} 2 & -1 & 1 \\ 3 & 0 & -1 \\ 0 & 4 & 3 \end{bmatrix}, X = \begin{bmatrix} x \\ y \\ z \end{bmatrix}, B = \begin{bmatrix} 3 \\ 15 \\ -1 \end{bmatrix}$$

$$X = A^{-1}B = \begin{bmatrix} \frac{4}{29} & \frac{7}{29} & \frac{1}{29} \\ -\frac{9}{29} & \frac{6}{29} & \frac{5}{29} \\ \frac{12}{29} & -\frac{8}{29} & \frac{3}{29} \end{bmatrix}\begin{bmatrix} 3 \\ 15 \\ -1 \end{bmatrix} = \begin{bmatrix} 4 \\ 2 \\ -3 \end{bmatrix}$$

Thus, $x = 4$, $y = 2$, and $z = -3$.

In Exercises 17–20, a hang glider manufacturer has the labor-hour and wage requirements indicated below.

$$L = \begin{bmatrix} 1.0 & 0.6 & 0.2 \\ 2.4 & 1.0 & 0.2 \\ 2.8 & 2.0 & 0.5 \end{bmatrix} \begin{matrix} \text{Model } A \\ \text{Model } B \\ \text{Model } C \end{matrix}$$

(Columns: Assembly, Finishing, Packaging)

Labor Requirements (in hours)

$$W = \begin{bmatrix} \$12 & \$10 \\ \$\ 8 & \$\ 9 \\ \$\ 5 & \$\ 6 \end{bmatrix} \begin{matrix} \text{Assembly} \\ \text{Finishing} \\ \text{Packaging} \end{matrix}$$

(Columns: Plant 1, Plant 2)

Wage Requirements (in dollars per hour)

17. What is the labor cost for model A?

Solution

$$[1.0 \quad 0.6 \quad 0.2]\begin{bmatrix} 12 & 10 \\ 8 & 9 \\ 5 & 6 \end{bmatrix} = [17.8 \quad 16.6]$$

For model A, add the costs for plant 1 and plant 2.

$\$17.80 + \$16.60 = \$34.40$

18. What is the labor cost for model B?

Solution

$$[2.4 \quad 1.0 \quad 0.2]\begin{bmatrix} 12 & 10 \\ 8 & 9 \\ 5 & 6 \end{bmatrix} = [37.8 \quad 34.2]$$

For model B, add the costs for plant 1 and plant 2.

$\$37.80 + \$34.20 = \$72.00$

19. What is the labor cost for model C?

Solution

$$[2.8 \quad 2.0 \quad 0.5]\begin{bmatrix} 12 & 10 \\ 8 & 9 \\ 5 & 6 \end{bmatrix} = [52.1 \quad 49]$$

For model C, add the costs for plant 1 and plant 2.

$\$52.10 + \$49.00 = \$101.10$

20. Find LW and interpret the result.

Solution

$$LW = \begin{bmatrix} 17.8 & 16.6 \\ 37.8 & 34.2 \\ 52.1 & 49.0 \end{bmatrix} \begin{matrix} \text{Model } A \\ \text{Model } B \\ \text{Model } C \end{matrix}$$

(Columns: Plant 1, Plant 2)

The total labor cost for each model is listed by plant.

7.5 Applications of Determinants and Matrices

- You should be able to find the area of a triangle with vertices (x_1, y_1), (x_2, y_2), and (x_3, y_3).

$$\text{Area} = \pm\frac{1}{2}\begin{vmatrix} x_1 & y_1 & 1 \\ x_2 & y_2 & 1 \\ x_3 & y_3 & 1 \end{vmatrix}$$

  The $\pm$ symbol indicates that the appropriate sign should be chosen so that the area is positive.

- You should be able to test to see if three points, (x_1, y_1), (x_2, y_2), and (x_3, y_3), are collinear.

$$\begin{vmatrix} x_1 & y_1 & 1 \\ x_2 & y_2 & 1 \\ x_3 & y_3 & 1 \end{vmatrix} = 0, \text{ if and only if they are collinear.}$$

- You should be able to find the equation of the line through (x_1, y_1) and (x_2, y_2) by evaluating.

$$\begin{vmatrix} x & y & 1 \\ x_1 & y_1 & 1 \\ x_2 & y_2 & 1 \end{vmatrix} = 0$$

- You should be able to encode and decode messages by using an invertible $n \times n$ matrix.

SOLUTIONS TO SELECTED EXERCISES

3. Use a determinant to find the area of the triangle with the vertices $(-2, -3)$, $(2, -3)$, $(0, 4)$. (See graph in textbook.)

Solution

$$\text{Area} = \pm\frac{1}{2}\begin{vmatrix} -2 & -3 & 1 \\ 2 & -3 & 1 \\ 0 & 4 & 1 \end{vmatrix} = \pm\frac{1}{2}\left(-2\begin{vmatrix} -3 & 1 \\ 4 & 1 \end{vmatrix} - 2\begin{vmatrix} -3 & 1 \\ 4 & 1 \end{vmatrix}\right)$$

$$= \pm\frac{1}{2}(14 + 14) = 14 \text{ square units}$$

7. Use a determinant to find the area of the triangle with the vertices $(-2, 4)$, $(2, 3)$, $(-1, 5)$.

Solution

$$\text{Area} = \pm\frac{1}{2}\begin{vmatrix} -2 & 4 & 1 \\ 2 & 3 & 1 \\ -1 & 5 & 1 \end{vmatrix} = \pm\frac{1}{2}\begin{vmatrix} -2 & 4 & 1 \\ 4 & -1 & 0 \\ 1 & 1 & 0 \end{vmatrix} = \pm\frac{1}{2}\begin{vmatrix} 4 & -1 \\ 1 & 1 \end{vmatrix} = \frac{5}{2} \text{ square units}$$

11. Find a value of x so that the triangle with vertices $(-5, 1)$, $(0, 2)$, and $(-2, x)$ has an area of 4.

Solution

$$4 = \pm\frac{1}{2}\begin{vmatrix} -5 & 1 & 1 \\ 0 & 2 & 1 \\ -2 & x & 1 \end{vmatrix}$$

$$\pm 8 = -5\begin{vmatrix} 2 & 1 \\ x & 1 \end{vmatrix} - 2\begin{vmatrix} 1 & 1 \\ 2 & 1 \end{vmatrix}$$

$$\pm 8 = -5(2 - x) - 2(-1)$$

$$\pm 8 = 5x - 8$$

$$8 \pm 8 = 5x$$

$$x = \frac{8 \pm 8}{5}$$

$$x = \frac{16}{5} \text{ or } x = 0$$

13. *Area of a Region* A large region of forest has been infested by gypsy moths. The region is roughly triangular, as shown the the figure in the textbook. From the northernmost vertex A of the region, the distance to the other vertices is 25 miles south and 10 miles east (for vertex B), and 20 miles south and 28 miles east (for vertex C). Use a graphing utility to approximate the number of square miles in this region.

Solution

Vertices: $(0, 25)$, $(10, 0)$, $(28, 5)$

$$\text{Area} = \pm\tfrac{1}{2}\begin{vmatrix} 0 & 25 & 1 \\ 10 & 0 & 1 \\ 28 & 5 & 1 \end{vmatrix} = 250 \text{ square miles}$$

17. Use a determinant to ascertain whether the following points are collinear.

$\left(2, -\tfrac{1}{2}\right)$, $(-4, 4)$, $(6, -3)$

Solution

$$\begin{vmatrix} 2 & -\frac{1}{2} & 1 \\ -4 & 4 & 1 \\ 6 & -3 & 1 \end{vmatrix} = -3 \neq 0$$

The points are not collinear.

21. Use a determinant to find an equation of the line through the points $(0, 0)$ and $(5, 3)$.

Solution

$$\text{Equation: } \begin{vmatrix} x & y & 1 \\ 0 & 0 & 1 \\ 5 & 3 & 1 \end{vmatrix} = 5y - 3x = 0 \quad \text{or} \quad 3x - 5y = 0$$

25. Use a determinant to find an equation of the line through the points $\left(-\tfrac{1}{2}, 3\right)$ and $\left(\tfrac{5}{2}, 1\right)$.

Solution

$$\text{Equation: } \begin{vmatrix} x & y & 1 \\ -\frac{1}{2} & 3 & 1 \\ \frac{5}{2} & 1 & 1 \end{vmatrix} = 3x + \tfrac{5}{2}y - \tfrac{1}{2} - \tfrac{15}{2} + \tfrac{1}{2}y - x = 0 \quad \text{or} \quad 2x + 3y - 8 = 0$$

27. Find x so that the points $(2, -5)$, $(4, x)$, and $(5, -2)$ are collinear.

Solution

$$\begin{vmatrix} 2 & -5 & 1 \\ 4 & x & 1 \\ 5 & -2 & 1 \end{vmatrix} = 0$$

$$2\begin{vmatrix} x & 1 \\ -2 & 1 \end{vmatrix} + 5\begin{vmatrix} 4 & 1 \\ 5 & 1 \end{vmatrix} + 1\begin{vmatrix} 4 & x \\ 5 & -2 \end{vmatrix} = 0$$

$$2(x+2) + 5(-1) + (-8 - 5x) = 0$$

$$-3x - 9 = 0$$

$$x = -3$$

31. Find the uncoded row matrices of order 1×3 for the message, TROUBLE IN RIVER CITY. Then encode the message using the matrix.

$$A = \begin{bmatrix} 1 & -1 & 0 \\ 1 & 0 & -1 \\ -6 & 2 & 3 \end{bmatrix}$$

Solution

The uncoded row matrices are the rows of the 7×3 matrix on the left.

$$\begin{matrix} \text{T R O} \\ \text{U B L} \\ \text{E \_ I} \\ \text{N \_ R} \\ \text{I V E} \\ \text{R \_ C} \\ \text{I T Y} \end{matrix} \begin{bmatrix} 20 & 18 & 15 \\ 21 & 2 & 12 \\ 5 & 0 & 9 \\ 14 & 0 & 18 \\ 9 & 22 & 5 \\ 18 & 0 & 3 \\ 9 & 20 & 25 \end{bmatrix} \begin{bmatrix} 1 & -1 & 0 \\ 1 & 0 & -1 \\ -6 & 2 & 3 \end{bmatrix} = \begin{bmatrix} -52 & 10 & 27 \\ -49 & 3 & 34 \\ -49 & 13 & 27 \\ -94 & 22 & 54 \\ 1 & 1 & -7 \\ 0 & -12 & 9 \\ -121 & 41 & 55 \end{bmatrix}$$

Answer: -52 10 27 -49 3 34 -49 13 27 -94 22 54 1 1 -7 0 -12 9 -121 41 55

35. Write a cryptogram for the message, HAPPY BIRTHDAY, using the matrix.

$$A = \begin{bmatrix} 1 & 2 & 2 \\ 3 & 7 & 9 \\ -1 & -4 & -7 \end{bmatrix}$$

Solution

H	A	P	P	Y	–	B	I	R	T	H	D	A	Y	–
[8	1	16]	[16	25	0]	[2	9	18]	[20	8	4]	[1	25	0]

$$[\,8 \quad 1 \quad 16]A = [\,-5 \quad -41 \quad -87]$$

$$[\,16 \quad 25 \quad 0]A = [\,91 \quad 207 \quad 257]$$

$$[\,2 \quad 9 \quad 18]A = [\,11 \quad -5 \quad -41]$$

$$[\,20 \quad 8 \quad 4]A = [\,40 \quad 80 \quad 84]$$

$$[\,1 \quad 25 \quad 0]A = [\,76 \quad 177 \quad 227]$$

Cryptogram: -5 -41 -87 91 207 257 11 -5 -41 40 80 84 76 177 227

39. Use A^{-1} to decode the following cryptogram.

$$A = \begin{bmatrix} 4 & 2 & 1 \\ -3 & -3 & -1 \\ 3 & 2 & 1 \end{bmatrix}$$

33 9 9 55 28 14 95 50 25 99 53 29 −22 −32 −9

Solution

$$A^{-1} = \begin{bmatrix} 4 & 2 & 1 \\ -3 & -3 & -1 \\ 3 & 2 & 1 \end{bmatrix}^{-1} = \begin{bmatrix} 1 & 0 & -1 \\ 0 & -1 & -1 \\ -3 & 2 & 6 \end{bmatrix}$$

$$\begin{bmatrix} 33 & 9 & 9 \\ 55 & 28 & 14 \\ 95 & 50 & 25 \\ 99 & 53 & 29 \\ -22 & -32 & -9 \end{bmatrix} \begin{bmatrix} 1 & 0 & -1 \\ 0 & -1 & -1 \\ -3 & 2 & 6 \end{bmatrix} = \begin{bmatrix} 6 & 9 & 12 \\ 13 & 0 & 1 \\ 20 & 0 & 5 \\ 12 & 5 & 22 \\ 5 & 14 & 0 \end{bmatrix} \begin{matrix} \text{F I L} \\ \text{M \quad A} \\ \text{T \quad E} \\ \text{L E V} \\ \text{E N \quad} \end{matrix}$$

Message: FILM AT ELEVEN

43. The following cryptogram was encoded with a 2×2 matrix.

8 21 −15 −10 −13 −13 5 10 5 25 5 19 −1 6 20 40 −18 −18 1 16

The last word of the message is _RON. What is the message?

Solution

Let A be the 2×2 matrix needed to decode the message.

$$\begin{bmatrix} -18 & -18 \\ 1 & 16 \end{bmatrix} A = \begin{bmatrix} 0 & 18 \\ 15 & 14 \end{bmatrix} \begin{matrix} \quad \text{R} \\ \text{O N} \end{matrix}$$

$$A = \begin{bmatrix} -18 & -18 \\ 1 & 16 \end{bmatrix}^{-1} \begin{bmatrix} 0 & 18 \\ 15 & 14 \end{bmatrix} = \begin{bmatrix} -\frac{8}{135} & -\frac{1}{15} \\ \frac{1}{270} & \frac{1}{15} \end{bmatrix} \begin{bmatrix} 0 & 18 \\ 15 & 14 \end{bmatrix} = \begin{bmatrix} -1 & -2 \\ 1 & 1 \end{bmatrix}$$

$$\begin{bmatrix} 8 & 21 \\ -15 & -10 \\ -13 & -13 \\ 5 & 10 \\ 5 & 25 \\ 5 & 19 \\ -1 & 6 \\ 20 & 40 \\ -18 & -18 \\ 1 & 16 \end{bmatrix} \begin{bmatrix} -1 & -2 \\ 1 & 1 \end{bmatrix} = \begin{bmatrix} 13 & 5 \\ 5 & 20 \\ 0 & 13 \\ 5 & 0 \\ 20 & 15 \\ 14 & 9 \\ 7 & 8 \\ 20 & 0 \\ 0 & 18 \\ 15 & 14 \end{bmatrix} \begin{matrix} \text{M E} \\ \text{E T} \\ \quad \text{M} \\ \text{E} \quad \\ \text{T O} \\ \text{N I} \\ \text{G H} \\ \text{T} \quad \\ \quad \text{R} \\ \text{O N} \end{matrix}$$

Message: MEET ME TONIGHT RON.

Review Exercises for Chapter 7

SOLUTIONS TO SELECTED EXERCISES

3. Write the matrix in reduced row-echelon form.

$$\begin{bmatrix} 1 & 2 & 3 \\ -2 & 0 & 2 \\ 2 & 1 & 2 \end{bmatrix}$$

Solution

$$\begin{bmatrix} 1 & 2 & 3 \\ -2 & 0 & 2 \\ 2 & 1 & 2 \end{bmatrix} \quad \begin{matrix} \\ 2R_1 + R_2 \rightarrow \\ -2R_1 + R_3 \rightarrow \end{matrix} \begin{bmatrix} 1 & 2 & 3 \\ 0 & 4 & 8 \\ 0 & -3 & -4 \end{bmatrix}$$

$$\begin{matrix} -2R_2 + R_1 \rightarrow \\ \frac{1}{4}R_2 \rightarrow \\ 3R_2 + R_3 \rightarrow \end{matrix} \begin{bmatrix} 1 & 0 & -1 \\ 0 & 1 & 2 \\ 0 & 0 & 2 \end{bmatrix}$$

$$\begin{matrix} R_3 + R_1 \rightarrow \\ -2R_3 + R_2 \rightarrow \\ \frac{1}{2}R_3 \rightarrow \end{matrix} \begin{bmatrix} 1 & 0 & 0 \\ 0 & 1 & 0 \\ 0 & 0 & 1 \end{bmatrix}$$

7. Use matrices to solve the system.

$$4x - 3y = 18$$

$$x + y = 1$$

Solution

$$\begin{bmatrix} 1 & 1 & \vdots & 1 \\ 4 & -3 & \vdots & 18 \end{bmatrix} \quad -4R_1 + R_2 \rightarrow \begin{bmatrix} 1 & 1 & \vdots & 1 \\ 0 & -7 & \vdots & 14 \end{bmatrix}$$

$-7y = 14 \Rightarrow y = -2$

$x + (-2) = 1 \Rightarrow x = 3$

Answer: $(3, -2)$

11. Use matrices to solve the system.

$$x + 2y + 2z = 10$$

$$2x + 3y + 5z = 20$$

Solution

$$\begin{bmatrix} 1 & 2 & 2 & \vdots & 10 \\ 2 & 3 & 5 & \vdots & 20 \end{bmatrix} \quad -2R_1 + R_2 \rightarrow \begin{bmatrix} 1 & 2 & 2 & \vdots & 10 \\ 0 & -1 & 1 & \vdots & 0 \end{bmatrix}$$

$$\begin{matrix} 2R_2 + R_1 \rightarrow \\ -R_2 \rightarrow \end{matrix} \begin{bmatrix} 1 & 0 & 4 & \vdots & 10 \\ 0 & 1 & -1 & \vdots & 0 \end{bmatrix}$$

$x + 4z = 10$

$y - z = 0$

Let $z = a$, then $y = a$ and $x = 10 - 4a$.

Answer: $(10 - 4a, a, a)$

15. *Borrowing Money* A company borrowed \$600,000 to expand its product line. Some of the money was borrowed at 9%, some at 10%, and some at 12%. How much was borrowed at each rate if the annual interest was \$63,000 and the amount borrowed at 10% was three times the amount borrowed at 9%?

Solution

Let x = amount at 9%, y = amount at 10%, and z = amount at 12%.

$$\begin{aligned} x + \quad y + \quad z &= 600{,}000 \\ 0.09x + 0.10y + 0.12z &= 63{,}000 \\ 3x - \quad y \quad\quad &= 0 \end{aligned}$$

$$\left[\begin{array}{ccc:c} 1 & 1 & 1 & 600{,}000 \\ 0.09 & 0.10 & 0.12 & 63{,}000 \\ 3 & -1 & 0 & 0 \end{array}\right]$$

$$\begin{array}{r} \\ -0.09R_1 + R_2 \rightarrow \\ -3R_1 + R_3 \rightarrow \end{array} \left[\begin{array}{ccc:c} 1 & 1 & 1 & 600{,}000 \\ 0 & 0.01 & 0.03 & 9{,}000 \\ 0 & -4 & -3 & -1{,}800{,}000 \end{array}\right]$$

$$\begin{array}{r} -R_2 + R_1 \rightarrow \\ 100R_2 \rightarrow \\ 4R_2 + R_3 \rightarrow \end{array} \left[\begin{array}{ccc:c} 1 & 0 & -2 & -300{,}000 \\ 0 & 1 & 3 & 900{,}000 \\ 0 & 0 & 9 & 1{,}800{,}000 \end{array}\right]$$

$$\begin{array}{r} 2R_3 + R_1 \rightarrow \\ -3R_3 + R_2 \rightarrow \\ \frac{1}{9}R_3 \rightarrow \end{array} \left[\begin{array}{ccc:c} 1 & 0 & 0 & 100{,}000 \\ 0 & 1 & 0 & 300{,}000 \\ 0 & 0 & 1 & 200{,}000 \end{array}\right]$$

$x = 100{,}000$

$y = 300{,}000$

$z = 200{,}000$

Answer: \$100,000 at 9%, \$300,000 at 10%, and \$200,000 at 12%

19. Find (a) $A + B$, (b) $A - B$, (c) $4A$, and (d) $4A - 3B$.

$$A = \begin{bmatrix} 1 & 3 & -2 & 6 \\ 0 & 1 & 3 & 2 \end{bmatrix}, \quad B = \begin{bmatrix} 2 & 1 & 4 & -5 \\ 3 & -6 & 3 & -2 \end{bmatrix}$$

Solution

(a) $A + B = \begin{bmatrix} 1 & 3 & -2 & 6 \\ 0 & 1 & 3 & 2 \end{bmatrix} + \begin{bmatrix} 2 & 1 & 4 & -5 \\ 3 & -6 & 3 & -2 \end{bmatrix} = \begin{bmatrix} 3 & 4 & 2 & 1 \\ 3 & -5 & 6 & 0 \end{bmatrix}$

(b) $A - B = \begin{bmatrix} 1 & 3 & -2 & 6 \\ 0 & 1 & 3 & 2 \end{bmatrix} - \begin{bmatrix} 2 & 1 & 4 & -5 \\ 3 & -6 & 3 & -2 \end{bmatrix} = \begin{bmatrix} -1 & 2 & -6 & 11 \\ -3 & 7 & 0 & 4 \end{bmatrix}$

(c) $4A = 4\begin{bmatrix} 1 & 3 & -2 & 6 \\ 0 & 1 & 3 & 2 \end{bmatrix} = \begin{bmatrix} 4 & 12 & -8 & 24 \\ 0 & 4 & 12 & 8 \end{bmatrix}$

(d) $4A - 3B = \begin{bmatrix} 4 & 12 & -8 & 24 \\ 0 & 4 & 12 & 8 \end{bmatrix} - \begin{bmatrix} 6 & 3 & 12 & -15 \\ 9 & -18 & 9 & -6 \end{bmatrix} = \begin{bmatrix} -2 & 9 & -20 & 39 \\ -9 & 22 & 3 & 14 \end{bmatrix}$

23. Find (a) AB, (b) BA, and (c) A^2 (if possible).

(*Note:* $A^2 = AA$.)

$$A = \begin{bmatrix} 1 & 0 & 2 \\ 3 & 1 & -2 \\ 1 & 1 & 1 \end{bmatrix}, \quad B = \begin{bmatrix} 2 & 0 & 0 \\ 1 & -2 & 1 \\ 5 & 4 & -2 \end{bmatrix}$$

Solution

$$\text{(a) } AB = \begin{bmatrix} 1 & 0 & 2 \\ 3 & 1 & -2 \\ 1 & 1 & 1 \end{bmatrix}\begin{bmatrix} 2 & 0 & 0 \\ 1 & -2 & 1 \\ 5 & 4 & -2 \end{bmatrix} = \begin{bmatrix} 12 & 8 & -4 \\ -3 & -10 & 5 \\ 8 & 2 & -1 \end{bmatrix}$$

$$\text{(b) } BA = \begin{bmatrix} 2 & 0 & 0 \\ 1 & -2 & 1 \\ 5 & 4 & -2 \end{bmatrix}\begin{bmatrix} 1 & 0 & 2 \\ 3 & 1 & -2 \\ 1 & 1 & 1 \end{bmatrix} = \begin{bmatrix} 2 & 0 & 4 \\ -4 & -1 & 7 \\ 15 & 2 & 0 \end{bmatrix}$$

$$\text{(c) } A^2 = \begin{bmatrix} 1 & 0 & 2 \\ 3 & 1 & -2 \\ 1 & 1 & 1 \end{bmatrix}\begin{bmatrix} 1 & 0 & 2 \\ 3 & 1 & -2 \\ 1 & 1 & 1 \end{bmatrix} = \begin{bmatrix} 3 & 2 & 4 \\ 4 & -1 & 2 \\ 5 & 2 & 1 \end{bmatrix}$$

27. Find AB (if possible).

$$A = \begin{bmatrix} 4 & 0 & 0 \\ 0 & 3 & 0 \\ 0 & 0 & -2 \end{bmatrix}, \quad B = \begin{bmatrix} \frac{1}{4} & 0 & 0 \\ 0 & \frac{1}{3} & 0 \\ 0 & 0 & -\frac{1}{2} \end{bmatrix}$$

Solution

$$AB = \begin{bmatrix} 4 & 0 & 0 \\ 0 & 3 & 0 \\ 0 & 0 & -2 \end{bmatrix}\begin{bmatrix} \frac{1}{4} & 0 & 0 \\ 0 & \frac{1}{3} & 0 \\ 0 & 0 & -\frac{1}{2} \end{bmatrix} = \begin{bmatrix} 1 & 0 & 0 \\ 0 & 1 & 0 \\ 0 & 0 & 0 \end{bmatrix}$$

31. Solve for $X = 5A - 3B$ given

$$A = \begin{bmatrix} 1 & -2 \\ 0 & 1 \\ 2 & 3 \end{bmatrix} \text{ and } B = \begin{bmatrix} 0 & 1 \\ 1 & 1 \\ 3 & 5 \end{bmatrix}.$$

Solution

$$X = 5A - 3B = 5\begin{bmatrix} 1 & -2 \\ 0 & 1 \\ 2 & 3 \end{bmatrix} - 3\begin{bmatrix} 0 & 1 \\ 1 & 1 \\ 3 & 5 \end{bmatrix}$$

$$= \begin{bmatrix} 5 & -10 \\ 0 & 5 \\ 10 & 15 \end{bmatrix} - \begin{bmatrix} 0 & 3 \\ 3 & 3 \\ 9 & 15 \end{bmatrix} = \begin{bmatrix} 5 & -13 \\ -3 & 2 \\ 1 & 0 \end{bmatrix}$$

35. *Inventory Levels* A company sells four different models of car sound systems through three retail outlets. The inventory of each model in the three outlets is given by the matrix S shown in the textbook. The wholesale and retail price for each model is given by the matrix T, also shown in the textbook.

Use a graphing utility to compute ST and interpret the result.

Solution

$$ST = \begin{bmatrix} 2 & 3 & 2 & 1 \\ 0 & 4 & 3 & 3 \\ 4 & 0 & 1 & 2 \end{bmatrix}\begin{bmatrix} 300 & 500 \\ 400 & 650 \\ 200 & 350 \\ 800 & 1200 \end{bmatrix} = \begin{array}{c} \begin{array}{cc} \text{Wholesale} & \text{Retail} \end{array} \\ \begin{bmatrix} 3000 & 4850 \\ 4600 & 7250 \\ 3000 & 4750 \end{bmatrix} \end{array} \begin{array}{c} \text{A} \\ \text{B} \\ \text{C} \end{array}$$

This represents the wholesale and retail price of the inventory at each outlet.

39. Find the inverse of the matrix (if it exists).

$$\begin{bmatrix} 1 & 3 \\ 2 & 5 \end{bmatrix}$$

Solution

$$[A \;\vdots\; I] = \left[\begin{array}{cc|cc} 1 & 3 & 1 & 0 \\ 2 & 5 & 0 & 1 \end{array}\right] \Rightarrow \left[\begin{array}{cc|cc} 1 & 3 & 1 & 0 \\ 0 & -1 & -2 & 1 \end{array}\right]$$

$$\Rightarrow \left[\begin{array}{cc|cc} 1 & 0 & -5 & 3 \\ 0 & 1 & 2 & -1 \end{array}\right] = [I \;\vdots\; A^{-1}]$$

$$A^{-1} = \begin{bmatrix} -5 & 3 \\ 2 & -1 \end{bmatrix}$$

41. Find the inverse of the following matrix (if it exists).

$$\begin{bmatrix} -1 & 0 & 0 & 0 \\ 0 & 2 & 0 & 0 \\ 0 & 0 & 4 & 0 \\ 0 & 0 & 0 & 6 \end{bmatrix}$$

Solution

Since the matrix only has nonzero entries on the main diagonal,

$$\begin{bmatrix} -1 & 0 & 0 & 0 \\ 0 & 2 & 0 & 0 \\ 0 & 0 & 4 & 0 \\ 0 & 0 & 0 & 6 \end{bmatrix}^{-1} = \begin{bmatrix} -1 & 0 & 0 & 0 \\ 0 & \frac{1}{2} & 0 & 0 \\ 0 & 0 & \frac{1}{4} & 0 \\ 0 & 0 & 0 & \frac{1}{6} \end{bmatrix}.$$

45. Use an inverse matrix to solve the following system of linear equations. (Use the inverse matrix found in Exercise 42, Section 7–R.)

$$\begin{aligned} 3x + 2y + 2z &= 13 \\ 2y + z &= 4 \\ x + z &= 5 \end{aligned}$$

Solution

$$\begin{bmatrix} 3 & 2 & 2 \\ 0 & 2 & 1 \\ 1 & 0 & 1 \end{bmatrix} \begin{bmatrix} x \\ y \\ z \end{bmatrix} = \begin{bmatrix} 13 \\ 4 \\ 5 \end{bmatrix}$$

$$\begin{bmatrix} x \\ y \\ z \end{bmatrix} = \frac{1}{4} \begin{bmatrix} 2 & -2 & -2 \\ 1 & 1 & -3 \\ -2 & 2 & 6 \end{bmatrix} \begin{bmatrix} 13 \\ 4 \\ 5 \end{bmatrix} = \begin{bmatrix} 2 \\ \frac{1}{2} \\ 3 \end{bmatrix}$$

Answer: $x = 2,\ y = \frac{1}{2},\ z = 3$

49. *Investment Portfolio* Consider a person who invests in blue-chip stock, common stock, and municipal bonds. The average yield is 8% on the blue-chip stock, 6% on common stock, and 7% on municipal bonds. Twice as much is invested in municipal bonds as in common stock. Moreover, the total annual return for all three types of investment (stocks and bonds) is $2,400. A system of linear equations (where x, y, and z represent the amounts invested in blue-chip stock, common stock, and municipal bonds) is as follows.

$$\begin{aligned} x + y + z &= \text{(total investment)} \\ 0.08x + 0.06y + 0.07z &= 2400 \\ 2y - z &= 0 \end{aligned}$$

Use the inverse of the coefficient matrix of this system to find the amount invested in each type of bond when the total investment equals $32,000.

Solution

$$[A \ \vdots \ I] = \left[\begin{array}{ccc|ccc} 1 & 1 & 1 & 1 & 0 & 0 \\ 0.08 & 0.06 & 0.07 & 0 & 1 & 0 \\ 0 & 2 & -1 & 0 & 0 & 1 \end{array}\right]$$

$$-0.08R_1 + R_2 \rightarrow \left[\begin{array}{ccc|ccc} 1 & 1 & 1 & 1 & 0 & 0 \\ 0 & -0.02 & -0.01 & -0.08 & 1 & 0 \\ 0 & 2 & -1 & 0 & 0 & 1 \end{array}\right]$$

$$\begin{array}{r} -R_2 + R_1 \rightarrow \\ -50R_2 \rightarrow \\ -2R_2 + R_3 \rightarrow \end{array} \left[\begin{array}{ccc|ccc} 1 & 0 & 0.5 & -3 & 50 & 0 \\ 0 & 1 & 0.5 & 4 & -50 & 0 \\ 0 & 0 & -2 & -8 & 100 & 1 \end{array}\right]$$

$$\begin{array}{r} -0.5R_3 + R_1 \rightarrow \\ -0.5R_3 + R_2 \rightarrow \\ -\frac{1}{2}R_3 \rightarrow \end{array} \left[\begin{array}{ccc|ccc} 1 & 0 & 0 & -5 & 75 & 0.25 \\ 0 & 1 & 0 & 2 & -25 & 0.25 \\ 0 & 0 & 1 & 4 & -50 & -0.50 \end{array}\right]$$

$$= [I \ \vdots \ A^{-1}]$$

$$A^{-1} = \begin{bmatrix} -5 & 75 & 0.25 \\ 2 & -25 & 0.25 \\ 4 & -50 & -0.50 \end{bmatrix}$$

$$\begin{bmatrix} x \\ y \\ z \end{bmatrix} = \begin{bmatrix} -5 & 75 & 0.25 \\ 2 & -25 & 0.25 \\ 4 & -50 & -0.50 \end{bmatrix} \begin{bmatrix} 32{,}000 \\ 2{,}400 \\ 0 \end{bmatrix} = \begin{bmatrix} 20{,}000 \\ 4{,}000 \\ 8{,}800 \end{bmatrix}$$

Answer: $20,000 at 8%
$4,000 at 6%
$8,000 at 7%

53. Find the determinant of the matrix.

$$\begin{bmatrix} 5 & 2 \\ 0 & 0 \end{bmatrix}$$

Solution

$$\begin{vmatrix} 5 & 2 \\ 0 & 0 \end{vmatrix} = 0 - 0 = 0$$

59. Find the determinant of the matrix.

$$\begin{bmatrix} 3 & 0 & 0 & 0 \\ 0 & 2 & 0 & 0 \\ 0 & 0 & -1 & 0 \\ 0 & 0 & 0 & -10 \end{bmatrix}$$

Solution

$$\begin{vmatrix} 3 & 0 & 0 & 0 \\ 0 & 2 & 0 & 0 \\ 0 & 0 & -1 & 0 \\ 0 & 0 & 0 & -10 \end{vmatrix} = (3)(2)(-1)(-10) = 60$$

63. Use a determinant to find the area of a triangle with the vertices $(-1, 1)$, $(2, 3)$, and $(4, -1)$.

Solution

$$\text{Area} = \pm\tfrac{1}{2}\begin{vmatrix} -1 & 1 & 1 \\ 2 & 3 & 1 \\ 4 & -1 & 1 \end{vmatrix} = \pm\tfrac{1}{2}[-1(4) - 1(-2) + 1(-14)] = 8 \text{ square units}$$

67. Use a determinant to find an equation of the line through the points $(-5, 2)$ and $(3, 0)$.

Solution

$$\text{Equation: } \begin{vmatrix} x & y & 1 \\ -5 & 2 & 1 \\ 3 & 0 & 1 \end{vmatrix} = x(2) - y(-8) + 1(-6) = 0 \quad \text{or} \quad 2x + 8y = 6$$

$$x + 4y - 3 = 0$$

69. Use the matrix to encode the message, HAPPY ANNIVERSARY.

$$A = \begin{bmatrix} 2 & 3 \\ 3 & 4 \end{bmatrix}$$

Solution

The uncoded row matrices are the rows of the 9×2 matrix on the left.

$$\begin{matrix} \text{H A} \\ \text{P P} \\ \text{Y} \\ \text{A N} \\ \text{N I} \\ \text{V E} \\ \text{R S} \\ \text{A R} \\ \text{Y} \end{matrix} \begin{bmatrix} 8 & 1 \\ 16 & 16 \\ 25 & 0 \\ 1 & 14 \\ 14 & 9 \\ 22 & 5 \\ 18 & 19 \\ 1 & 18 \\ 25 & 0 \end{bmatrix} \begin{bmatrix} 2 & 3 \\ 3 & 4 \end{bmatrix} = \begin{bmatrix} 19 & 28 \\ 80 & 112 \\ 50 & 75 \\ 44 & 59 \\ 55 & 78 \\ 59 & 86 \\ 93 & 130 \\ 56 & 75 \\ 50 & 75 \end{bmatrix}$$

Encoded: 19 28 80 112 50 75 44 59 55 78 59 86 93
130 56 75 50 75

Test for Chapter 7

1. Write an augmented matrix that represents the system.

$$\begin{aligned} 2x - y &= -3 \\ x + y &= 2 \end{aligned}$$

Solution

$$\left[\begin{array}{cc:c} 2 & -1 & -3 \\ 1 & 1 & 2 \end{array}\right]$$

2. Write an augmented matrix that represents the system.

$$\begin{aligned} 3x + 2y + 2z &= 17 \\ 2y + z &= -1 \\ x + z &= 8 \end{aligned}$$

Solution

$$\left[\begin{array}{ccc:c} 3 & 2 & 2 & 17 \\ 0 & 2 & 1 & -1 \\ 1 & 0 & 1 & 8 \end{array}\right]$$

3. Use matrices to solve the system.

$$\begin{aligned} 3x + 4y &= -6 \\ -x + 3y &= -11 \end{aligned}$$

Solution

$$\left[\begin{array}{cc:c} -1 & 3 & -11 \\ 3 & 4 & -6 \end{array}\right] \quad \begin{array}{r} -R_1 \rightarrow \\ -3R_1 + R_2 \rightarrow \end{array} \left[\begin{array}{cc:c} 1 & -3 & 11 \\ 0 & 13 & -39 \end{array}\right] \quad \tfrac{1}{13}R_2 \rightarrow \left[\begin{array}{cc:c} 1 & 0 & 2 \\ 0 & 1 & -3 \end{array}\right]$$

$x = 2$

$y = -3$

Answer: $(2, -3)$

4. Use matrices to solve the system.

$$\begin{aligned} x - 2y + z &= 14 \\ y - 3z &= 2 \\ z &= -6 \end{aligned}$$

Solution

$$\left[\begin{array}{ccc:c} 1 & -2 & 1 & 14 \\ 0 & 1 & -3 & 2 \\ 0 & 0 & 1 & -6 \end{array}\right] \quad 2R_2 + R_1 \rightarrow \left[\begin{array}{ccc:c} 1 & 0 & -5 & 18 \\ 0 & 1 & -3 & 2 \\ 0 & 0 & 1 & -6 \end{array}\right]$$

$$\begin{array}{r} 5R_3 + R_1 \rightarrow \\ 3R_3 + R_2 \rightarrow \\ \end{array} \left[\begin{array}{ccc:c} 1 & 0 & 0 & -12 \\ 0 & 1 & 0 & -16 \\ 0 & 0 & 1 & -6 \end{array}\right]$$

$x = -12, y = -16, z = -6$

Answer: $(-12, -16, -6)$

5. Use matrices to solve the system.

$$\begin{aligned} 2x - 3y + z &= 14 \\ x + 2y \quad &= -4 \\ y - z &= -4 \end{aligned}$$

Solution

$$\left[\begin{array}{rrr|r} 2 & -3 & 1 & 14 \\ 1 & 2 & 0 & -4 \\ 0 & 1 & -1 & -4 \end{array}\right] \quad \begin{array}{r} -R_2 + R_1 \rightarrow \\ \\ \\ \end{array} \left[\begin{array}{rrr|r} 1 & -5 & 1 & 18 \\ 1 & 2 & 0 & -4 \\ 0 & 1 & -1 & -4 \end{array}\right]$$

$$\begin{array}{r} \\ -R_1 + R_2 \rightarrow \\ \\ \end{array} \left[\begin{array}{rrr|r} 1 & -5 & 1 & 18 \\ 0 & 7 & -1 & -22 \\ 0 & 1 & -1 & -4 \end{array}\right]$$

$$\begin{array}{r} \\ R_3 \\ R_2 \end{array} \left[\begin{array}{rrr|r} 1 & -5 & 1 & 18 \\ 0 & 1 & -1 & -4 \\ 0 & 7 & -1 & -22 \end{array}\right]$$

$$\begin{array}{r} 5R_2 + R_1 \rightarrow \\ \\ -7R_2 + R_3 \rightarrow \end{array} \left[\begin{array}{rrr|r} 1 & 0 & -4 & -2 \\ 0 & 1 & -1 & -4 \\ 0 & 0 & 6 & 6 \end{array}\right]$$

$$\begin{array}{r} 4R_3 + R_1 \rightarrow \\ R_3 + R_2 \rightarrow \\ \frac{1}{6}R_3 \rightarrow \end{array} \left[\begin{array}{rrr|r} 1 & 0 & 0 & 2 \\ 0 & 1 & 0 & -3 \\ 0 & 0 & 1 & 1 \end{array}\right]$$

$x = 2,\ y = -3,\ z = 1$

Answer: $(2, -3, 1)$

In Exercises 6–9, use the matrices to find the indicated matrix.

$$A = \begin{bmatrix} 1 & 3 \\ 2 & 4 \end{bmatrix} \qquad B = \begin{bmatrix} 2 & -1 & 3 \\ 4 & 0 & 1 \end{bmatrix}$$

$$C = \begin{bmatrix} 0 & -2 \\ 3 & 5 \end{bmatrix} \qquad D = \begin{bmatrix} 3 \\ 2 \\ -1 \end{bmatrix}$$

6. Find $2A + C$.

Solution

$$2A + C = 2\begin{bmatrix} 1 & 3 \\ 2 & 4 \end{bmatrix} + \begin{bmatrix} 0 & -2 \\ 3 & 5 \end{bmatrix} = \begin{bmatrix} 2 & 6 \\ 4 & 8 \end{bmatrix} + \begin{bmatrix} 0 & -2 \\ 3 & 5 \end{bmatrix} = \begin{bmatrix} 2 & 4 \\ 7 & 13 \end{bmatrix}$$

7. Find AB.

Solution

$$AB = \begin{bmatrix} 1 & 3 \\ 2 & 4 \end{bmatrix}\begin{bmatrix} 2 & -1 & 3 \\ 4 & 0 & 1 \end{bmatrix} = \begin{bmatrix} 14 & -1 & 6 \\ 20 & -2 & 10 \end{bmatrix}$$

8. Find BD.

Solution

$$BD = \begin{bmatrix} 2 & -1 & 3 \\ 4 & 0 & 1 \end{bmatrix}\begin{bmatrix} 3 \\ 2 \\ -1 \end{bmatrix} = \begin{bmatrix} 1 \\ 11 \end{bmatrix}$$

9. Find A^2.

Solution

$$A^2 = \begin{bmatrix} 1 & 3 \\ 2 & 4 \end{bmatrix}\begin{bmatrix} 1 & 3 \\ 2 & 4 \end{bmatrix} = \begin{bmatrix} 7 & 15 \\ 10 & 22 \end{bmatrix}$$

10. Find the inverse of the matrix $A = \begin{bmatrix} 2 & -1 \\ -3 & 4 \end{bmatrix}$.

Solution

$$A^{-1} = \frac{1}{2(4) - (-1)(-3)}\begin{bmatrix} 4 & -(-1) \\ -(-3) & 2 \end{bmatrix} = \frac{1}{5}\begin{bmatrix} 4 & 1 \\ 3 & 2 \end{bmatrix} = \begin{bmatrix} 0.8 & 0.2 \\ 0.6 & 0.4 \end{bmatrix}$$

11. Find the inverse of the matrix $A = \begin{bmatrix} 1 & 0 \\ 0 & 1 \end{bmatrix}$.

Solution

Since $A = I$, $A^{-1} = A = \begin{bmatrix} 1 & 0 \\ 0 & 1 \end{bmatrix}$.

12. Find the inverse of the matrix $A = \begin{bmatrix} 3 & 2 & 2 \\ 0 & 2 & 1 \\ 1 & 0 & 1 \end{bmatrix}$

Solution

$$[A \ \vdots \ I] = \left[\begin{array}{ccc|ccc} 3 & 2 & 2 & 1 & 0 & 0 \\ 0 & 2 & 1 & 0 & 1 & 0 \\ 1 & 0 & 1 & 0 & 0 & 1 \end{array}\right] \Rightarrow \left[\begin{array}{ccc|ccc} 1 & 2 & 0 & 1 & 0 & -2 \\ 0 & 2 & 1 & 0 & 1 & 0 \\ 0 & -2 & 1 & -1 & 0 & 3 \end{array}\right]$$

$$\Rightarrow \left[\begin{array}{ccc|ccc} 1 & 0 & -1 & 1 & -1 & -2 \\ 0 & 1 & \frac{1}{2} & 0 & \frac{1}{2} & 0 \\ 0 & 0 & 2 & -1 & 1 & 3 \end{array}\right] \Rightarrow \left[\begin{array}{ccc|ccc} 1 & 0 & 0 & \frac{1}{2} & -\frac{1}{2} & -\frac{1}{2} \\ 0 & 1 & 0 & \frac{1}{4} & \frac{1}{4} & -\frac{3}{4} \\ 0 & 0 & 1 & -\frac{1}{2} & \frac{1}{2} & \frac{3}{2} \end{array}\right]$$

$$= [I \ \vdots \ A^{-1}]$$

$$A^{-1} = \begin{bmatrix} \frac{1}{2} & -\frac{1}{2} & -\frac{1}{2} \\ \frac{1}{4} & \frac{1}{4} & -\frac{3}{4} \\ -\frac{1}{2} & \frac{1}{2} & \frac{3}{2} \end{bmatrix} = \frac{1}{4}\begin{bmatrix} 2 & -2 & -2 \\ 1 & 1 & -3 \\ -2 & 2 & 6 \end{bmatrix} = \begin{bmatrix} 0.50 & -0.50 & -0.50 \\ 0.25 & 0.25 & -0.75 \\ -0.50 & 0.50 & 1.50 \end{bmatrix}$$

13. Find the determinant of the matrix.

$$\begin{bmatrix} 3 & -1 \\ 4 & 7 \end{bmatrix}$$

Solution

$$\begin{vmatrix} 3 & -1 \\ 4 & 7 \end{vmatrix} = 3(7) - (-1)(4) = 21 + 4 = 25$$

14. Find the determinant of the matrix.

$$\begin{bmatrix} 3 & 2 & -1 \\ 1 & 0 & 2 \\ 4 & 5 & 2 \end{bmatrix}$$

Solution

$$\begin{vmatrix} 3 & 2 & -1 \\ 1 & 0 & 2 \\ 4 & 5 & 2 \end{vmatrix} = -1\begin{vmatrix} 2 & -1 \\ 5 & 2 \end{vmatrix} - 2\begin{vmatrix} 3 & 2 \\ 4 & 5 \end{vmatrix} = -1(9) - 2(7) = -23$$

(Use Row 2)

15. Find the determinant of the matrix.

$$\begin{bmatrix} 2 & 0 & 0 \\ 0 & 5 & 0 \\ 0 & 0 & -2 \end{bmatrix}$$

Solution

$$\begin{vmatrix} 2 & 0 & 0 \\ 0 & 5 & 0 \\ 0 & 0 & -2 \end{vmatrix} = 2(5)(-2) = -20$$

16. Use the inverse of the matrix A in Exercise 12 to solve the system in Exercise 2.

$$\begin{aligned} 3x + 2y + 2z &= 17 \\ 2y + z &= -1 \\ x + z &= 8 \end{aligned}$$

Solution

$$\begin{bmatrix} 3 & 2 & 2 \\ 0 & 2 & 1 \\ 1 & 0 & 1 \end{bmatrix}\begin{bmatrix} x \\ y \\ z \end{bmatrix} = \begin{bmatrix} 17 \\ -1 \\ 8 \end{bmatrix}$$

$$\begin{bmatrix} x \\ y \\ z \end{bmatrix} = \frac{1}{4}\begin{bmatrix} 2 & -2 & -2 \\ 1 & 1 & -3 \\ -2 & 2 & 6 \end{bmatrix}\begin{bmatrix} 17 \\ -1 \\ 8 \end{bmatrix} = \begin{bmatrix} 5 \\ -2 \\ 3 \end{bmatrix}$$

Answer: $(5, -2, 3)$

17. Find two nonzero matrices whose product is a zero matrix.

Solution

Answers will vary. One possible solution is

$$A = \begin{bmatrix} 2 & -2 \\ -2 & 2 \end{bmatrix} \text{ and } B = \begin{bmatrix} 3 & 3 \\ 3 & 3 \end{bmatrix}.$$

18. Use a determinant to decide whether $(2, 9)$, $(-2, 1)$, $(3, 11)$ are collinear.

Solution

$$\begin{vmatrix} 2 & 9 & 1 \\ -2 & 1 & 1 \\ 3 & 11 & 1 \end{vmatrix} = 2(-10) - 9(-5) + 1(-25) = 0$$

The points are collinear.

19. Find the area of the triangle whose vertices are $(-2, 4)$, $(0, 5)$, and $(3, -1)$.

Solution

$$A = \pm\frac{1}{2}\begin{vmatrix} -2 & 4 & 1 \\ 0 & 5 & 1 \\ 3 & -1 & 1 \end{vmatrix} = -\frac{1}{2}(-15) = \frac{15}{2}$$

20. A manufacturer produces three models of a product, which are shipped to two warehouses. The number of units i that are shipped to warehouse j is represented by a_{ij} in the matrix

$$A = \begin{bmatrix} 1000 & 3000 \\ 2000 & 4000 \\ 5000 & 8000 \end{bmatrix}.$$

The price per unit is represented by the matrix $B = [\$25 \quad \$20 \quad \$32]$. Find the product BA and interpret the result.

Solution

$$BA = [25 \quad 20 \quad 32]\begin{bmatrix} 1000 & 3000 \\ 2000 & 4000 \\ 5000 & 8000 \end{bmatrix} = [225{,}000 \quad 411{,}000]$$

\$225,000 represents the value of the inventory at one warehouse and \$411,000 represents the value of the inventory at the other warehouse, or the total expense of stocking each warehouse.

Practice Test for Chapter 7

1. Put the matrix in reduced echelon form.

$$\begin{bmatrix} 1 & -2 & 4 \\ 3 & -5 & 9 \end{bmatrix}$$

For Exercises 2–4, use matrices to solve the system of equations.

2. $3x + 5y = 3$
$2x - y = -11$

3. $2x + 3y = -3$
$3x + 2y = 8$
$x + y = 1$

4. $x + 3z = -5$
$2x + y = 0$
$3x + y - z = 3$

5. Multiply $\begin{bmatrix} 1 & 4 & 5 \\ 2 & 0 & -3 \end{bmatrix}\begin{bmatrix} 1 & 6 \\ 0 & -7 \\ -1 & 2 \end{bmatrix}$.

6. Given $A = \begin{bmatrix} 9 & 1 \\ -4 & 8 \end{bmatrix}$ and $B = \begin{bmatrix} 6 & -2 \\ 3 & 5 \end{bmatrix}$, find $3A - 5B$.

7. Using the matrices in Exercise 6, find AB and BA.

8. True or false: $(A + B)(A + 3B) = A^2 + 4AB + 3B^2$ where A and B are matrices.

For Exercises 9–10, find the inverse of the matrix, if it exists.

9. $\begin{bmatrix} 1 & 2 \\ 3 & 5 \end{bmatrix}$

10. $\begin{bmatrix} 1 & 1 & 1 \\ 3 & 6 & 5 \\ 6 & 10 & 8 \end{bmatrix}$

11. Use an inverse matrix to solve the systems:

(a) $x + 2y = 4$
$3x + 5y = 1$

(b) $x + 2y = 3$
$3x + 5y = -2$

For Exercises 12–14, find the determinant of the matrix.

12. $\begin{bmatrix} 6 & -1 \\ 3 & 4 \end{bmatrix}$

13. $\begin{bmatrix} 1 & 3 & -1 \\ 5 & 9 & 0 \\ 6 & 2 & -5 \end{bmatrix}$

14. $\begin{bmatrix} 1 & 4 & 2 & 3 \\ 0 & 1 & -2 & 0 \\ 3 & 5 & -1 & 1 \\ 2 & 0 & 6 & 1 \end{bmatrix}$

15. True or false:

$$\begin{vmatrix} 3 & 0 & 0 \\ 0 & 3 & 0 \\ 0 & 0 & 3 \end{vmatrix} = -3^3 \begin{vmatrix} 1 & 0 & 0 \\ 0 & 0 & 1 \\ 0 & 1 & 0 \end{vmatrix}$$

16. Evaluate the determinant of the matrix.

$$\begin{bmatrix} 6 & 4 & 3 & 0 & 6 \\ 0 & 5 & 1 & 4 & 8 \\ 0 & 0 & 2 & 7 & 3 \\ 0 & 0 & 0 & 9 & 2 \\ 0 & 0 & 0 & 0 & 1 \end{bmatrix}$$

17. Use a determinant to find the area of the triangle with vertices $(6, 3)$, $(-1, 4)$, and $(2, -5)$.

18. Use a determinant to determine if the points $(0, -7)$, $(3, 2)$, and $(-4, -19)$ are collinear.

19. Use a determinant to find the equation of the line through $(4,\ -3)$ and $(\frac{1}{2},\ 7)$.

20. Encode the message, STAND FIRM, using the matrix $\begin{bmatrix} 3 & -1 \\ 7 & 5 \end{bmatrix}$.

CHAPTER EIGHT
Sequences and Probability

8.1 Sequences and Summation Notation

- Given the general nth term in a sequence, you should be able to find, or list, the terms.
- You should be able to find an expression for the nth term of a sequence.
- Be able to evaluate terms of a sequence involving factorials.
- You should be able to use sigma notation for a sum.
- Know the properties of sums.

(a) $\sum_{i=1}^{n} ca_i = c\sum_{i=1}^{n} a_i$, c a constant

(b) $\sum_{i=1}^{n} (a_i + b_i) = \sum_{i=1}^{n} a_i + \sum_{i=1}^{n} b_i$

(c) $\sum_{i=1}^{n} (a_i - b_i) = \sum_{i=1}^{n} a_i - \sum_{i=1}^{n} b_i$

SOLUTIONS TO SELECTED EXERCISES

5. Write the first five terms of $a_n = (-2)^n$. (Assume n begins with 1.)

Solution

$a_n = (-2)^n$

$a_1 = (-2)^1 = -2$

$a_2 = (-2)^2 = 4$

$a_3 = (-2)^3 = -8$

$a_4 = (-2)^4 = 16$

$a_5 = (-2)^5 = -32$

9. Write the first five terms of $a_n = 1 + (-1)^n$. (Assume n begins with 1.)

Solution

$a_n = 1 + (-1)^n$

$a_1 = 1 + (-1)^1 = 0$

$a_2 = 1 + 1 = 2$

$a_3 = 1 - 1 = 0$

$a_4 = 2$

$a_5 = 0$

13. Write the first five terms of the following sequence. (Assume n begins with 1.)

$$a_n = \frac{3^n}{n!}$$

Solution

$$a_n = \frac{3^n}{n!}$$

$$a_1 = \frac{3^1}{1!} = \frac{3}{1} = 3$$

$$a_2 = \frac{3^2}{2!} = \frac{9}{2}$$

$$a_3 = \frac{27}{6} = \frac{9}{2}$$

$$a_4 = \frac{81}{24} = \frac{27}{8}$$

$$a_5 = \frac{243}{120} = \frac{81}{40}$$

17. Write the first five terms of the following sequence. (Assume n begins with 1.)

$$a_n = \frac{(-1)^n}{n^2}.$$

Solution

$$a_n = \frac{(-1)^n}{n^2}$$

$$a_1 = \frac{-1}{1} = -1$$

$$a_2 = \frac{1}{4}$$

$$a_3 = \frac{-1}{9}$$

$$a_4 = \frac{1}{16}$$

$$a_5 = \frac{-1}{25}$$

23. Evaluate $\frac{6!}{4!}$.

Solution

$$\frac{6!}{4!} = \frac{6 \cdot 5 \cdot 4!}{4!} = 30$$

27. Simplify the expression $\frac{(n+1)!}{n!}$. Then evaluate the expression when $n = 0$ and $n = 1$.

Solution

$$\frac{(n+1)!}{n!} = \frac{(n+1)n!}{n!} = n+1$$

$n = 0$: $n + 1 = 1$

$n = 1$: $n + 1 = 2$

31. Write an expression for the *most apparent* nth term of 1, 4, 7, 10, 13, ... (Assume n begins with 1.)

Solution

$$a_n = 1 + (n-1)3 = 3n - 2$$

35. Write an expression for the *most apparent* nth term of $\frac{2}{3}, \frac{3}{4}, \frac{4}{5}, \frac{5}{6}, \frac{6}{7}, \ldots$ (Assume n begins with 1.)

Solution

$$a_n = \frac{n+1}{n+2}$$

39. Write an expression for the *most apparent* nth term of the following. (Assume n begins with 1.)

$$1 + \frac{1}{1},\ 1 + \frac{1}{2},\ 1 + \frac{1}{3},\ 1 + \frac{1}{4},\ 1 + \frac{1}{5},\ \ldots$$

Solution

$$a_n = 1 + \frac{1}{n}$$

43. Write an expression for the *most apparent* nth term of $1, -1, 1, -1, 1, \ldots$ (Assume n begins with 1.)

Solution

$a_n = (-1)^{n+1}$

47. Find the sum.

$$\sum_{i=1}^{5}(2i+1)$$

Solution

$$\sum_{i=1}^{5}(2i+1) = (2+1)+(4+1)+(6+1)+(8+1)+(10+1) = 35$$

51. Find the sum.

$$\sum_{i=0}^{4} i^2$$

Solution

$$\sum_{i=0}^{4} i^2 = 0^2+1^2+2^2+3^2+4^2 = 30$$

55. Find the sum.

$$\sum_{i=1}^{4}(i-1)^2$$

Solution

$$\sum_{i=1}^{4}(i-1)^2 = (1-1)^2+(2-1)^2+(3-1)^2+(4-1)^2 = 14$$

59. Use summation notation to write the sum.

$$\frac{1}{3(1)}+\frac{1}{3(2)}+\frac{1}{3(3)}+\cdots+\frac{1}{3(9)}$$

Solution

$$\frac{1}{3(1)}+\frac{1}{3(2)}+\frac{1}{3(3)}+\cdots+\frac{1}{3(9)} = \sum_{i=1}^{9}\frac{1}{3i}$$

63. Use summation notation to write the sum.

$$3-9+27-81+243-729$$

Solution

$$3-9+27-81+243-729 = \sum_{i=1}^{6}(-1)^{i+1}3^i$$

67. Use summation notation to write the sum.

$$\frac{1}{4}+\frac{3}{8}+\frac{7}{16}+\frac{15}{32}+\frac{31}{64}$$

Solution

$$\frac{1}{4}+\frac{3}{8}+\frac{7}{16}+\frac{15}{32}+\frac{31}{64} = \sum_{i=1}^{5}\frac{2^i-1}{2^{i+1}}$$

71. *Ratio of Men to Women* Until the mid-1940s, the population of the United States had more men than women. After that, there were more women than men. The ratio of men to women is approximately given by the model

$$a_n = 1.09 - 0.027n + 0.0012n^2, \qquad n = 1, 2, \ldots, 9$$

where a_n is the ratio of men to women and n is the year with $n = 1, 2, 3, \ldots, 9$ corresponding to 1910, 1920, 1930, ..., 1990. (*Source: U.S. Bureau of Census.*)

(a) Use a graphing utility to find the terms of this finite sequence.

(b) Construct a bar graph that represents the sequence.

(c) In 1990, the population of the United States was 250 million. How many of these were women? How many were men?

Solution

$a_n = 1.09 - 0.027n + 0.0012n^2,\ n = 1,\ 2,\ \ldots,\ 9$

(a) $a_1 = 1.064$

$a_2 = 1.041$

$a_3 = 1.020$

$a_4 = 1.001$

$a_5 = 0.985$

$a_6 = 0.971$

$a_7 = 0.960$

$a_8 = 0.951$

$a_9 = 0.944$

(b)

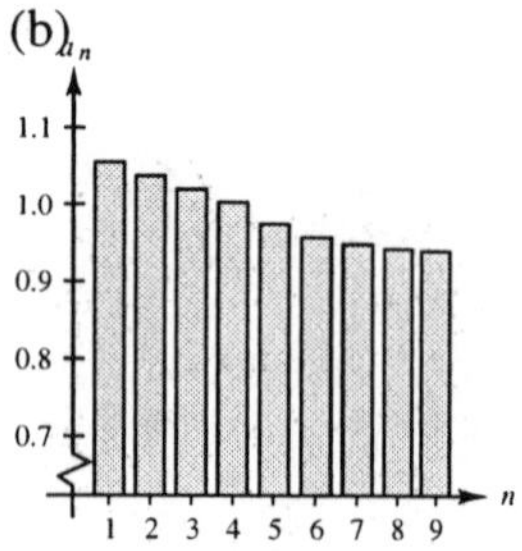

(c) Let $x =$ number of men

$$\frac{x}{250 - x} = 0.944$$

$$x = 121.4 \text{ million men}$$

$$250 - x = 128.6 \text{ million women}$$

75. *Dividends* A company declares dividends per share of common stock that can be approximated by the model $a_n = 0.35n + 1.25$, $n = 0,\ 1,\ 2,\ 3,\ 4$, where a_n is the dividend in dollars and $n = 0$ is the year corresponding to 1990. Approximate the sum of the dividends per share of common stock for the years 1990 through 1994 by evaluating

$$\sum_{n=0}^{4}(0.35n + 1.25).$$

Solution

$\sum_{n=0}^{4}(0.35n + 1.25) = 1.25 + 1.60 + 1.95 + 2.30 + 2.65 = \$9.75$

8.2 Arithmetic Sequences

- You should be able to recognize an arithmetic sequence, find its common difference, and find its nth term. $a_n = dn + c$
- You should be able to find the nth partial sum of an arithmetic sequence with common difference d using the formula

$$S_n = \frac{n}{2}(a_1 + a_n).$$

SOLUTIONS TO SELECTED EXERCISES

5. Determine whether the sequence $\frac{9}{4}, 2, \frac{7}{4}, \frac{3}{2}, \frac{5}{4}, 1, \ldots$ is arithmetic. If it is, find the common difference.

Solution

Arithmetic sequence, $d = -\frac{1}{4}$

9. Determine whether the sequence $5.3, 5.7, 6.1, 6.5, 6.9, \ldots$ is arithmetic. If it is, find the common difference.

Solution

Arithmetic sequence, $d = 0.4$

13. Write the first five terms of the specified sequence. Determine whether the sequence is arithmetic. If it is, find the common difference.

$$a_n = \frac{1}{n+1}$$

Solution

$\frac{1}{2}, \frac{1}{3}, \frac{1}{4}, \frac{1}{5}, \frac{1}{6}$

Not an arithmetic sequence

17. Write the first five terms of the sequence $a_n = (2+n) - (1+n)$. Determine whether the sequence is arithmetic. If it is, find the common difference.

Solution

$a_n = (2+n) - (1+n) = 1$

$1, 1, 1, 1, 1$

Arithmetic sequence, $d = 0$

21. Find a formula for a_n for $a_1 = 100,\ d = -8$.

Solution

$a_n = a_1 + (n-1)d = 100 + (n-1)(-8) = -8n + 108$

25. Find a formula for a_n.
$a_4 = 12,\ a_{10} = 30.$

Solution

$a_4 = 12,\ a_{10} = 30$

$a_{10} = a_4 + 6d$

$30 = 12 + 6d$

$18 = 6d$

$d = 3$

$a_n = dn + c$

$a_4 = 3(4) + c$

$12 = 12 + c$

$c = 0$

$a_n = 3n$

29. Write the first five terms.
$a_1 = 5,\ d = 6.$

Solution

$a_1 = 5$

$a_2 = 5 + 6 = 11$

$a_3 = 11 + 6 = 17$

$a_4 = 17 + 6 = 23$

$a_5 = 23 + 6 = 29$

33. Write the first five terms.
$a_1 = 2,\ a_{12} = 46.$

Solution

$46 = 2 + (12 - 1)d$

$44 = 11d$

$4 = d$

$a_1 = 2$

$a_2 = 2 + 4 = 6$

$a_3 = 6 + 4 = 10$

$a_4 = 10 + 4 = 14$

$a_5 = 14 + 4 = 18$

37. Find the sum of the first n terms.
$8,\ 20,\ 32,\ 44,\ \ldots,\ n = 10.$

Solution

$a_1 = 8,\ d = 12,\ a_{10} = 8 + 9(12) = 116$

$S_{10} = \frac{10}{2}(8 + 116) = 620$

41. Find the sum of the first n terms.
$40,\ 37,\ 34,\ 31,\ \ldots,\ n = 10.$

Solution

$a_1 = 40,\ d = -3,\ a_{10} = 40 + 9(-3) = 13$

$S_{10} = \frac{10}{2}(40 + 13) = 265$

45. Find the sum.

$$\sum_{n=1}^{50} n$$

Solution

$a_1 = 1,\ d = 1,\ a_{50} = 50$

$$\sum_{n=1}^{50} n = \frac{50}{2}(1 + 50) = 1275$$

49. Find the sum.

$$\sum_{n=1}^{500}(n+3)$$

Solution

$a_1 = 4, \quad d = 1, \quad a_{500} = 503$

$$\sum_{n=1}^{500}(n+3) = \tfrac{500}{2}(4+503) = 126{,}750$$

53. Find the sum.

$$\sum_{n=0}^{50}(100-5n)$$

Solution

$a_0 = 1000, \quad d = -5, \quad a_{50} = 750$

$$\sum_{n=0}^{50}(100-5n) = \tfrac{51}{2}(1000+750)$$

$$= 44{,}625$$

57. *Job Offer* A person accepts a position with a company and will receive a salary of $27,500 for the first year. The person is guaranteed a raise of $1,500 per year for the first five years.

(a) What will the person's salary be during the sixth year of employment?

(b) Use a graphing utility to find how much the company paid the person at the end of 6 years.

Solution

(a) $a_1 = \$27{,}500, \quad d = \$1{,}500$

$a_6 = a_1 + 5d = 27{,}500 + 5(1500) = \$35{,}000$

(b) $S_6 = \tfrac{6}{2}[27{,}500 + 35{,}000] = \$187{,}500$

61. *Total Sales* The annual sales for Campbell Soup Company from 1984 to 1994 can be approximated by the model

$$a_n = 327.0n + 2421.1, \qquad n = 4, 5, 6, 7, \ldots, 14$$

where a_n is the total annual sales (in millions of dollars) and $n = 4$ represents 1984.

(a) Sketch a bar graph showing the annual sales for Campbell Soup Company from 1984 to 1994.

(b) Find the total sales from 1984 to 1994. (*Source: Campbell Soup Company.*)

Solution

(a)

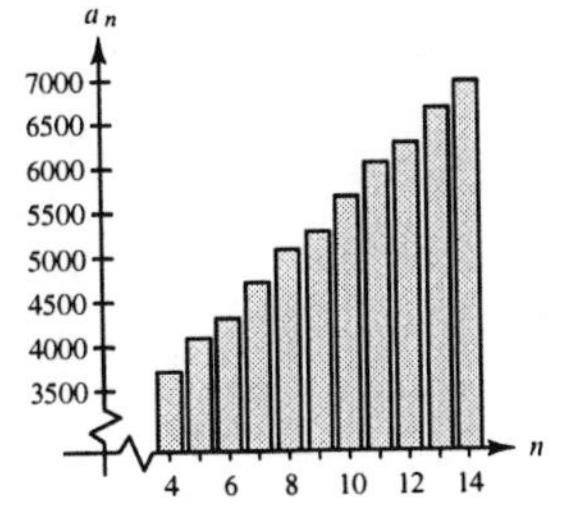

(b) $\displaystyle\sum_{n=4}^{14}(327.0n + 2421.1) = \tfrac{11}{2}(3729.1 + 6999.1) = \$59{,}005.10$

65. *Number of Logs* Logs are stacked in a pile, as shown in the figure in the textbook. The top row has 15 logs and the bottom row has 21 longs. How many logs are in the stack?

Solution

$a_1 = 15, \quad a_7 = 21$

$S_7 = \tfrac{7}{2}(15 + 21) = 126$ logs

8.3 Geometric Sequences

- You should be able to identify a geometric sequence, find its common ratio, and find the nth term. $a_n = a_1 r^{n-1}$
- You should be able to find the nth partial sum of a geometric sequence with common ratio r using the formula

$$S_n = \frac{a_1(1 - r^n)}{1 - r}, \quad r \neq 1.$$

- You should know that if $|r| < 1$, then

$$\sum_{n=0}^{\infty} a_1 r^n = \sum_{n=1}^{\infty} a_1 r^{n-1} = \frac{a_1}{1 - r}.$$

SOLUTIONS TO SELECTED EXERCISES

5. Determine whether the sequence is geometric. If it is, find its common ratio and write a formula for a_n.

$$1, -\tfrac{1}{2}, \tfrac{1}{4}, -\tfrac{1}{8}, \ldots$$

Solution

Geometric sequence, $r = -\frac{1}{2}$

$a_n = \left(-\frac{1}{2}\right)^{n-1}$

11. Write the first five terms.

$$a_1 = 2, \quad r = 3.$$

Solution

$a_1 = 2$

$a_2 = 2(3) = 6$

$a_3 = 6(3) = 18$

$a_4 = 18(3) = 54$

$a_5 = 54(3) = 162$

15. Write the first five terms.

$$a_1 = 5, \quad r = -\tfrac{1}{10}$$

Solution

$a_1 = 5$

$a_2 = 5\left(-\frac{1}{10}\right) = -\frac{1}{2}$

$a_3 = \left(-\frac{1}{2}\right)\left(-\frac{1}{10}\right) = \frac{1}{20}$

$a_4 = \frac{1}{20}\left(-\frac{1}{10}\right) = -\frac{1}{200}$

$a_5 = \left(-\frac{1}{200}\right)\left(-\frac{1}{10}\right) = \frac{1}{2{,}000}$

19. Find the indicated term of the geometric sequence.

$$a_1 = 4, \quad r = \tfrac{1}{2}, \quad \text{10th term}$$

Solution

$a_n = a_1 r^{n-1}$

$a_{10} = 4\left(\frac{1}{2}\right)^9 = \left(\frac{1}{2}\right)^7 = \frac{1}{128}$

23. Find the indicated term of the geometric sequence.

$$a_1 = 100, \quad r = e, \quad \text{9th term.}$$

Solution

$a_n = a_1 r^{n-1}$

$a_9 = 100(e)^8 = 100e^8$

27. Find the indicated term of the geometric sequence.

$a_1 = 16, \ a_4 = \frac{27}{4}$, 3rd term

Solution

$\frac{27}{4} = 16r^3 \Rightarrow r = \frac{3}{4}$

$a_n = a_1 r^{n-1}$

$a_3 = 16\left(\frac{3}{4}\right)^2 = 9$

31. Find the sum.

$$\sum_{n=1}^{10} 8(2^n)$$

Solution

$$\sum_{n=1}^{10} 8(2^n) = a_1\left(\frac{1-r^n}{1-r}\right)$$

$$= 8(2)\left[\frac{1-2^{10}}{1-2}\right] = 16{,}368$$

35. Find the sum.

$$\sum_{n=0}^{8} 2^n$$

Solution

$$\sum_{n=0}^{8} 2^n = \sum_{n=1}^{9} 2^{n-1} \Rightarrow a_1 = 1, \ r = 2$$

$$S_9 = \frac{1(1-2^9)}{1-2} = 511$$

39. Find the sum.

$$\sum_{n=0}^{\infty}\left(\frac{1}{2}\right)^n = 1 + \frac{1}{2} + \frac{1}{4} + \frac{1}{8} + \cdots$$

Solution

$a_1 = 1, \ r = \frac{1}{2}$

$$\sum_{n=0}^{\infty}\left(\frac{1}{2}\right)^n = \frac{a_1}{1-r} = \frac{1}{1-(1/2)} = 2$$

43. Find the sum.

$$\sum_{n=0}^{\infty} 4\left(\frac{1}{4}\right)^n = 4 + 1 + \frac{1}{4} + \frac{1}{16} + \cdots$$

Solution

$a_1 = 4, \ r = \frac{1}{4}$

$$\sum_{n=0}^{\infty} 4\left(\frac{1}{4}\right)^n = \frac{a_1}{1-r} = \frac{4}{1-(1/4)} = \frac{16}{3}$$

47. *Compound Interest* Suppose you deposited $75 in an account at the beginning of each month for 10 years. The account pays 8% compounded monthly. Use a graphing utility to find your balance at the end of 10 years. If the interest were compounded continuously, what would the balance be?

Solution

Compounded monthly: $A = \sum_{n=1}^{120} 75\left(1 + \frac{0.08}{12}\right)^n \approx \$13{,}812.43$

Compounded continuously: $A = \sum_{n=1}^{120} 75e^{(0.08n)/12} \approx \$13{,}833.34$

51. *Sales* The annual sales for the La-Z-Boy Chair Company from 1980 to 1989 can be approximated by the model

$$a_n = 153e^{0.159n}, \qquad n = 0, 1, 2, 3, \ldots, 9$$

where a_n is the annual sales (in millions of dollars) and n represents the year with $n = 0$ corresponding to 1980. Use the formula for the sum of a geometric sequence to approximate the total sales earned during this 10-year period. (*Source: La-Z-Boy Chair Company.*)

Solution

$$S = \sum_{n=0}^{9} 153e^{0.159n} \approx 3465.71 \text{ million dollars}$$

55. *Area* The sides of a square are 16 inches in length. A new square is formed by connecting the midpoints of the sides of the original square, and two of the triangles are shaded. If this process is repeated five more times, what will the total area of the shaded region be? (See figure in textbook.)

Solution

$$\sum_{n=1}^{6} \frac{1}{4}\left(\frac{1}{2}\right)^{n-1} \approx 0.4921875$$

Total area of shaded region: $0.4921875(16)^2 = 126$ sq. in.

Mid-Chapter Quiz for Chapter 8

1. Write the first five terms of the sequence. (Begin with $n = 1$.)

$$a_n = 3n - 1$$

Solution

$a_1 = 3(1) - 1 = 2$

$a_2 = 3(2) - 1 = 5$

$a_3 = 3(3) - 1 = 8$

$a_4 = 3(4) - 1 = 11$

$a_5 = 3(5) - 1 = 14$

2. Write the first five terms of the sequence. (Begin with $n = 1$.)

$$a_n = \frac{n}{n+1}$$

Solution

$a_1 = 3(1)\frac{1}{1+1} = \frac{1}{2}$

$a_2 = 3(2)\frac{2}{2+1} = \frac{2}{3}$

$a_3 = 3(3)\frac{3}{3+1} = \frac{3}{4}$

$a_4 = 3(4)\frac{4}{4+1} = \frac{4}{5}$

$a_5 = 3(5)\frac{5}{5+1} = \frac{5}{6}$

3. Evaluate $\frac{7!}{4!}$.

Solution

$$\frac{7!}{4!} = \frac{7 \cdot 6 \cdot 5 \cdot \cancel{4!}}{\cancel{4!}}$$
$$= 7 \cdot 6 \cdot 5$$
$$= 210$$

4. Evaluate $\frac{27!}{24!}$.

Solution

$$\frac{27!}{24!} = \frac{27 \cdot 26 \cdot 25 \cdot \cancel{24!}}{\cancel{24!}}$$
$$= 27 \cdot 26 \cdot 25$$
$$= 17{,}550$$

5. Find an expression for a_n. (Begin with $n = 1$.)

$$2, \frac{1}{2}, \frac{2}{9}, \frac{1}{8}, \frac{2}{25}, \ldots$$

Solution

Write the sequence as

$$\frac{2}{1^2}, \frac{2}{2^2}, \frac{2}{3^2}, \frac{2}{4^2}, \frac{2}{5^2}, \ldots$$

$$a_n = \frac{2}{n^2}$$

6. Find an expression for a_n. (Begin with $n = 1$.)

$$\frac{1}{2}, \frac{1}{3}, \frac{1}{4}, \frac{1}{5}, \frac{1}{6}, \ldots$$

Solution

$$a_n = \frac{1}{n+1}$$

7. Find an expression for a_n. (Begin with $n = 1$.)

Arithmetic: $a_n = 2,\ d = 5$

Solution

$$a_n = a_1 + (n-1)d$$
$$= 2 + (n-1)(5)$$
$$= 5n - 3$$

8. Find an expression for a_n. (Begin with $n = 1$.)

Arithmetic: $a_1 = 7,\ d = 3$

Solution

$$a_n = a_1 + (n-1)d$$
$$= 7 + (n-1)(3)$$
$$= 3n + 4$$

In Exercises 9–12, decide whether the sequence is arithmetic, geometric, or neither. Explain your reasoning.

9. $a_n = 2 + 3(n-1)$

Solution

The first five terms of the sequence are

$$2, 5, 8, 11, 14, \ldots$$

The common difference is 3. The sequence is arithmetic.

10. $1, -2, 4, -8, \ldots$

Solution

$$1, -2, 4, -8, \ldots$$

The common ratio is -2. The sequence is geometric.

11. $1, \frac{1}{2}, \frac{1}{4}, \frac{1}{8}, \ldots$

Solution

$$1, \frac{1}{2}, \frac{1}{4}, \frac{1}{8}, \ldots$$

The common ratio is $\frac{1}{2}$. The sequence is geometric.

12. $a_n = 2^n$

Solution

The first five terms of the sequence are

$$2, 4, 8, 16, 32, \ldots$$

The common ratio is 2. The sequence is geometric.

13. Evaluate $\sum_{i=1}^{5}(2i-1)$.

Solution

$$\sum_{i=1}^{5}(2i-1) = 1+3+5+7+9 = 25$$

14. Evaluate $\sum_{i=1}^{4} 2i^2$.

Solution

$$\sum_{i=1}^{4} 2i^2 = 2+8+18+32 = 60$$

15. Evaluate $\sum_{n=1}^{25} 2n$.

Solution

$a_n = 2n$ is an arithmetic sequence.

$a_1 = 2,\ a_{25} = 50$

$$\sum_{n=1}^{25} 2n = \tfrac{25}{2}(2+50) = 650$$

16. Evaluate $\sum_{n=1}^{20}(3n+1)$

Solution

$a_n = 3n+1$ is an arithmetic sequence.

$a_1 = 4,\ a_{20} = 61$

$$\sum_{n=1}^{20}(3n+1) = \tfrac{20}{2}(4+61) = 650$$

17. Evaluate $\sum_{n=0}^{\infty}\left(\frac{1}{3}\right)^n$

Solution

$$\sum_{n=0}^{\infty}\left(\tfrac{1}{3}\right)^n = 1 + \tfrac{1}{3} + \tfrac{1}{9} + \tfrac{1}{27} + \tfrac{1}{81} + \dots$$

$$= \frac{1}{1-(1/3)} = \frac{3}{2}$$

This is the sum of a geometric sequence with $a_1 = 1$ and $r = \frac{1}{3}$.

18. Evaluate $\sum_{n=0}^{\infty} 5\left(\frac{1}{5}\right)^n$

Solution

$$\sum_{n=0}^{\infty} 5\left(\tfrac{1}{5}\right)^n = 5 + 1 + \tfrac{1}{5} + \tfrac{1}{25} + \tfrac{1}{125} + \dots$$

$$= \frac{5}{1-(1/5)} = \frac{25}{4}$$

This is the sum of a geometric sequence with $a_1 = 5$ and $r = \frac{1}{5}$.

19. *Job Offer* A person accepts a position with a company and will receive a salary of $28,500 for the first year. The person is guaranteed a raise of $1,500 per year for the first five years.

(a) What will the salary be during the sixth year of employment?

(b) How much will the company have paid the person by the end of the sixth year?

Solution

(a) $a_6 = 28{,}500 + 5(1500) = \$36{,}000$

(b) $S_6 = \frac{6}{2}(28{,}500 + 36{,}000) = \$193{,}500$

20. *Compound Interest* Suppose that you deposit $50 in an account at the beginning of each month for ten years. The account pays 6% compounded monthly. What would your balance be at the end of 10 years?

Solution

$$\sum_{n=1}^{120} 50\left(1+\frac{0.06}{12}\right)^n \approx \$8{,}234.94$$

8.4 The Binomial Theorem

- You should be able to use the formula

$$(x + y)^n = x^n + nx^{n-1}y + \frac{n(n-1)}{2!}x^{n-2}y^2 + \cdots +_n C_r x^{n-r}y^r + \cdots + y^n$$

where ${}_nC_r = \dfrac{n!}{(n-r)!r!}$, to expand $(x + y)^n$.

- You should be able to use Pascal's Triangle in binomial expansion.

```
                1
              1   1
            1   2   1
          1   3   3   1
        1   4   6   4   1
      1   5  10  10   5   1
    1   6  15  20  15   6   1
  1   7  21  35  35  21   7   1
```

SOLUTIONS TO SELECTED EXERCISES

3. Evaluate ${}_{12}C_0$.

Solution

$${}_{12}C_0 = \frac{12!}{0!12!} = 1$$

9. Evaluate ${}_{100}C_2$.

Solution

$${}_{100}C_2 = \frac{100!}{2!98!} = \frac{100 \cdot 99}{2 \cdot 1} = 4950$$

13. Use the Binomial Theorem to expand $(x + 2)^3$. Simplify your answer.

Solution

$$\begin{aligned}(x + 2)^3 &= {}_3C_0x^3 + {}_3C_1x^2(2) + {}_3C_2x(2)^2 + {}_3C_3(2)^3 \\ &= x^3 + 3x^2(2) + 3x(2)^2 + (2)^3 \\ &= x^3 + 6x^2 + 12x + 8\end{aligned}$$

17. Use the Binomial Theorem to expand $(x + y)^5$. Simplify your answer.

Solution

$$\begin{aligned}(x + y)^5 &= {}_5C_0x^5 + {}_5C_1x^4y + {}_5C_2x^3y^2 + {}_5C_3x^2y^3 + {}_5C_4xy^4 + {}_5C_5y^5 \\ &= x^5 + 5x^4y + 10x^3y^2 + 10x^2y^3 + 5xy^4 + y^5\end{aligned}$$

21. Use the Binomial Theorem to expand $(2 + 3s)^6$. Simplify your answer.

Solution

$$\begin{aligned}(2 + 3s)^6 &= {}_6C_0(2)^6 + {}_6C_1(2)^5(3s) + {}_6C_2(2)^4(3s)^2 + {}_6C_3(2)^3(3s)^3 + {}_6C_4(2)^2(3s)^4 \\ &\quad + {}_6C_5(2)(3s)^5 + {}_6C_6(3s)^6 \\ &= 64 + 6(32)(3s) + 15(16)(9s^2) + 20(8)(27s^3) + 15(4)(81s^4) + 6(2)(243s^5) + 729s^6 \\ &= 64 + 576s + 2160s^2 + 4320s^3 + 4860s^4 + 2916s^5 + 729s^6\end{aligned}$$

25. Use the Binomial Theorem to expand $(1 - 2x)^3$. Simplify your answer.

Solution

$$\begin{aligned}(1 - 2x)^3 &= {}_3C_0 1^3 - {}_3C_1 1^2(2x) + {}_3C_2 1(2x)^2 - {}_3C_3(2x)^3 \\ &= 1 - 3(2x) + 3(2x)^2 - (2x)^3 \\ &= 1 - 6x + 12x^2 - 8x^3\end{aligned}$$

29. Use the Binomial Theorem to expand the expression. Simplify your answer.

$$\left(\frac{1}{x} + y\right)^5$$

Solution

$$\begin{aligned}\left(\frac{1}{x} + y\right)^5 &= {}_5C_0\left(\frac{1}{x}\right)^5 + {}_5C_1\left(\frac{1}{x}\right)^4 y + {}_5C_2\left(\frac{1}{x}\right)^3 y^2 + {}_5C_3\left(\frac{1}{x}\right)^2 y^3 + {}_5C_4\left(\frac{1}{x}\right) y^4 + {}_5C_5 y^5 \\ &= \frac{1}{x^5} + \frac{5y}{x^4} + \frac{10y^2}{x^3} + \frac{10y^3}{x^2} + \frac{5y^4}{x} + y^5\end{aligned}$$

33. Use the Binomial Theorem to expand $(2 - 3i)^6$. Simplify your answer by using the fact that $i^2 = -1$.

Solution

$$\begin{aligned}(2 - 3i)^6 &= {}_6C_0 2^6 - {}_6C_1 2^5(3i) + {}_6C_2 2^4(3i)^2 - {}_6C_3 2^3(3i)^3 + {}_6C_4 2^2(3i)^4 \\ &\quad - {}_6C_5 2(3i)^5 + {}_6C_6(3i)^6 \\ &= 64 - 576i - 2160 + 4320i + 4860 - 2916i - 729 \\ &= 2035 + 828i\end{aligned}$$

37. Expand $(2t - 1)^5$ using Pascal's Triangle to determine the coefficients.

Solution

5th Row of Pascal's Triangle: 1 5 10 10 5 1

$$\begin{aligned}(2t - 1)^5 &= 1(2t)^5 + 5(2t)^4(-1) + 10(2t)^3(-1)^2 + 10(2t)^2(-1)^3 + 5(2t)(-1)^4 + 1(-2)^5 \\ &= 32t^5 - 80t^4 + 80t^3 - 40t^2 + 10t - 1\end{aligned}$$

41. Find the indicated term in the expansion of the binomial expression $(x+3)^{12}$, ax^5.

Solution

The term involving x^5 in the expansion of $(x+3)^{12}$ is

$$_{12}C_7 x^5(3)^7 = \frac{12!}{7!5!} \cdot 3^7 x^5 = 1{,}732{,}104x^5.$$

45. Find the indicated term in the expansion of the binomial expression $(3x-2y)^9$, ax^4y^5.

Solution

The coefficient of x^4y^5 in the expansion of $(3x-2y)^9$ is

$$_9C_5(3)^4(-2)^5 = \frac{9!}{5!4!}(81)(-32) = -326{,}592.$$

The term is $-326{,}592x^4y^5$.

49. Use the Binomial Theorem to expand the expression. In the study of probability, it is sometimes necessary to use the expansion of $(p+q)^n$, where $p+q=1$.

$$\left(\tfrac{1}{2}+\tfrac{1}{2}\right)^7$$

Solution

7th Row of Pascal's Triangle: 1, 7, 21, 35, 35, 21, 7, 1

$$\left(\tfrac{1}{2}+\tfrac{1}{2}\right)^7 = \left(\tfrac{1}{2}\right)^7 + 7\left(\tfrac{1}{2}\right)^6\left(\tfrac{1}{2}\right) + 21\left(\tfrac{1}{2}\right)^5\left(\tfrac{1}{2}\right)^2 + 35\left(\tfrac{1}{2}\right)^4\left(\tfrac{1}{2}\right)^3 + 35\left(\tfrac{1}{2}\right)^3\left(\tfrac{1}{2}\right)^4$$
$$+ 21\left(\tfrac{1}{2}\right)^2\left(\tfrac{1}{2}\right)^5 + 7\left(\tfrac{1}{2}\right)\left(\tfrac{1}{2}\right)^6 + \left(\tfrac{1}{2}\right)^7$$
$$= \tfrac{1}{128} + \tfrac{7}{128} + \tfrac{21}{128} + \tfrac{35}{128} + \tfrac{35}{128} + \tfrac{21}{128} + \tfrac{7}{128} + \tfrac{1}{128} = 1$$

53. Use the Binomial Theorem to expand $(0.6+0.4)^5$. In the study of probability, it is sometimes necessary to use the expansion of $(p+q)^n$, where $p+q=1$.

Solution

$$(0.6+0.4)^5 = (0.6)^5 + 5(0.6)^4(0.4) + 10(0.6)^3(0.4)^2 + 10(0.6)^2(0.4)^3$$
$$+ 5(0.6)(0.4)^4 + (0.4)^5$$
$$= 0.07776 + 0.25920 + 0.34560 + 0.23040 + 0.07680 + 0.01024 = 1$$

55. *Finding a Pattern* Describe the pattern formed by the sums of the numbers along the diagonal segments of Pascal's Triangle.

Solution

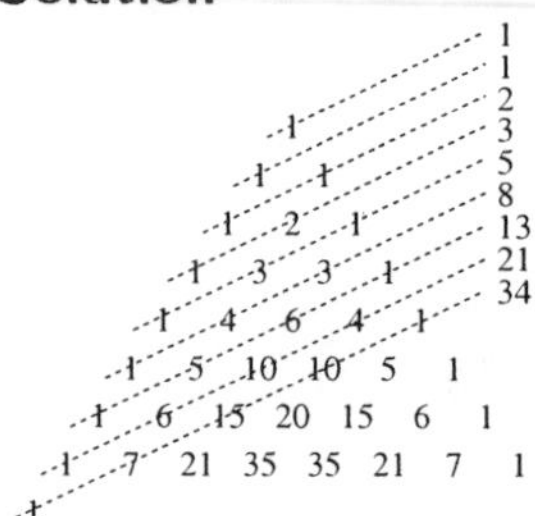

The first nine terms of the sequence are

$$1, 1, 2, 3, 5, 8, 13, 21, 34, \ldots$$

After the first two terms, the next terms are formed by adding the previous two terms.

$a_1 = 1,\ a_2 = 1$

$a_3 = a_1 + a_2 = 1 + 1 = 2$

$a_4 = a_2 + a_3 = 1 + 2 = 3$

$a_5 = a_3 + a_4 = 2 + 3 = 5$

$a_6 = a_4 + a_5 = 3 + 5 = 8$

$a_7 = a_5 + a_6 = 5 + 8 = 13$

$\vdots$

This is called the Fibonacci sequence.

8.5 Counting Principles

- You should know The Fundamental Principle of Counting.
  If E_1 and E_2 are two events such that E_1 can occur in m_1 different ways and E_2 can occur in m_2 different ways, then the number of ways that both events can occur is m_1m_2.
- $_nP_r = \dfrac{n!}{(n-r)!}$ is the number of permutations of n elements taken r at a time.
- Given a set of n objects that has n_1 of one kind, n_2 of a second kind, and so on, the number of distinguishable permutations is

$$\frac{n!}{n_1!n_2!\dots n_k!}.$$

- $_nC_r = \dfrac{n!}{(n-r)!r!}$ is the number of combinations of n elements taken r at a time.

SOLUTIONS TO SELECTED EXERCISES

1. *Job Applicants* A small college needs two additional faculty members: a chemist and a statistician. In how many ways can these positions be filled if there are three applicants for the chemistry position and four for the position in statistics?

Solution

Chemist: 3 choices

Statistician: 4 choices

Total: $3 \cdot 4 = 12$ ways

5. *License Plate Numbers* In a certain state, the automobile license plates consist of two letters followed by a four-digit number. How many distinct license plate numbers can be formed?

Solution

$26 \cdot 26 \cdot 10 \cdot 10 \cdot 10 \cdot 10 = 6{,}760{,}000$

9. *Three-Digit Numbers* How many three-digit numbers can be formed under the following conditions?

(a) Leading digits cannot be zero.

(b) Leading digits cannot be zero and no repetition of digits is allowed.

(c) Leading digits cannot be zero and the number must be a multiple of 5.

Solution

(a) $9 \cdot 10 \cdot 10 = 900$ (b) $9 \cdot 9 \cdot 8 = 648$ (c) $9 \cdot 10 \cdot 2 = 180$

13. *Concert Seats* Three couples have reserved seats in a given row for a concert. In how many different ways can they be seated, given the following conditions?

(a) There are no restrictions.

(b) The two members of each couple wish to sit together.

Solution

(a) $6 \cdot 5 \cdot 4 \cdot 3 \cdot 2 \cdot 1 = 720$

(b) $6 \cdot 1 \cdot 4 \cdot 1 \cdot 2 \cdot 1 = 48$

17. Evaluate ${}_8P_3$.

Solution

$${}_8P_3 = \frac{8!}{5!} = 8 \cdot 7 \cdot 6 = 336$$

21. Evaluate ${}_{100}P_2$.

Solution

$${}_{100}P_2 = \frac{100!}{98!} = 100 \cdot 99 = 9900$$

25. *Posing for a Photograph* In how many ways can five children line up in one row to have their picture taken?

Solution

$${}_5P_5 = \frac{5!}{1!} = 120$$

29. *Forming an Experimental Group* In order to conduct a certain experiment, four students are randomly selected from a class of 20. How many different groups of four students are possible?

Solution

$${}_{20}C_4 = \frac{20!}{16!4!} = \frac{20 \cdot 19 \cdot 18 \cdot 17}{4 \cdot 3 \cdot 2 \cdot 1} = 4845$$

33. *Number of Subsets* How many different subsets of 4 elements can be formed from a set of 100 elements?

Solution

$${}_{100}C_4 = \frac{100!}{96!4!} = \frac{100 \cdot 99 \cdot 98 \cdot 97}{4 \cdot 3 \cdot 2 \cdot 1} = 3{,}921{,}225$$

37. *Job Applicants* An employer interviews eight people for four openings in the company. Three of the eight people are women. If all eight are qualified, in how many ways could the employer fill the four positions if (a) the selection is random and (b) exactly two are women?

Solution

(a) $${}_8C_4 = \frac{8!}{4!4!} = \frac{8 \cdot 7 \cdot 6 \cdot 5}{4 \cdot 3 \cdot 2 \cdot 1} = 70$$

(b) $${}_3C_2 \cdot {}_5C_2 = \frac{3!}{2!1!} \cdot \frac{5!}{2!3!} = 3 \cdot 10 = 30$$

41. *Diagonals of a Polygon* Find the number of diagonals of a pentagon. (A line segment connecting any two nonadjacent vertices is called a *diagonal* of the polygon.)

Solution

$${}_5C_2 - 5 = \frac{5!}{2!3!} - 5 = 10 - 5 = 5$$

45. Find the number of distinguishable permutations of A, A, G, E, E, E, M.

Solution

$$\frac{7!}{2!3!} = 420$$

49. Find the number of distinguishable permutations of A, L, G, E, B, R, A.

Solution

$$\frac{7!}{2!} = 7 \cdot 6 \cdot 5 \cdot 4 \cdot 3 = 2520$$

8.6 Probability

You should know the following basic principles of probability.

- If an event E has $n(E)$ equally likely outcomes and its sample space S has $n(S)$ equally likely outcomes, then the probability of event E is

$$P(E) = \frac{n(E)}{n(S)}.$$

- Know the properties of the probability of an event.
  (a) $0 \le P(E) \le 1$
  (b) If $P(E) = 0$, then E is an impossible event.
  (c) If $P(E) = 1$, then E is a certain event.
- If A and B are mutually exclusive events, then $P(A \cup B) = P(A) + P(B)$.
  If A and B are not mutually exclusive events, then

$$P(A \cup B) = P(A) + P(B) - P(A \cap B).$$

- If A and B are independent events, then the probability that both A and B will occur is $P(A)P(B)$.
- The complement of an event A is $P(A') = 1 - P(A)$.

SOLUTIONS TO SELECTED EXERCISES

3. A coin is tossed three times. Find the probability of getting at least one head.

Solution

$E = \{HHH, HHT, HTH, THH, HTT, THT, TTH\}$

$S = \{HHH, HHT, HTH, THH, HTT, THT, TTH, TTT\}$

$$P(E) = \frac{n(E)}{n(S)} = \frac{7}{8}$$

7. Two six-sided dice are tossed. Find the probability that the sum is at least 7.

Solution

$E = \{(1,6), (2,5), (3,4), (4,3), (5,2), (6,1), (2,6), (3,5), (4,4), (5,3),$
$(6,2), (3,6), (4,5), (5,4), (6,3), (4,6), (5,5), (6,4), (5,6), (6,5), (6,6)\}$

$n(S) = 6 \cdot 6 = 36$

$$P(E) = \frac{n(E)}{n(S)} = \frac{21}{36} = \frac{7}{12}$$

13. A card is selected from a standard deck of 52 cards. Find the probability of getting a black card that is not a face card.

Solution

$E = \{1C, 2C, 3C, 4C, 5C, 6C, 7C, 8C, 9C, 10C,$
$1S, 2S, 3S, 4S, 5S, 6S, 7S, 8S, 9S, 10S\}$

where C stands for clubs and S stands for spades.

$$P(E) = \frac{n(E)}{n(S)} = \frac{20}{52} = \frac{5}{13}$$

17. Two marbles are drawn from a bag containing one green, two yellow, and three red marbles. Find the probability of drawing neither of the yellow marbles.

Solution

$P(RR) = \frac{3}{6} \cdot \frac{2}{5} = \frac{6}{30}$

$P(RG) = \frac{3}{6} \cdot \frac{1}{5} = \frac{3}{30}$

$P(GR) = \frac{1}{6} \cdot \frac{3}{5} = \frac{3}{30}$

$$\begin{aligned} P(Y') &= P(RR) + P(RG) + P(GR) \\ &= \tfrac{6}{30} + \tfrac{3}{30} + \tfrac{3}{30} \\ &= \tfrac{12}{30} = \tfrac{2}{5} \end{aligned}$$

19. You are given the probability, $p = 0.7$, that an event *will* happen. Find the probability that the event *will not* happen.

Solution

$1 - p = 1 - 0.7 = 0.3$

23. *Winning an Election* Three people have been nominated for president of a college class. From a small poll, it is estimated that the probability of Jane winning the election is 0.37, and the probability of Larry winning the election is 0.44. What is the probability of the third candidate winning the election?

Solution

$1 - (0.37 + 0.44) = 0.19$

27. *Random Number Generator* Two integers (between 1 and 30 inclusive) are chosen by a random number generator on a computer. What is the probability that (a) the numbers are both even, (b) one number is even and one is odd, (c) both numbers are less than 10, and (d) the same number is chosen twice?

Solution

(a) $\left(\frac{1}{2}\right)\left(\frac{1}{2}\right) = \frac{1}{4}$

(b) $1 - p\,(\text{both even}) - p\,(\text{both odd}) = 1 - \frac{1}{4} - \frac{1}{4} = \frac{1}{2}$

(c) $\left(\frac{9}{30}\right)\left(\frac{9}{30}\right) = \frac{9}{100}$

(d) $(1)\left(\frac{1}{30}\right) = \frac{1}{30}$

31. *Drawing Cards from a Deck* Two cards are selected at random from an ordinary deck of 52 playing cards. Find the probability that two aces are selected, given the following conditions.

(a) The cards are drawn in sequence, with the first card being replaced and the deck reshuffled prior to the second drawing.

(b) The two cards are drawn consecutively, without replacement.

Solution

(a) $\frac{4}{52} \cdot \frac{4}{52} = \frac{1}{169}$

(b) $\frac{4}{52} \cdot \frac{3}{51} = \frac{1}{221}$

35. *Letter Mix-Up* Four letters and envelopes are addressed to four different people. If the letters are randomly inserted into the envelopes, what is the probability that (a) exactly one will be inserted in the correct envelope and (b) at least one will be inserted in the correct envelope?

Solution

Total ways to insert letters: $4! = 24$ ways

4 correct: 1 way

3 correct: not possible

2 correct: 6 ways

1 correct: 8 ways

0 correct: 9 ways

(a) $\dfrac{8}{24} = \dfrac{1}{3}$

(b) $\dfrac{8+6+1}{24} = \dfrac{15}{24} = \dfrac{5}{8}$

39. *Defective Units* A shipment of 1,000 compact disc players contains 4 defective units. A retail outlet has ordered 20 units. (a) What is the probability that all 20 units are good? (b) What is the probability that at least 1 unit is defective?

Solution

(a) $\dfrac{{}_{996}C_{20}}{{}_{1,000}C_{20}} \approx 0.923$

(b) $1 - p = 0.077$

43. *Making a Sale* A salesperson makes a sale at approximately one-third of the offices she calls on. If, on a given day, she goes to four offices, what is the probability that she will make a sale at (a) all four offices, (b) none of the offices, and (c) at least one office?

Solution

(a) $\frac{1}{3} \cdot \frac{1}{3} \cdot \frac{1}{3} \cdot \frac{1}{3} = \frac{1}{81}$

(b) $\left(\frac{2}{3}\right)^4 = \frac{16}{81}$

(c) $1 - \frac{16}{81} = \frac{65}{81}$

47. *Is That Cash or Charge?* According to a survey by *USA Today*, the method used by Christmas shoppers to pay for gifts is as shown in the pie chart in the textbook. Suppose two Christmas shoppers are chosen at random. What is the probability that both shoppers will pay for their gifts in cash only?

Solution

$(0.32)(0.32) = 0.1024$

49. *Number of Telephones* According to a survey by *Maritz Ameripoll*, the number of telephones in an American household is as shown in the figure in the textbook. Suppose three households are chosen at random. What is the probability that all three households have at least four telephones?

Solution

$(0.10 + 0.06)^3 \approx 0.004$

Review Exercises for Chapter 8

SOLUTIONS TO SELECTED EXERCISES

3. Write the first five terms of $a_n = \dfrac{2n}{2n+1}$. (Assume n begins with 1.)

Solution

$$a_1 = \frac{2(1)}{2(1)+1} = \frac{2}{3}$$

$$a_2 = \frac{2(2)}{2(2)+1} = \frac{4}{5}$$

$$a_3 = \frac{2(3)}{2(3)+1} = \frac{6}{7}$$

$$a_4 = \frac{2(4)}{2(4)+1} = \frac{8}{9}$$

$$a_5 = \frac{2(5)}{2(5)+1} = \frac{10}{11}$$

7. Simplify the fraction $\dfrac{15!}{13!}$.

Solution

$$\frac{15!}{13!} = \frac{15 \cdot 14 \cdot 13!}{13!} = 15 \cdot 14 = 210$$

11. Write an expression for the *most* apparent nth term of $1, \frac{1}{2}, \frac{1}{3}, \frac{1}{4}, \frac{1}{5}, \ldots$ (Assume n begins with 1.)

Solution

$$a_n = \frac{1}{n}$$

15. Find the sum $\displaystyle\sum_{i=1}^{5}(2i + 3)$.

Solution

$$\sum_{i=1}^{5}(2i+3) = [2(1)+3] + [2(2)+3] + [2(3)+3] + [2(4)+3] + [2(5)+3]$$

$$= 5 + 7 + 9 + 11 + 13 = 45$$

19. Use summation notation to write $\frac{3}{1+1}+\frac{3}{1+2}+\frac{3}{1+3}+\frac{3}{1+4}+\dots+\frac{3}{1+12}$.

Solution

$$\frac{3}{1+1}+\frac{3}{1+2}+\frac{3}{1+3}+\frac{3}{1+4}+\dots+\frac{3}{1+12}=\sum_{k=1}^{12}\frac{3}{1+k}$$

23. *Compound Interest* Suppose on your next birthday you deposit $2,000 in an account that earns 8% compounded quarterly. The balance in the account after n quarters is given by

$$A_n = 2,000\left(1+\frac{0.08}{4}\right)^n \qquad n = 1, 2, 3, \dots$$

(a) Compute the first eight terms of the sequence.

(b) Find the balance in this account 10 years from the date of deposit by computing the fortieth term of the sequence.

(c) Assuming you do not withdraw money from the account, find the balance 30 years from the date of deposit by computing the one hundred twentieth term of the sequence.

Solution

(a) $A_1 = \$2,040.00$

$A_2 = \$2,080.80$

$A_3 \approx \$2,122.42$

$A_4 \approx \$2,164.86$

$A_5 \approx \$2,208.16$

$A_6 \approx \$2,252.32$

$A_7 \approx \$2,297.37$

$A_8 \approx \$2,343.32$

(b) $A_{40} \approx \$4,416.08$

(c) $A_{120} \approx \$21,530.33$

27. Determine whether $4, \frac{7}{2}, 3, \frac{5}{2}, 2, \dots$ is arithmetic. If it is, find the common difference.

Solution

Arithmetic, $d = -\frac{1}{2}$

31. Find a formula for the arithmetic sequence.

$$a_1 = 3,\ d = 5$$

Solution

$a_1 = 3,\ d = 5$

$a_n = a_1 + (n-1)d = 3 + (n-1)5 = 5n - 2$

35. Find the sum of the first 10 terms of the arithmetic sequence.

$$3,\ 9,\ 15,\ 21,\ 27,\ \dots$$

Solution

$a_n = 6n - 3$

$a_1 = 3,\ a_{10} = 57$

$S_{10} = \frac{10}{2}(a_1 + a_{10}) = 5(3 + 57) = 300$

39. *Job Offers* A person is offered a job by two companies. The position with Company A has a salary of \$24,500 for the first year with a guaranteed annual raise of \$1,500 per year for the first five years. The position with Company B has a salary of \$22,000 for the first year with a guaranteed annual raise of \$2,400 per year for the first five years.

(a) What will the salary be during the sixth year of employment at Company A? Company B?

(b) How much will Company A have paid the person at the end of six years?

(c) How much will Company B have paid the person at the end of six years?

(d) Which job should the person accept and why?

Solution

$$A_n = 24{,}500 + (n-1)1500 = 1500n + 23{,}000$$

$$B_n = 22{,}000 + (n-1)2400 = 2400n + 19{,}600$$

(a) $A_6 = \$32{,}000$

$B_6 = \$34{,}000$

(b) $\sum_{i=1}^{6} A_i = \$169{,}500$

(c) $\sum_{i=1}^{6} B_i = \$168{,}000$

(d) For the short term, A pays better; for the long term B pays more.

43. Determine whether 16, 8, 4, 2, 1, $\frac{1}{2}$, $\frac{1}{4}$, $\frac{1}{8}$, ... is geometric. If it is, find its common ratio.

Solution

Geometric, $r = \frac{1}{2}$

49. Find the indicated term of the geometric sequence.

$$a_1 = 8,\ r = \tfrac{1}{2},\ n = 10.$$

Solution

$$a_{10} = a_1 r^9 = 8\left(\tfrac{1}{2}\right)^9 = \tfrac{1}{64}$$

53. Find the sum of $\sum_{n=1}^{10} 4(2^n)$.

Solution

$$\sum_{n=1}^{10} 4(2^n) = a_1\left(\frac{1-r^n}{1-r}\right) = 8\left(\frac{1-2^{10}}{1-2}\right) = 8,184$$

55. Find the sum of $\sum_{n=0}^{\infty}\left(\frac{1}{3}\right)^n = 1 + \frac{1}{3} + \frac{1}{9} + \frac{1}{27} + \cdots$.

Solution

$$\sum_{n=0}^{\infty}\left(\frac{1}{3}\right)^n = \frac{a}{1-r} = \frac{1}{1-\frac{1}{3}} = \frac{3}{2} = 1.5$$

59. *Profit* The annual profit for a company from 1980 to 1994 can be approximated by the model $a_n = 156.4e^{0.15n}$ where a_n is the annual sales (in millions of dollars) and n represents the year with $n = 0$ corresponding to 1980.

(a) Sketch a bar graph that represents this company's profit during the 15-year period.

(b) Find the total profit during the 15-year period.

Solution

(a) $a_0 \approx 156.40$ $\quad a_8 \approx 519.27$

$a_1 \approx 181.71$ $\quad a_9 \approx 603.30$

$a_2 \approx 211.12$ $\quad a_{10} \approx 700.94$

$a_3 \approx 245.28$ $\quad a_{11} \approx 814.37$

$a_4 \approx 284.98$ $\quad a_{12} \approx 946.16$

$a_5 \approx 331.10$ $\quad a_{13} \approx 1099.29$

$a_6 \approx 384.68$ $\quad a_{14} \approx 1277.19$

$a_7 \approx 446.94$

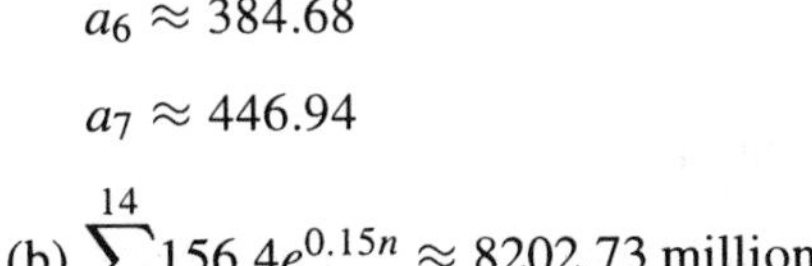

(b) $\displaystyle\sum_{n=0}^{14} 156.4e^{0.15n} \approx 8202.73$ million

63. Evaluate ${}_{30}C_{30}$.

Solution

$${}_{30}C_{30} = \frac{30!}{30!0!} = 1$$

67. Use the Binomial Theorem to expand $(x+y)^{10}$. Simplify your answer.

Solution

$$\begin{aligned}(x+y)^{10} &= {}_{10}C_0x^{10} + {}_{10}C_1x^9y + {}_{10}C_2x^8y^2 + {}_{10}C_3x^7y^3 + {}_{10}C_4x^6y^4 + {}_{10}C_5x^5y^5 \\ &\quad + {}_{10}C_6x^4y^6 + {}_{10}C_7x^3y^7 + {}_{10}C_8x^2y^8 + {}_{10}C_9xy^9 + {}_{10}C_{10}y^{10} \\ &= x^{10} + 10x^9y + 45x^8y^2 + 120x^7y^3 + 210x^6y^4 + 252x^5y^5 \\ &\quad + 210x^4y^6 + 120x^3y^7 + 45x^2y^8 + 10xy^9 + y^{10}\end{aligned}$$

71. Expand $(3-2y)^4$ using Pascal's Triangle to determine the coefficients.

Solution

4th Row of Pascal's Triangle: 1 4 6 4 1

$$\begin{aligned}(3-2y)^4 &= 1(3)^4 - 4(3)^3(2y) + 6(3)^2(2y)^2 - 4(3)(2y)^3 + 1(2y)^4 \\ &= 81 - 216y + 216y^2 - 96y^3 + 16y^4\end{aligned}$$

75. Use the Binomial Theorem to expand $\left(\frac{1}{3}+\frac{2}{3}\right)^5$. In the study of probability, it is sometimes necessary to use the expansion of $(p+q)^n$ where $p+q=1$.

Solution

$$\begin{aligned}\left(\tfrac{1}{3}+\tfrac{2}{3}\right)^5 &= 1\left(\tfrac{1}{3}\right)^5 + 5\left(\tfrac{1}{3}\right)^4\left(\tfrac{2}{3}\right) + 10\left(\tfrac{1}{3}\right)^3\left(\tfrac{2}{3}\right)^2 + 10\left(\tfrac{1}{3}\right)^2\left(\tfrac{2}{3}\right)^3 + 5\left(\tfrac{1}{3}\right)\left(\tfrac{2}{3}\right)^4 + \left(\tfrac{2}{3}\right)^5 \\ &= \tfrac{1}{243} + \tfrac{10}{243} + \tfrac{40}{243} + \tfrac{80}{243} + \tfrac{80}{243} + \tfrac{32}{243}\end{aligned}$$

81. *True-False Exam* In how many ways can a 20-question true-false exam be answered? (Assume that no questions are omitted.)

Solution

$(2)^{20} = 1{,}048{,}576$ ways

85. Evaluate ${}_{30}P_1$.

Solution

$${}_{30}P_1 = \frac{30!}{29!} = 30$$

89. *Starting Lineup* In how many ways can the starting lineup for a coed, six-member flag football team be selected from a team of eight men and eight women? League rules require that every team have three men and three women on the field.

Solution

${}_8C_3 \cdot {}_8C_3 = 3136$

93. Find the probability of getting a red card that is not a face card in the experiment of selecting one card from a standard deck of 52 playing cards.

Solution

$\frac{20}{52} = \frac{5}{13}$

97. You are given the probability, $p = 0.48$, that an event will happen. Find the probability that the event will not happen.

Solution

$1 - p = 1 - 0.48 = 0.52$

101. *Poker Hand* Five cards are drawn from an ordinary deck of 52 playing cards. What is the probability of getting a full house? (*Note:* A full house consists of three of one kind and two of another. For example, Q-Q-Q-2-2 and A-A-A-6-6 are full houses.)

Solution

$$\frac{{}_{13}C_1 \cdot {}_4C_3 \cdot {}_{12}C_1 \cdot {}_4C_2}{{}_{52}C_5} = \frac{3744}{2{,}598{,}960} = \frac{6}{4165}$$

103. *A Boy or a Girl?* Assume that the probability of the birth of a child of a particular sex is 50%. In a family with 10 children, what is the probability that

(a) All the children are girls?

(b) All the children are the same sex?

(c) There is at least one girl?

Solution

(a) $\left(\frac{1}{2}\right)^{10} = \frac{1}{1024}$

(b) $2\left(\frac{1}{2}\right)^{10} = \frac{1}{512}$

(c) $1 - \frac{1}{1024} = \frac{1023}{1024}$

Test for Chapter 8

1. Write out the first five terms of the sequence. (Begin with $n = 1$.)

$$a_n = 3n + 1$$

Solution

$a_1 = 3(1) + 1 = 4$
$a_2 = 3(2) + 1 = 7$
$a_3 = 3(3) + 1 = 10$
$a_4 = 3(4) + 1 = 13$
$a_5 = 3(5) + 1 = 16$

2. Write out the first five terms of the sequence. (Begin with $n = 1$.)

$$a_n = (-1)^n n^2$$

Solution

$a_1 = (-1)^1(1)^2 = -1$
$a_2 = (-1)^2(2)^2 = 4$
$a_3 = (-1)^3(3)^2 = -9$
$a_4 = (-1)^4(4)^2 = 16$
$a_5 = (-1)^5(5)^2 = -25$

3. Write out the first five terms of the sequence. (Begin with $n = 1$.)

$$a_n = n!$$

Solution

$a_1 = 1! = 1$
$a_2 = 2! = 2$
$a_3 = 3! = 6$
$a_4 = 4! = 24$
$a_5 = 5! = 120$

4. Write out the first five terms of the sequence. (Begin with $n = 1$.)

$$a_n = \left(\tfrac{1}{2}\right)^n$$

Solution

$a_1 = \left(\tfrac{1}{2}\right)^1 = \tfrac{1}{2}$
$a_2 = \left(\tfrac{1}{2}\right)^2 = \tfrac{1}{4}$
$a_3 = \left(\tfrac{1}{2}\right)^3 = \tfrac{1}{8}$
$a_4 = \left(\tfrac{1}{2}\right)^4 = \tfrac{1}{16}$
$a_5 = \left(\tfrac{1}{2}\right)^5 = \tfrac{1}{32}$

In Exercises 5–8, decide whether the sequence is arithmetic, geometry, or neither. If possible, find its common difference or ratio.

5. $a_n = 5n - 2$

Solution

3, 8, 13, 18, ... is arithmetic. The common difference is 5.

6. $a_n = 2^n + 2$

Solution

4, 6, 10, 18, 34, ... is neither arithmetic nor geometric.

7. $a_n = 5(2^n)$

Solution

10, 20, 40, 80, 160, ... is geometric with a common ratio of 2.

8. $a_n = 5n^2 - 2$

Solution

3, 18, 43, 78, 123, ... is neither arithmetic nor geometric.

9. Evaluate $\sum_{n=1}^{50}(2n+1)$

Solution

Arithmetic

$a_1 = 3,\ a_{50} = 101$

$$\sum_{n=1}^{50}(2n+1) = \tfrac{50}{2}(3+101)$$
$$= 2600$$

10. Evaluate $\sum_{n=1}^{20} 3\left(\tfrac{3}{2}\right)^n$

Solution

Geometric

$a_1 = \tfrac{9}{2},\ r = \tfrac{3}{2}$

$$\sum_{n=1}^{20} 3\left(\tfrac{3}{2}\right)^n = \frac{\tfrac{9}{2}\left[1-\left(\tfrac{3}{2}\right)^{20}\right]}{1-\tfrac{3}{2}}$$
$$\approx 29{,}918.311$$

11. Evaluate $\sum_{n=1}^{\infty}\left(\tfrac{1}{3}\right)^n$

Solution

Geometric

$a_1 = \tfrac{1}{3},\ r = \tfrac{1}{3}$

$$\sum_{n=1}^{\infty}\left(\tfrac{1}{3}\right)^n = \frac{\tfrac{1}{3}}{1-\tfrac{1}{3}} = \tfrac{1}{2}$$

12. A deposit of \$10,000 is made in an account that earns 8% compounded monthly. The balance in the account after n months is given by

$$A_n = 10{,}000\left(1+\frac{0.08}{12}\right)^n, \qquad n = 1, 2, 3, \ldots$$

Find the balance in this account after 10 years.

Solution

$$A_{120} = 10{,}000\left(1+\frac{0.08}{12}\right)^{120}$$
$$\approx \$22{,}196.40$$

13. You deposited \$100 in an account at the beginning of each month for 10 years. The account pays 6% compounded monthly. What is the balance at the end of 10 years?

Solution

Compounded monthly: $A = \sum_{i=1}^{120} 100\left(1+\frac{0.06}{12}\right)^i \approx \$16{,}469.87$

14. Use the Binomial Theorem to expand $(x+2)^7$.

Solution

$$\begin{aligned}(x+2)^7 &= {}_7C_0x^7 + {}_7C_1x^6(2) + {}_7C_2x^5(2)^2 + {}_7C_3x^4(2)^3 + {}_7C_4x^3(2)^4 + {}_7C_5x^2(2)^5 \\ &\quad +_7 C_6x(2)^6 +_7 C_7(2)^7 \\ &= x^7 + 7x^6(2) + 21x^5(4) + 35x^4(8) + 35x^3(16) + 21x^2(32) + 7x(64) + 128 \\ &= x^7 + 14x^6 + 84x^5 + 280x^4 + 560x^3 + 672x^2 + 448x + 128\end{aligned}$$

15. Determine the numerical coefficient of the x^7 term of $(x-3)^{12}$.

Solution

The coefficient of x^7 is ${}_{12}C_5(-3)^5 = -192,456$.

16. A customer in an electronics store can choose one of five CD players, one of six speaker systems, and one of four radio/ cassette players to design a sound system. How many different systems can be designed?

Solution

$(5)(6)(4) = 120$ different systems

17. In how many ways can a 15-question true-false exam be answered? (Assume that no question is omitted.) Explain your reasoning.

Solution

Since each question has two possible answers, we have

$2^{15} = 32{,}768$ ways.

18. How many five-digit numbers can be formed if the leading digit cannot be zero and the number must be odd?

Solution

If the number is odd, it must end in 1, 3, 5, 7, or 9.

$(9)(10)(10)(10)(5) = 45{,}000$

19. On a game show you are given five different digits to arrange in the proper order to represent the price of a car. You only know the first digit. What is the probability that you will arrange the other four digits correctly?

Solution

$\left(\frac{1}{4}\right)\left(\frac{1}{3}\right)\left(\frac{1}{2}\right)(1) = \frac{1}{24}$

20. Five coins are tossed. What is the probability that they are all heads?

Solution

$\left(\frac{1}{2}\right)^5 = \frac{1}{32}$

Practice Test for Chapter 8

1. Write out the first five terms of the sequence $a_n = \dfrac{2n}{(n+2)!}$.

2. Write an expression for the nth term of the sequence $\{\frac{4}{3}, \frac{5}{9}, \frac{9}{27}, \frac{7}{81}, \frac{8}{243}, \ldots\}$.

3. Find the sum $\sum_{i=1}^{6}(2i-1)$.

4. Write out the first five terms of the arithmetic sequence where $a_1 = 23$ and $d = -2$.

5. Find a_n for the arithmetic sequence with $a_1 = 12$, $d = 3$, and $n = 50$.

6. Find the sum of the first 200 positive integers.

7. Write out the first five terms of the geometric sequence with $a_1 = 7$ and $r = 2$.

8. Evaluate $\sum_{n=0}^{9}\left(\frac{2}{3}\right)^n$.

9. Evaluate $\sum_{n=0}^{\infty}(0.03)^n$.

10. Simplify $\dfrac{(n+3)!}{n!}$.

11. A deposit of \$50 is made at the beginning of each month for 20 years in an account that pays 7% compounded monthly. What is the balance at the end of 20 years?

12. Evaluate ${}_{13}C_4$.

13. Expand $(x+3)^5$.

14. Find the term involving x^7 in $(x-2)^{12}$.

15. Evaluate ${}_{30}P_4$.

16. How many ways can six people sit at a table with six chairs?

17. Twelve cars run in a race. How many different ways can they come in first, second, and third places? (Assume that there are no ties.)

18. Two six-sided dice are tossed. Find the probability that the total of the two dice is less than 5.

19. Two cards are selected at random from a deck of 52 playing cards. Find the probability that the first card is a King and the second card is a black ten.

20. A manufacturer has determined that for every 1,000 units it produces, 3 will be faulty. What is the probability that an order of 50 units will have one or more faulty units?

Cumulative Test for Chapters 6–8

1. Solve the system by substitution.

$$x + 2y = 22$$
$$-x + 4y = 20$$

Solution

$x + 2y = 22 \Rightarrow x = 22 - 2y$

$-x + 4y = 20 \Rightarrow -(22 - 2y) + 4y = 20$

$6y - 22 = 20$

$6y = 42$

$y = 7$

$x = 22 - 2(7) = 8$

Answer: $x = 8,\ y = 7$

2. Solve the system by graphing.

$$2x - 3y = 0$$
$$4x + y = 14$$

Solution

$2x - 3y = 0 \Rightarrow y = \frac{2}{3}x$

$4x + y = 14 \Rightarrow y = -4x + 14$

Point of intersection: (3, 2)

Answer: $x = 3,\ y = 2$

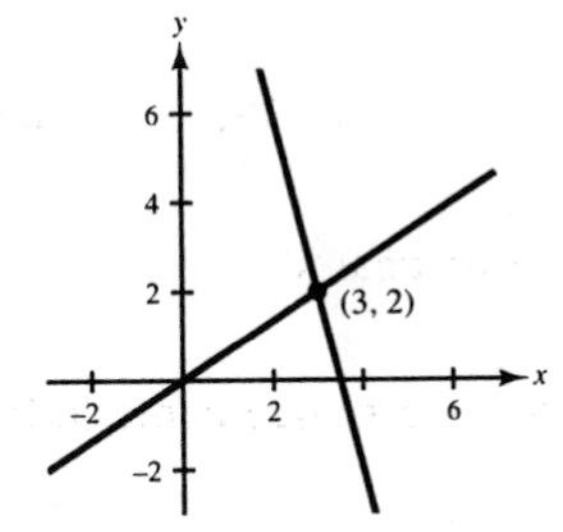

3. Solve the system by elimination.

$$\begin{aligned} 2x - 3y + z &= 18 \\ 3x \qquad - 2z &= -4 \\ x - y + 3z &= 20 \end{aligned}$$

Solution

$$\begin{aligned} x - y + 3z &= 20 \quad \text{Interchange the equations} \\ 2x - 3y + z &= 18 \\ 3x \qquad - 2z &= -4 \end{aligned}$$

$$\begin{aligned} x - y + 3z &= 20 \\ -y - 5z &= -22 \quad -2 \text{ Equation 1} + \text{Equation 2} \\ 3y - 11z &= -64 \quad -3 \text{ Equation 1} + \text{Equation 3} \end{aligned}$$

$$\begin{aligned} x - y + 3z &= 20 \\ y + 5z &= -130 \quad -1\text{Equation 2} \\ -26z &= -130 \; 3 \text{ Equation 2} + \text{Equation 3} \\ z &= 5 \; \text{Solve and use back-substitution.} \\ y + 5(5) &= 22 \Rightarrow y = -3 \\ x - (-3) + 3(5) &= 20 \Rightarrow x = 2 \end{aligned}$$

Answer: $x = 2,\ y = -3,\ z = 5$

4. Solve the system by matrices.

$$\begin{aligned} 3x - 4y + 2z &= -32 \\ 2x + 3y \qquad &= 8 \\ y - 3z &= 19 \end{aligned}$$

Solution

$$\begin{bmatrix} 3 & -4 & 2 & \vdots & -32 \\ 2 & 3 & 0 & \vdots & 8 \\ 0 & 1 & -3 & \vdots & 19 \end{bmatrix} \Rightarrow \begin{bmatrix} 1 & -7 & 2 & \vdots & -40 \\ 0 & 17 & -4 & \vdots & 88 \\ 0 & 1 & -3 & \vdots & 19 \end{bmatrix}$$

$$\Rightarrow \begin{bmatrix} 1 & 0 & 310 & \vdots & -1,552 \\ 0 & 1 & 44 & \vdots & -216 \\ 0 & 0 & -47 & \vdots & 235 \end{bmatrix} \Rightarrow \begin{bmatrix} 1 & 0 & 0 & \vdots & -2 \\ 0 & 1 & 0 & \vdots & 4 \\ 0 & 0 & 1 & \vdots & -5 \end{bmatrix}$$

Answer: $x = -2,\ y = 4,\ z = -5$

5. Use the matrices shown in the textbook to find $2A - C$.

Solution

$$2A - C = 2\begin{bmatrix} 3 & 2 & 2 \\ 1 & 2 & 2 \\ 1 & 0 & 1 \end{bmatrix} - \begin{bmatrix} 5 & 0 & -4 \\ 3 & 0 & 1 \\ 2 & -1 & -3 \end{bmatrix} = \begin{bmatrix} 6 & 4 & 4 \\ 2 & 4 & 4 \\ 2 & 0 & 2 \end{bmatrix} - \begin{bmatrix} 5 & 0 & -4 \\ 3 & 0 & 1 \\ 2 & -1 & -3 \end{bmatrix}$$

$$= \begin{bmatrix} 1 & 4 & 8 \\ -1 & 4 & 3 \\ 0 & 1 & 5 \end{bmatrix}$$

6. Use the matrices shown in the textbook to find AB.

Solution

$$AB = \begin{bmatrix} 3 & 2 & 2 \\ 1 & 2 & 2 \\ 1 & 0 & 1 \end{bmatrix} \begin{bmatrix} 4 & 1 \\ -1 & 2 \\ 3 & 1 \end{bmatrix} = \begin{bmatrix} 16 & 9 \\ 8 & 7 \\ 7 & 2 \end{bmatrix}$$

7. Use the matrices shown in the textbook to find BD.

Solution

$$BD = \begin{bmatrix} 4 & 1 \\ -1 & 2 \\ 3 & 1 \end{bmatrix} \begin{bmatrix} 1 & 3 \\ -2 & 0 \end{bmatrix} = \begin{bmatrix} 2 & 12 \\ -5 & -3 \\ 1 & 9 \end{bmatrix}$$

8. Use the matrices shown in the textbook to find A^{-1}.

Solution

$$[A \;\vdots\; I] = \left[\begin{array}{ccc|ccc} 3 & 2 & 2 & 1 & 0 & 0 \\ 1 & 2 & 2 & 0 & 1 & 0 \\ 1 & 0 & 1 & 0 & 0 & 1 \end{array}\right] \Rightarrow \left[\begin{array}{ccc|ccc} 1 & -2 & -2 & 1 & -2 & 0 \\ 0 & 4 & 4 & -1 & 3 & 0 \\ 0 & 2 & 3 & -1 & 2 & 1 \end{array}\right]$$

$$\Rightarrow \left[\begin{array}{ccc|ccc} 1 & 0 & 1 & 0 & 0 & 1 \\ 0 & 1 & 1 & -\frac{1}{4} & \frac{3}{4} & 0 \\ 0 & 0 & 1 & -\frac{1}{2} & \frac{1}{2} & 1 \end{array}\right] \Rightarrow \left[\begin{array}{ccc|ccc} 1 & 0 & 0 & \frac{1}{2} & -\frac{1}{2} & 0 \\ 0 & 1 & 0 & \frac{1}{4} & \frac{1}{4} & -1 \\ 0 & 0 & 1 & -\frac{1}{2} & \frac{1}{2} & 1 \end{array}\right]$$

$$= [I \;\vdots\; A^{-1}]$$

$$A^{-1} = \begin{bmatrix} \frac{1}{2} & -\frac{1}{2} & 0 \\ \frac{1}{4} & \frac{1}{4} & -1 \\ -\frac{1}{2} & \frac{1}{2} & 1 \end{bmatrix} = \frac{1}{4}\begin{bmatrix} 2 & -2 & 0 \\ 1 & 1 & -4 \\ -2 & 2 & 4 \end{bmatrix}$$

9. Evaluate the determinant of the matrix.

$$\begin{bmatrix} 1 & -2 & 4 \\ 3 & 7 & -5 \\ 6 & 1 & 4 \end{bmatrix}$$

Solution

$$\begin{vmatrix} 1 & -2 & 4 \\ 3 & 7 & -5 \\ 6 & 1 & 4 \end{vmatrix} = \begin{vmatrix} 7 & -5 \\ 1 & 4 \end{vmatrix} + 2\begin{vmatrix} 3 & -5 \\ 6 & 4 \end{vmatrix} + 4\begin{vmatrix} 3 & 7 \\ 6 & 1 \end{vmatrix} = 33 + 2(42) + 4(-39) = -39$$

10. Evaluate the determinant of the matrix.

$$\begin{bmatrix} 1 & 0 & 0 \\ 0 & 7 & 0 \\ 0 & 0 & -1 \end{bmatrix}$$

Solution

$$\begin{vmatrix} 1 & 0 & 0 \\ 0 & 7 & 0 \\ 0 & 0 & -1 \end{vmatrix} = (1)(7)(-1) = -7 \text{ (Diagonal matrix)}$$

11. Find the point of equilibrium for a system with demand function $p = 75 - 0.0005x$ and supply function $p = 30 + 0.002x$.

Solution

Demand = Supply

$$75 - 0.0005x = 30 + 0.002x$$
$$45 = 0.0025x$$
$$x = 18{,}000 \text{ units}$$
$$p = \$66.00$$

Point of equilibrium: (18,000, 66)

12. A total of \$45,000 is invested at 7.5% and 8.5% simple interest. If the yearly interest is \$3,625, how much of the \$45,000 is invested at each rate?

Solution

Let x = amount at 7.5% and y = amount at 8.5%.

$$x + y = 45{,}000 \Rightarrow y = 45{,}000 - x$$
$$0.075x + 0.085y = 3625$$
$$0.075x + 0.085(45{,}000 - x) = 3625$$
$$0.075x + 3825 - 0.085x = 3625$$
$$-0.01x = -200$$
$$x = 20{,}000$$
$$y = 45{,}000 - 20{,}000 = 25{,}000$$

Answer: \$20,000 at 7.5%, \$25,000 at 8.5%

13. A factory produces three different models of a product, which are shipped to three different warehouses. The number of units i that are shipped to warehouse j is represented by the a_{ij} in the matrix.

$$A = \begin{bmatrix} 1000 & 3000 & 2000 \\ 2000 & 4000 & 5000 \\ 3000 & 1000 & 1000 \end{bmatrix}$$

The price per unit is represented by the matrix $B = [\,\$20 \quad \$30 \quad \$25\,]$. Find the product BA and state what each entry of the product represents.

Solution

$$BA = [\,20 \quad 30 \quad 25\,]\begin{bmatrix} 1000 & 3000 & 2000 \\ 2000 & 4000 & 5000 \\ 3000 & 1000 & 1000 \end{bmatrix} = [\,155{,}000 \quad 205{,}000 \quad 215{,}000\,]$$

Each entry represents the value of the inventory at each warehouse.

14. Write the first five terms of the sequence given by $a_n = 3 + 4n$.

Solution

$a_1 = 3 + 4(1) = 7$

$a_2 = 3 + 4(2) = 11$

$a_3 = 3 + 4(3) = 15$

$a_4 = 3 + 4(4) = 19$

$a_5 = 3 + 4(5) = 23$

15. Write the first five terms of the geometric sequence with $a_1 = 3$ and $r = \frac{1}{2}$.

Solution

$a_1 = 3,\ r = \frac{1}{2}$

$a_1 = 3$

$a_2 = 3\left(\frac{1}{2}\right) = \frac{3}{2}$

$a_3 = \left(\frac{3}{2}\right)\left(\frac{1}{2}\right) = \frac{3}{4}$

$a_4 = \left(\frac{3}{4}\right)\left(\frac{1}{2}\right) = \frac{3}{8}$

$a_5 = \left(\frac{3}{8}\right)\left(\frac{1}{2}\right) = \frac{3}{16}$

16. Use the Binomial Theorem to expand $(y - 4)^5$.

Solution

$$\begin{aligned}(y-4)^5 &= {}_5C_0y^5 + {}_5C_1y^4(-4) + {}_5C_2y^3(-4)^2 + {}_5C_3y^2(-4)^3 + {}_5C_4y(-4)^4 + {}_5C_5(-4)^5 \\ &= y^5 + 5y^4(-4) + 10y^3(16) + 10y^2(-64) + 5y(256) - 1,024 \\ &= y^5 - 20y^4 + 160y^3 - 640y^2 + 1,280y - 1,024\end{aligned}$$

17. How many ways can the letters A, B, C, D, and E be arranged?

Solution

$(5)(4)(3)(2)(1) = 120$ ways

18. How many five-digit numbers are multiples of 5?

Solution

The number cannot begin with 0.

The number must end with 0 or 5.

$(9)(10)(10)(10)(2) = 18{,}000$ numbers

19. A shipment of 2000 microwave ovens contains 10 defective units. You order three units from the shipment. What is the probability that all three are good?

Solution

$\left(\frac{1990}{2000}\right)\left(\frac{1989}{1999}\right)\left(\frac{1988}{1998}\right) \approx 0.985.$

20. In how many ways can a 10-question true-false exam be answered? (Assume that no questions are omitted.)

Solution

$2^{10} = 1024$

APPENDIX A
Conic Sections

A.1 Conic Sections

3. Match $y^2 = 4x$ with its graph. [The graphs are labeled (a), (b), (c), (d), (e), (f), (g), and (h) in the textbook.]

Solution

$y^2 = 4x$

Parabola opening to the right

Graph (d)

7. Match $\dfrac{x^2}{1} - \dfrac{y^2}{4} = 1$ with its graph. [The graphs are labeled (a), (b), (c), (d), (e), (f), (g), and (h) in the textbook.]

Solution

$\dfrac{x^2}{1} - \dfrac{y^2}{4} = 1$

Hyperbola with horizontal transverse axis

Graph (e)

11. Find the vertex and focus of the parabola $y^2 = -6x$ and sketch its graph.

Solution

$y^2 = 4\left(-\frac{3}{2}\right)x;\ p = -\frac{3}{2}$

Vertex: $(0, 0)$

Focus: $\left(-\frac{3}{2}, 0\right)$

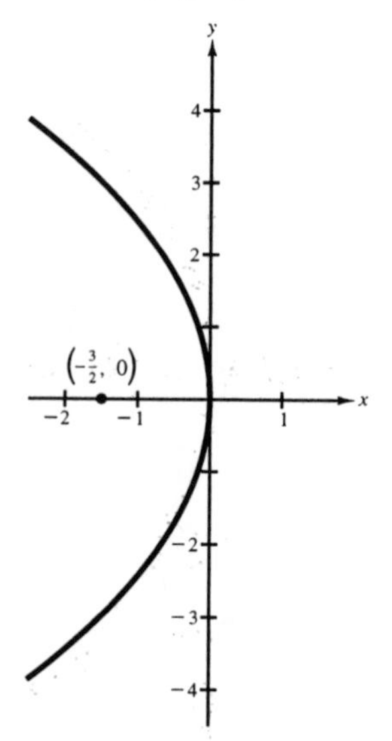

15. Find the vertex and focus of the parabola $y^2 - 8x = 0$ and sketch its graph.

Solution

$y^2 = 4(2)x;\ p = 2$

Vertex: $(0, 0)$

Focus: $(2, 0)$

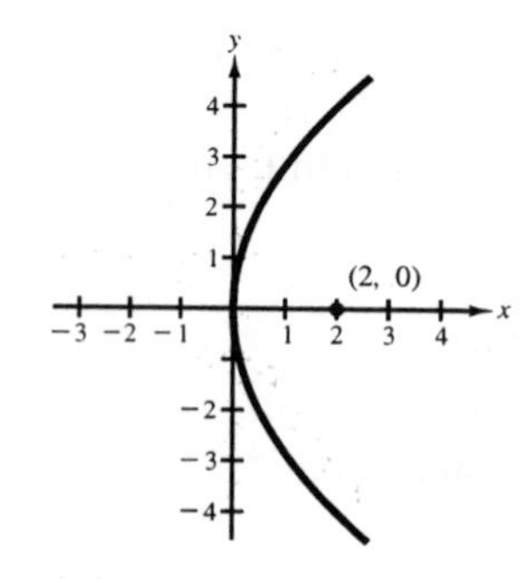

19. Find the equation of the parabola.

Vertex: $(0, 0)$

Focus: $(-2, 0)$

Solution

Focus: $(-2, 0) \Rightarrow p = -2$

$y^2 = 4(-2)x$

$y^2 = -8x$

23. Find the equation of the parabola.

Vertex: $(0, 0)$

Directrix: $y = 2$

Solution

Directrix: $y = 2 \Rightarrow p = -2$

$x^2 = 4(-2)y$

$x^2 = -8y$

27. Find the center and vertices of the ellipse $\dfrac{x^2}{25} + \dfrac{y^2}{16} = 1$ and sketch its graph.

Solution

$\dfrac{x^2}{25} + \dfrac{y^2}{16} = 1$

Horizontal major axis

$a = 5, \ b = 4$

Center: $(0, 0)$

Vertices: $(\pm 5, 0)$

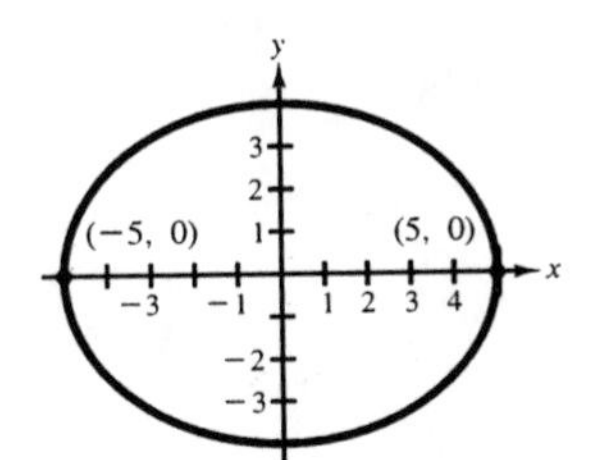

31. Find the center and vertices of the ellipse $\dfrac{x^2}{9} + \dfrac{y^2}{5} = 1$ and sketch its graph.

Solution

$\dfrac{x^2}{9} + \dfrac{y^2}{5} = 1$

Horizontal major axis

$a = 3, \ b = \sqrt{5}$

Center: $(0, 0)$

Vertices: $(\pm 3, 0)$

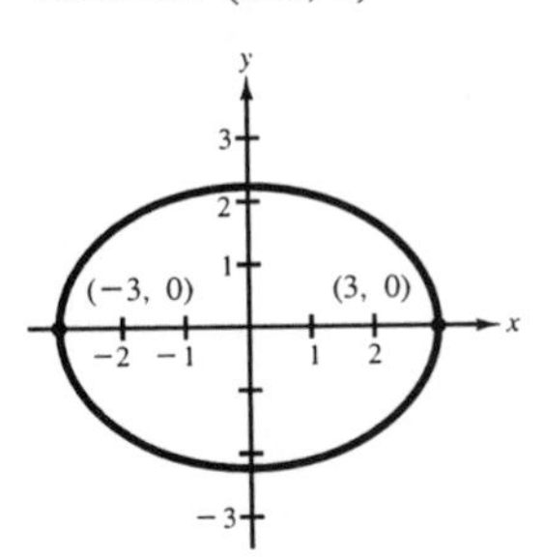

35. Find the equation of the ellipse.

Vertices: $(0, \pm 2)$

Minor axis of length 2

Solution

Vertices: $(0, \pm 2) \ \Rightarrow \ a = 2$

Minor axis of length 2 $\Rightarrow \ b = 1$

Vertical major axis

$\dfrac{x^2}{b^2} + \dfrac{y^2}{a^2} = 1$

$\dfrac{x^2}{1} + \dfrac{y^2}{4} = 1$

39. Find the equation of the ellipse.

Foci: $(\pm 5, 0)$

Major axis of length 12

Solution

Foci: $(\pm 5, 0) \ \Rightarrow \ c = 5$

Major axis of length 12 $\Rightarrow \ a = 6$

$b = \sqrt{6^2 - 5^2} = \sqrt{11}$

Horizontal major axis

$\dfrac{x^2}{a^2} + \dfrac{y^2}{b^2} = 1$

$\dfrac{x^2}{36} + \dfrac{y^2}{11} = 1$

43. Find the center and vertices of the hyperbola $x^2 - y^2 = 1$ and sketch its graph.

Solution

$x^2 - y^2 = 1$

$a = 1,\ b = 1$

Center: $(0, 0)$

Vertices $(\pm 1, 0)$

Asymptotes: $y = \pm x$

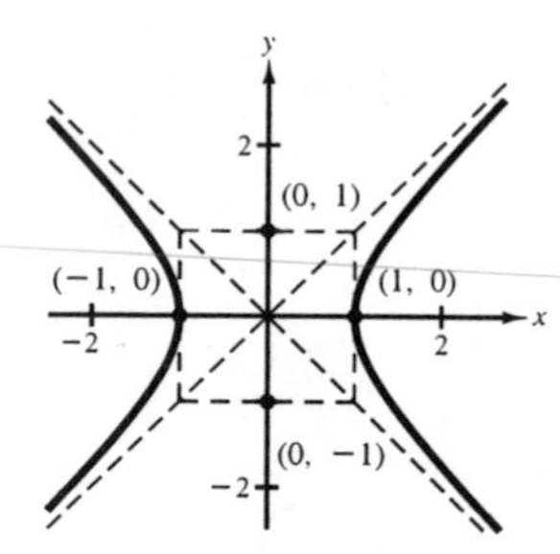

47. Find the center and vertices of the hyperbola $\frac{y^2}{25} - \frac{x^2}{144} = 1$ and sketch its graph.

Solution

$$\frac{y^2}{25} - \frac{x^2}{144} = 1$$

$a = 5,\ b = 12$

Center: $(0, 0)$

Vertices: $(0, \pm 5)$

Asymptotes: $y = \pm \frac{5}{12}x$

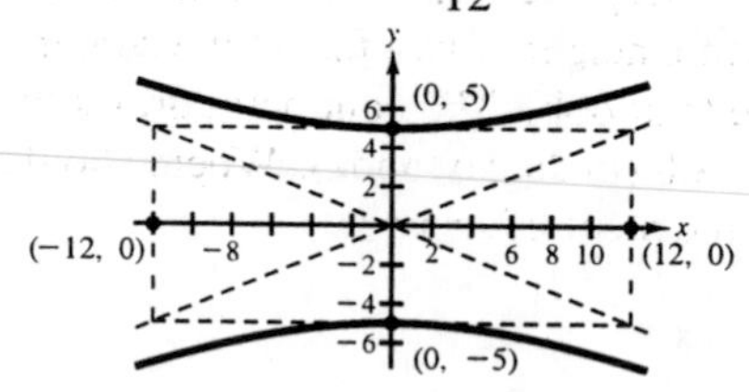

51. Find the equation of the hyperbola.

Vertices: $(0, \pm 2)$

Foci: $(0, \pm 4)$

Solution

Vertices: $(0, \pm 2) \Rightarrow a = 2$

Foci: $(0, \pm 4) \Rightarrow c = 4$

$b^2 = c^2 - a^2 = 12$

Vertical transverse axis

$$\frac{y^2}{a^2} - \frac{x^2}{b^2} = 1$$

$$\frac{y^2}{4} - \frac{x^2}{12} = 1$$

55. Find the equation of the hyperbola.

Foci: $(0, \pm 8)$

Asymptotes: $y = \pm 4x$

Solution

Foci: $(0, \pm 8) \Rightarrow c = 8$

Asymptotes: $y = \pm 4x$

Vertical transverse axis

$4 = \frac{a}{b} \Rightarrow a = 4b$

$16b^2 + b^2 = (8)^2$

$$b^2 = \frac{64}{17} \Rightarrow a^2 = \frac{1024}{17}$$

$$\frac{y^2}{a^2} - \frac{x^2}{b^2} = 1$$

$$\frac{17y^2}{1024} - \frac{17x^2}{64} = 1$$

59. *Satellite Antenna* The receiver in a parabolic television dish antenna is three feet from the vertex and is located at the focus (see figure in the textbook). Find an equation of a cross section of the reflector. (Assume that the dish is directed upward and the vertex is at the origin.)

Solution

Focus: $(0, 3)$

$x^2 = 4(3)y$

$x^2 = 12y$

61. *Fireplace Arch* A fireplace arch is to be constructed in the shape of a semi-ellipse. The opening is to have a height of two feet at the center and a width of 5 feet along the base (see figure in the textbook). The contractor draws the outline of the ellipse by the method shown in Figure A.8 (in the textbook). Where should the tacks be placed and what should be the length of the piece of string?

Solution

$2a = 5 \Rightarrow a = 2.5$

$b = 2$

$c^2 = (2.5)^2 - (2)^2$

$c = \sqrt{2.25} = 1.5$

The tacks should be placed at $(\pm 1.5, 0)$. The length of the string should be $2a = 5$ feet.

65. Sketch the graph of the ellipse $\dfrac{x^2}{4} + \dfrac{y^2}{1} = 1$, making use of the latus recta (see Exercise 64).

Solution

$\dfrac{x^2}{4} + \dfrac{y^2}{1} = 1$

$a = 2,\ b = 1,\ c = \sqrt{3}$

Points on the ellipse: $(\pm 2, 0),\ (0, \pm 1)$

Length of latus recta:

$\dfrac{2b^2}{a} = \dfrac{2(1)^2}{2} = 1$

Additional points: $\left(-\sqrt{3}, \pm\dfrac{1}{2}\right), \left(\sqrt{3}, \pm\dfrac{1}{2}\right)$

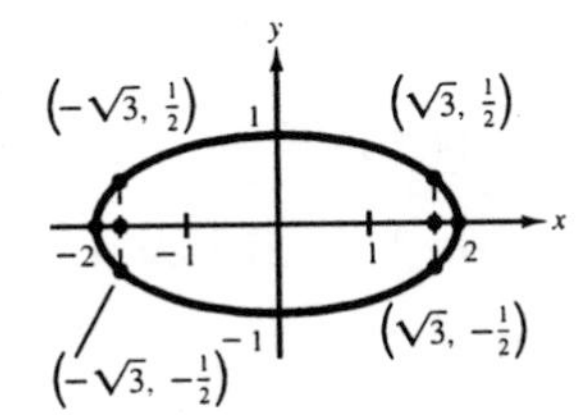

69. *LORAN* Long-distance radio navigation for aircraft and ships is accomplished by synchronized pulses transmitted by widely separated transmitting stations. These pulses travel at the speed of light (186,000 miles per second). The difference in the arrival times of these pulses at an aircraft or ship is constant on a hyperbola having the transmitting stations as foci. Assume that two stations, 300 miles apart, are positioned on the rectangular coordinate system at points with coordinates $(-150, 0)$ and $(150, 0)$ and that a ship is traveling on a path with coordinates $(x, 75)$ (see figure). Find the x-coordinate of the position of the ship if the time difference between the pulses from the transmitting stations is 1000 microseconds (0.001 second).

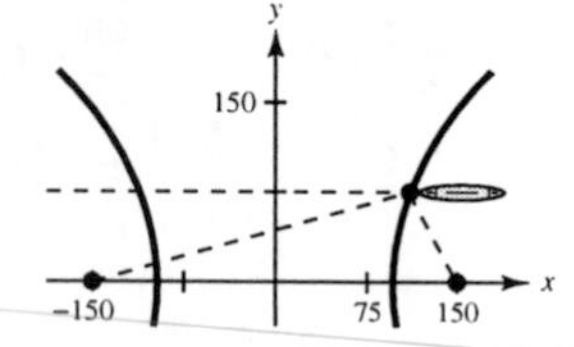

Solution

$r = 186{,}000$ miles per second

$t_1 = t_2 - 0.001$

$$d_1 = \sqrt{(150 - x)^2 + (0 - 75)^2} = \sqrt{x^2 - 300x + 28{,}125}$$

$$d_2 = \sqrt{(-150 - x)^2 + (0 - 75)^2} = \sqrt{x^2 + 300x + 28{,}125}$$

$$\frac{\sqrt{x^2 - 300x + 28{,}125}}{186{,}000} = \frac{\sqrt{x^2 + 300x + 28{,}125}}{186{,}000} - 0.001$$

$$\sqrt{x^2 - 300x + 28{,}125} = \sqrt{x^2 + 300x + 28{,}125} - 186$$

$$x^2 - 300x + 28{,}125 = x^2 + 300x + 28{,}125 - 372\sqrt{x^2 + 300x + 28{,}125} + 34{,}596$$

$$372\sqrt{x^2 + 300x + 28{,}125} = 600x + 34{,}596$$

$$138{,}384(x^2 + 300x + 28{,}125) = 360{,}000x^2 + 41{,}515{,}200x + 1{,}196{,}883{,}216$$

$$2{,}695{,}166{,}784 = 221{,}616x^2$$

$$x \approx 110.3 \text{ miles}$$

71. Use the definition of an ellipse to derive the standard form of the equation of an ellipse.

Solution

Let (x, y) be such that the sum of the distance from $(c, 0)$ and $(-c, 0)$ is $2a$. (Note that this is only deriving the standard form for the ellipse with horizontal major axis.)

$$
\begin{aligned}
2a &= \sqrt{(x-c)^2+y^2} + \sqrt{(x+c)^2+y^2} \\
2a - \sqrt{(x+c)^2+y^2} &= \sqrt{(x-c)^2+y^2} \\
4a^2 - 4a\sqrt{(x+c)^2+y^2} + (x+c)^2 + y^2 &= (x-c)^2 + y^2 \\
4a^2 + 4cx &= 4a\sqrt{(x+c)^2+y^2} \\
a^2 + cx &= a\sqrt{(x+c)^2+y^2} \\
a^4 + 2a^2cx + c^2x^2 &= a^2(x^2 + 2cx + c^2 + y^2) \\
a^4 + c^2x^2 &= a^2x^2 + a^2c^2 + a^2y^2 \\
a^2(a^2 - c^2) &= (a^2 - c^2)x^2 + a^2y^2
\end{aligned}
$$

Let $b^2 = a^2 - c^2$. Then we have

$$a^2b^2 = b^2x^2 + a^2y^2 \Rightarrow 1 = \frac{x^2}{a^2} + \frac{y^2}{b^2}.$$

A.2 Conic Sections and Translations

1. Find the vertex, focus, and directrix of the parabola $(x-1)^2+8(y+2)^2=0$ and sketch its graph.

Solution

$(x-1)^2+8(y+2)=0$

$(x-1)^2=4(-2)(y+2);\ p=-2$

Vertex: $(1,-2)$

Focus: $(1,-4)$

Directrix: $y=0$

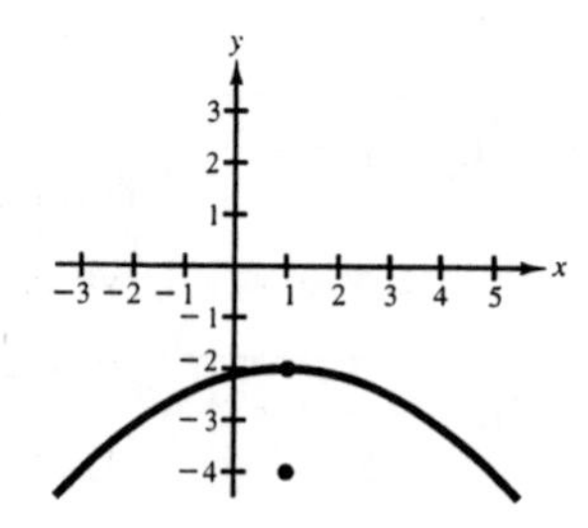

5. Find the vertex, focus, and directrix of the parabola $y=\frac{1}{4}(x^2-2x+5)$ and sketch its graph.

Solution

$y=\frac{1}{4}(x^2-2x+5)$

$y=\frac{1}{4}(x-1)^2+1$

$4(y-1)=(x-1)^2;\ p=1$

Vertex: $(1,1)$

Focus: $(1,2)$

Directrix: $y=0$

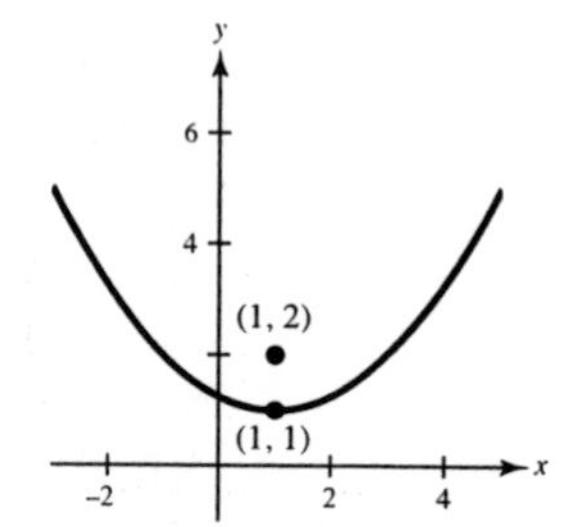

9. Find the vertex, focus, and directrix of the parabola $y^2+6y+8x+25=0$ and sketch its graph.

Solution

$y^2+6y+8x+25=0$

$(y+3)^2=4(-2)(x+2);\ p=-2$

Vertex: $(-2,-3)$

Focus: $(-4,-3)$

Directrix: $x=0$

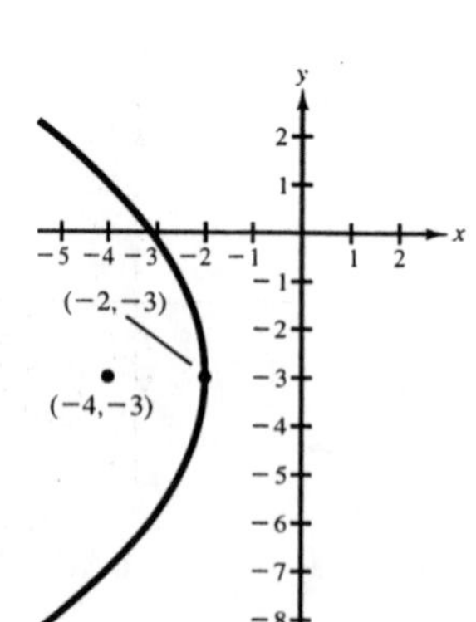

13. Find an equation of the parabola.

Vertex: $(3, 2)$

Focus: $(1, 2)$

Solution

Vertex: $(3, 2)$

Focus: $(1, 2)$

Horizontal axis

$p = 1 - 3 = -2$

$(y-2)^2 = 4(-2)(x-3)$

$(y-2)^2 = -8(x-3)$

17. Find an equation of the parabola.

Focus: $(2, 2)$

Directrix: $x = -2$

Solution

Focus: $(2, 2)$

Directrix: $x = -2$

Horizontal axis

Vertex: $(0, 2)$

$p = 2 - 0 = 2$

$(y-2)^2 = 4(2)(x-0)$

$(y-2)^2 = 8x$

21. Find the center, foci, and vertices of the ellipse $\dfrac{(x-1)^2}{9} + \dfrac{(y-5)^2}{25} = 1$ and sketch its graph.

Solution

$$\frac{(x-1)^2}{9} + \frac{(y-5)^2}{25} = 1$$

$a = 5,\ b = 3,\ c = \sqrt{a^2 - b^2} = 4$

Center: $(1, 5)$

Foci: $(1, 1), (1, 9)$

Vertices: $(1, 0), (1, 10)$

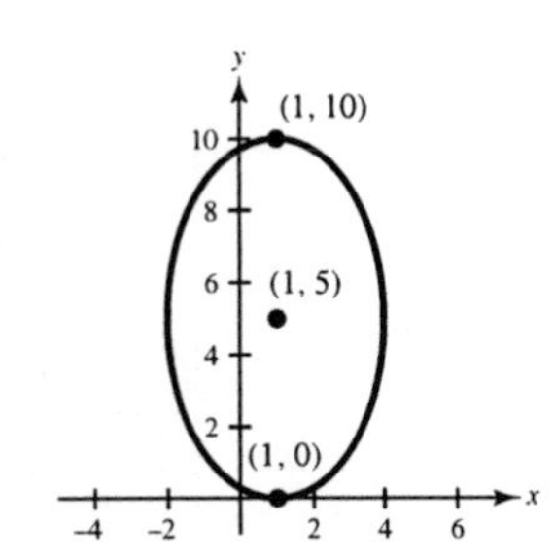

25. Find the center, foci, and vertices of the ellipse $16x^2 + 25y^2 - 32x + 50y + 16 = 0$ and sketch its graph.

Solution

$$16x^2 + 25y^2 - 32x + 50y + 16 = 0$$

$$16(x^2 - 2x + 1) + 25(y^2 + 2y + 1) = 25$$

$$\frac{(x-1)^2}{25/16} + (y+1)^2 = 1$$

$a = \dfrac{5}{4}, b = 1, c = \sqrt{a^2 - b^2} = \dfrac{3}{4}$

Horizontal major axis

Center: $(1, -1)$

Foci: $\left(\dfrac{1}{4}, -1\right), \left(\dfrac{7}{4}, -1\right)$

Vertices: $\left(-\dfrac{1}{4}, -1\right), \left(\dfrac{9}{4}, -1\right)$

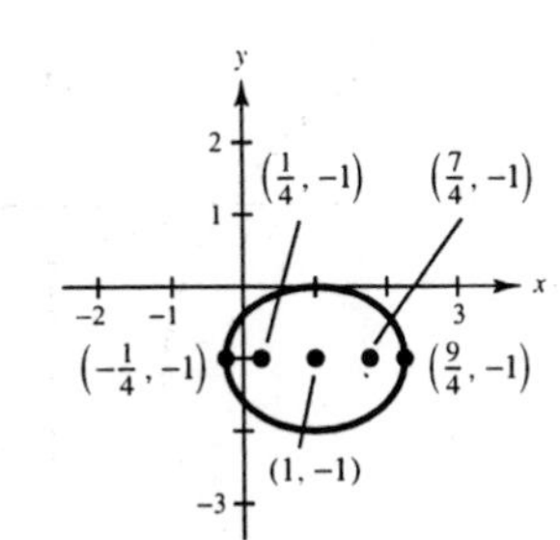

29. Find an equation for the ellipse.

Vertices: $(0, 2)$, $(4, 2)$

Minor axis of length 2

Solution

Vertices: $(0, 2)$, $(4, 2)$

Minor axis of length 2

$a = \dfrac{4-0}{2} = 2,\ b = \dfrac{2}{2} = 1$

Center: $(2, 2)$

Horizontal major axis

$$\frac{(x-2)^2}{4} + (y-2)^2 = 1$$

31. Find an equation for the ellipse.

Foci: $(0, 0)$, $(0, 8)$

Major axis of length 16

Solution

Foci: $(0, 0)$, $(0, 8)$

Major axis of length 16

$a = 8,\ c = 4,\ b^2 = 48$

Center: $(0, 4)$

Vertical major axis

$$\frac{x^2}{48} + \frac{(y-4)^2}{64} = 1$$

35. Find an equation for the ellipse.

Center: $(0, 4)$

$a = 2c$

Vertices: $(-4, 4)$, $(4, 4)$

Solution

Center: $(0, 4)$

$a = 2c$

Vertices: $(-4, 4)$, $(4, 4)$

$a = 4,\ c = 2,\ b^2 = 12$

Horizontal major axis

$$\frac{x^2}{16} + \frac{(y-4)^2}{12} = 1$$

39. Find the center, vertices, and foci of the hyperbola $(y+6)^2 - (x-2)^2 = 1$ and sketch its graph.

Solution

$(y+6)^2 - (x-2)^2 = 1$

$a = 1,\ b = 1,\ c = \sqrt{a^2 + b^2} = \sqrt{2}$

Center: $(2, -6)$

Vertical transverse axis

Vertices: $(2, -5)$, $(2, -7)$

Foci: $\left(2, -6 \pm \sqrt{2}\right)$

Asymptotes: $y = \pm(x-2) - 6$

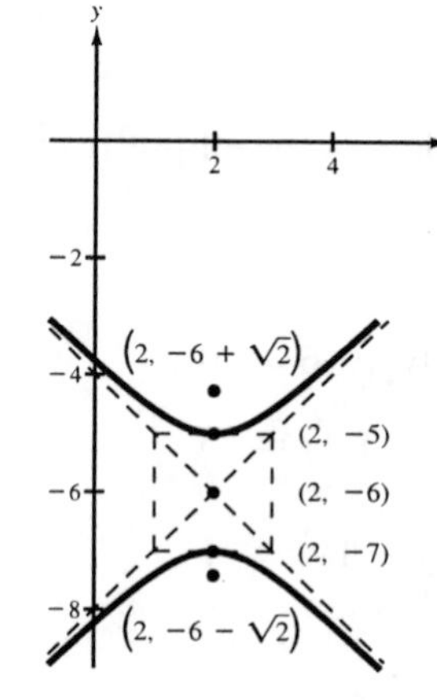

41. Find the center, vertices, and foci of the hyperbola $9x^2 - y^2 - 36x - 6y + 18 = 0$ and sketch its graph.

Solution

$$9x^2 - y^2 - 36x - 6y + 18 = 0$$
$$9(x^2 - 4x + 4) - (y^2 + 6y + 9) = 9$$
$$(x-2)^2 - \frac{(y+3)^2}{9} = 1$$

$a = 1,\ b = 3,\ c = \sqrt{a^2 + b^2} = \sqrt{10}$

Center: $(2, -3)$

Horizontal transverse axis

Vertices: $(1, -3), (3, -3)$

Foci: $(2 \pm \sqrt{10}, -3)$

Asymptotes: $y = \pm 3(x - 2) - 3$

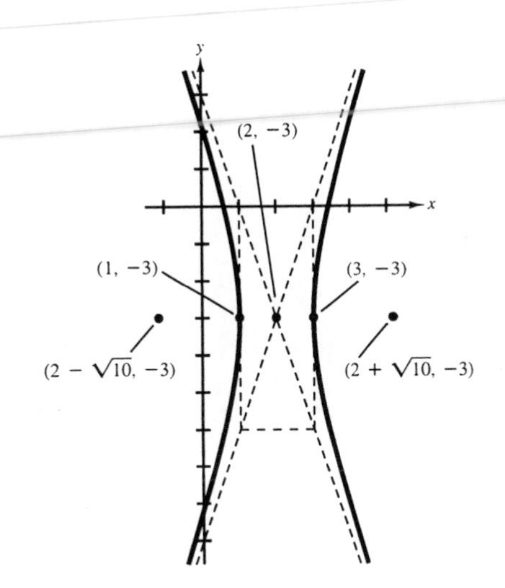

47. Find an equation for the hyperbola.

Vertices: $(2, 0), (6, 0)$

Foci: $(0, 0), (8, 0)$

Solution

Vertices: $(2, 0), (6, 0)$

Foci: $(0, 0), (8, 0)$

Center: $(4, 0)$

Horizontal transverse axis

$a = 2,\ c = 4,\ b^2 = c^2 - a^2 = 12$

$$\frac{(x-4)^2}{4} - \frac{y^2}{12} = 1$$

51. Find an equation for the hyperbola.

Vertices: $(2, 3), (2, -3)$

Passes through the point $(0, 5)$

Solution

Vertices: $(2, 3), (2, -3)$

Passes through the point $(0, 5)$

Center: $(2, 0)$

Vertical transverse axis

$a = 3$

$$\frac{y^2}{9} - \frac{(x-2)^2}{b^2} = 1$$
$$\frac{5^2}{9} - \frac{(0-2)^2}{b^2} = 1$$
$$b^2 = \frac{9}{4}$$
$$\frac{y^2}{9} - \frac{4(x-2)^2}{9} = 1$$

53. Find an equation for the hyperbola.

Vertices: $(0, 2)$, $(6, 2)$

Asymptotes: $y = \frac{2}{3}x,\ y = 4 - \frac{2}{3}x$

Solution

Vertices: $(0, 2)$, $(6, 2)$

Asymptotes: $y = \frac{2}{3}x,\ y = 4 - \frac{2}{3}x$

Center: $(3, 2)$

Horizontal transverse axis

$a = 3,\ \pm\frac{b}{a} = \pm\frac{2}{3} \Rightarrow b = 2$

$$\frac{(x-3)^2}{9} - \frac{(y-2)^2}{4} = 1$$

57. Classify the graph of the equation.

$$4x^2 - y^2 - 4x - 3 = 0$$

Solution

$$4x^2 - y^2 - 4x - 3 = 0$$

$$4\left(x - \tfrac{1}{2}\right)^2 - y^2 = 4$$

$$\left(x - \tfrac{1}{2}\right)^2 - \frac{y^2}{4} = 1$$

Hyperbola

61. Classify the graph of the equation.

$$25x^2 - 10x - 200y - 119 = 0$$

Solution

$$25x^2 - 10x - 200y - 119 = 0$$

$$25\left(x - \tfrac{1}{5}\right)^2 = 200\left(y + \tfrac{3}{5}\right)$$

$$\left(x - \tfrac{1}{5}\right)^2 = 8\left(y + \tfrac{3}{5}\right)$$

Parabola

65. Find an equation of the ellipse with vertices $(\pm 5, 0)$ and eccentricity $e = \frac{3}{5}$.

Solution

Vertices: $(\pm 5, 0)$

Center: $(0, 0)$

$e = \frac{c}{a} = \frac{3}{5}$

$a = 5,\ c = 3,\ b = 4$

$$\frac{x^2}{25} + \frac{y^2}{16} = 1$$

69. *Orbit of Saturn* The planet Saturn moves in an elliptical orbit with the sun at one of the foci (see figure in the textbook). The shortest distance and the greatest distance that the planet can get from the sun are 1.3495×10^9 kilometers and 1.5045×10^9 kilometers, respectively. Find the eccentricity of the orbit.

Solution

Least distance: $a - c = 1.3495 \times 10^9$

Greatest distance: $a + c = 1.5045 \times 10^9$

$$a = 1.3495 \times 10^9 + c$$

$$(1.3495 \times 10^9 + c) + c = 1.5045 \times 10^9$$

$$2c = 1.55 \times 10^8$$

$$c = 7.75 \times 10^7$$

$a = 1.3495 \times 10^9 + 7.75 \times 10^7$

$a = 1.427 \times 10^9$

$$e = \frac{c}{a} = \frac{7.75 \times 10^7}{1.427 \times 10^9} \approx 0.05431 = 5.431 \times 10^{-2}$$

71. Show that the equation of an ellipse can be written as

$$\frac{(x-h)^2}{a^2} + \frac{(y-k)^2}{a^2(1-e^2)} = 1.$$

Note that as e approaches zero, the ellipse approaches a circle of radius a.

Solution

$e = \frac{c}{a} \Rightarrow c = ae$

$b^2 = a^2 - c^2 = a^2 - (ae)^2 = a^2(1 - e^2)$

Thus, $\frac{(x-h)^2}{a^2} + \frac{(y-k)^2}{b^2} = 1$ can be written as $\frac{(x-h)^2}{a^2} + \frac{(y-k)^2}{a^2(1-e^2)} = 1.$

Practice Test Solutions for Chapter P

1. $-2(-2)^3 + 7(-2) - 4 = -2(-8) - 14 - 4 = -16 - 14 - 4 = -34$

2. $\frac{x}{z} - \frac{z}{y} = \frac{x}{z} \cdot \frac{y}{y} - \frac{z}{y} \cdot \frac{z}{z} = \frac{xy - z^2}{yz}$

3. $|x - 7| \le 4$

4. $10(-5)^3 = 10(-125) = -1250$

5. $(-4x^3)(-2x^{-5})\left(\frac{1}{16}x\right) = (-4)(-2)\left(\frac{1}{16}\right)x^{3+(-5)+1} = \frac{8}{16}x^{-1} = \frac{1}{2x}$

6. $0.0000412 = 4.12 \times 10^{-5}$

7. $125^{2/3} = (\sqrt[3]{125})^2 = (5)^2 = 25$

8. $\sqrt[4]{64x^7y^9} = \sqrt[4]{16 \cdot 4x^4x^3y^8y}$
$= 2xy^2\sqrt[4]{4x^3y}$

9. $\frac{6}{\sqrt{12}} = \frac{6}{2\sqrt{3}} \cdot \frac{\sqrt{3}}{\sqrt{3}} = \frac{6\sqrt{3}}{6} = \sqrt{3}$

10. $3\sqrt{80} - 7\sqrt{500} = 3(4\sqrt{5}) - 7(10\sqrt{5})$
$= 12\sqrt{5} - 70\sqrt{5} = -58\sqrt{5}$

11. $(8x^4 - 9x^2 + 2x - 1) - (3x^3 + 5x + 4) = 8x^4 - 3x^3 - 9x^2 - 3x - 5$

12. $(x - 3)(x^2 + x - 7) = x^3 + x^2 - 7x - 3x^2 - 3x + 21 = x^3 - 2x^2 - 10x + 21$

13. $[(x - 2) - y]^2 = (x - 2)^2 - 2y(x - 2) + y^2$
$= x^2 - 4x + 4 - 2xy + 4y + y^2 = x^2 + y^2 - 2xy - 4x + 4y + 4$

14. $16x^4 - 1 = (4x^2 + 1)(4x^2 - 1) = (4x^2 + 1)(2x + 1)(2x - 1)$

15. $6x^2 + 5x - 4 = (2x - 1)(3x + 4)$

16. $x^3 - 64 = x^3 - 4^3$
$= (x - 4)(x^2 + 4x + 16)$

17. $-\frac{3}{x} + \frac{x}{x^2 + 2} = \frac{-3(x^2 + 2) + x^2}{x(x^2 + 2)} = \frac{-2x^2 - 6}{x(x^2 + 2)} = -\frac{2(x^2 + 3)}{x(x^2 + 2)}$

18. $\frac{x - 3}{4x} \div \frac{x^2 - 9}{x^2} = \frac{x - 3}{4x} \cdot \frac{x^2}{(x + 3)(x - 3)} = \frac{x}{4(x + 3)}$

19. $\frac{1 - \frac{1}{x}}{1 - \frac{1}{1 - (1/x)}} = \frac{\frac{x - 1}{x}}{1 - \frac{1}{(x - 1)/x}} = \frac{\frac{x - 1}{x}}{1 - \frac{x}{x - 1}} = \frac{\frac{x - 1}{x}}{\frac{-1}{x - 1}} = \frac{x - 1}{x} \cdot \frac{x - 1}{-1} = \frac{-(x - 1)^2}{x}$

20. $B = 5000\left(1 + \frac{0.075}{4}\right)^{(4)(20)} \approx \$22{,}099.36$

Practice Test Solutions for Chapter 1

1. $5x + 4 = 7x - 8$

$4 + 8 = 7x - 5x$

$12 = 2x$

$x = 6$

2. $\frac{x}{3} - 5 = \frac{x}{5} + 1$

$15\left(\frac{x}{3} - 5\right) = 15\left(\frac{x}{5} + 1\right)$

$5x - 75 = 3x + 15$

$2x = 90$

$x = 45$

3. $\frac{3x + 1}{6x - 7} = \frac{2}{5}$

$5(3x + 1) = 2(6x - 7)$

$15x + 5 = 12x - 14$

$3x = -19$

$x = -\frac{19}{3}$

4. $(x - 3)^2 + 4 = (x + 1)^2$

$x^2 - 6x + 9 + 4 = x^2 + 2x + 1$

$-8x = -12$

$x = \frac{-12}{-8}$

$x = \frac{3}{2}$

5. $A = \frac{1}{2}(a + b)h$

$2A = ah + bh$

$2A - bh = ah$

$\frac{2A - bh}{h} = a$

6. $x + (x + 1) + (x + 2) = 132$

$3x + 3 = 132$

$3x = 129$

$x = 43$

$x + 1 = 44$

$x + 2 = 45$

7. Percent $= \frac{301}{4300} = 0.07 = 7\%$

8. Let x = number of quarters.
Then $53 - x$ = number of nickels.

$25x + 5(53 - x) = 605$

$20x + 265 = 605$

$20x = 340$

$x = 17$ quarters

$53 - x = 36$ nickels

9. Let x = amount in $9\frac{1}{2}\%$ fund.
Then $15{,}000 - x$ = amount in 11% fund.

$0.095x + 0.11(15{,}000 - x) = 1582.50$

$-0.015x + 1650 = 1582.50$

$-0.015x = -67.5$

$x = \$4500$ @ $9\frac{1}{2}\%$

$15{,}000 - x = \$10{,}500$ @ 11%

10. $28 + 5x - 3x^2 = 0$

$(4 - x)(7 + 3x) = 0$

$x = 4$ or $x = -\frac{7}{3}$

11. $(x+3)^2 = 49$

$$x + 3 = \pm\sqrt{49}$$

$$x = -3 \pm 7$$

$x = 4$ or $x = -10$

12. $(x-2)^2 = 24$

$$x - 2 = \pm\sqrt{24}$$

$$x - 2 = \pm 2\sqrt{6}$$

$$x = 2 \pm 2\sqrt{6}$$

13. $x^2 + 5x - 1 = 0$

$a = 1,\ b = 5,\ c = -1$

$$x = \frac{-5 \pm \sqrt{(5)^2 - 4(1)(-1)}}{2(1)}$$

$$= \frac{-5 \pm \sqrt{25+4}}{2}$$

$$= \frac{-5 \pm \sqrt{29}}{2}$$

14. $3x^2 - 2x + 4 = 0$

$a = 3,\ b = -2,\ c = 4$

$$x = \frac{-(-2) \pm \sqrt{(-2)^2 - 4(3)(4)}}{2(3)}$$

$$= \frac{2 \pm \sqrt{4 - 48}}{6}$$

$$= \frac{2 \pm \sqrt{-44}}{6}$$

No real solutions.

15.

$$60{,}000 = xy$$

$$y = \frac{60{,}000}{x}$$

$$2x + 2y = 1100$$

$$2x + 2\left(\frac{60{,}000}{x}\right) = 1100$$

$$x + \frac{60{,}000}{x} = 550$$

$$x^2 + 60{,}000 = 550x$$

$$x^2 - 550x + 60,000 = 0$$

$$(x - 150)(x - 400) = 0$$

$x = 150$ or $x = 400$

$y = 400$ $\quad y = 150$

Length: 400 feet, Width: 150 feet

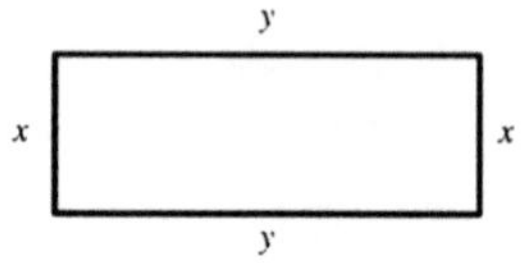

16.

$$x(x+2) = 624$$

$$x^2 + 2x - 624 = 0$$

$$(x - 24)(x + 26) = 0$$

$x = 24,\ x = -26,$ (extraneous solution)

$x + 2 = 26$

17. $x^3 - 10x^2 + 24x = 0$

$$x(x^2 - 10x + 24) = 0$$

$$x(x-4)(x-6) = 0$$

$$x = 0,\ x = 4,\ x = 6$$

18. $\sqrt[3]{6-x} = 4$

$6 - x = 64$

$-x = 58$

$x = -58$

19. $(x^2 - 8)^{2/5} = 4$

$x^2 - 8 = \pm 4^{5/2}$

$x^2 - 8 = \pm 32$

$x^2 = 40$ or $x^2 = -24$

$x = \pm\sqrt{40}$

$x = \pm 2\sqrt{10}$

No real solution

20. $x^4 - x^2 - 12 = 0$

$(x^2 - 4)(x^2 + 3) = 0$

$x^2 = \;\; 4$ or $x^2 = -3$

$x = \pm 2$

No real solutions

21. $4 - 3x > 16$

$-3x > 12$

$x < -4$

22. $4x^3 - 12x^2 \geq 0$

$4x^2(x - 3) \geq 0$

Critical numbers: $x = 0,\ 3$

Test intervals: $(-\infty, 0),\ (0, 3),\ (3, \infty)$

Solution set: $[3, \infty)$ and $x = 0$

-1 0 1 2 3 4 x

23. $\left|\frac{x-3}{2}\right| < 5$

$-5 < \frac{x-3}{2} < 5$

$-10 < x - 3 < 10$

$-7 < x < 13$

24. $\frac{x+1}{x-3} < 2$

$\frac{x+1}{x-3} - 2 < 0$

$\frac{x + 1 - 2(x - 3)}{x - 3} < 0$

$\frac{7 - x}{x - 3} < 0$

Solution intervals: $x < 3$ or $x \geq 7$

2 3 4 5 6 7 8 x

25. $|3x - 4| \geq 9$

$3x - 4 \leq -9$ or $3x - 4 \geq 9$

$3x \leq -5 \qquad 3x \geq 13$

$x \leq -\frac{5}{3} \qquad x \geq \frac{13}{3}$

Practice Test Solutions for Chapter 2

1. $d = \sqrt{(4-0)^2 + (-1-3)^2}$

$= \sqrt{16+16}$

$= \sqrt{32}$

$= 4\sqrt{2}$

2. Midpoint: $\left(\dfrac{4+0}{2}, \dfrac{-1+3}{2}\right) = (2, 1)$

3. $6 = \sqrt{(x-0)^2 + (-2-0)^2}$

$6 = \sqrt{x^2+4}$

$36 = x^2 + 4$

$x^2 = 32$

$x = \pm\sqrt{32}$

$x = \pm 4\sqrt{2}$

4. x-intercept: Let $y = 0$; $0 = \dfrac{x-2}{x+3}$

$0 = x - 2$

$x = 2 \quad (2, 0)$

y-intercept: Let $x = 0$; $y = \dfrac{0-2}{0+3}$

$y = -\dfrac{2}{3} \quad \left(0, -\dfrac{2}{3}\right)$

5. x-intercepts: Let $y = 0$; $0 = x\sqrt{9-x^2}$

$0 = x^2(9-x^2)$

$0 = x^2(3+x)(3-x)$

$x = 0, \pm 3, \quad (0, 0), (3, 0), (-3, 0)$

y-intercept: Let $x = 0$; $y = 0\sqrt{9-0^2} = 0(3) = 0, \quad (0, 0)$

6. $xy^2 = 6$

$x(-y)^2 = 6 \Rightarrow xy^2 = 6$ x-axis symmetry

$(-x)y^2 = 6 \Rightarrow xy^2 = -6$ No y-axis symmetry

$(-x)(-y)^2 = 6 \Rightarrow xy^2 = -6$ No origin symmetry

7. $y = \dfrac{x}{x^2+3}$

$-y = \dfrac{x}{x^2+3} \Rightarrow y = -\dfrac{x}{x^2+3}$ No x-axis symmetry

$y = \dfrac{-x}{(-x)^2+3} \Rightarrow y = -\dfrac{x}{x^2+3}$ No y-axis symmetry

$-y = \dfrac{-x}{(-x)^2+3} \Rightarrow y = \dfrac{x}{x^2+3}$ Origin symmetry

8. x-intercepts: $(0, 0)$, $(2, 0)$, $(-2, 0)$

Origin symmetry:

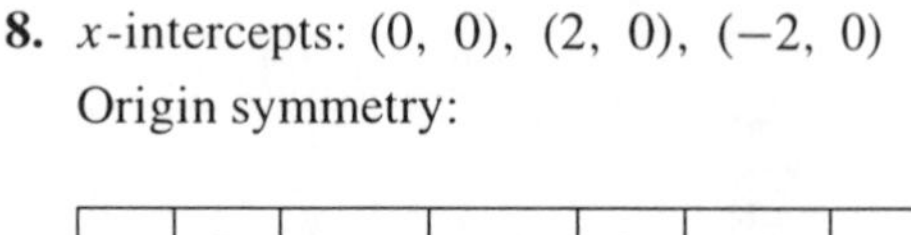

x	0	1	-1	2	-2	3
y	0	-3	3	0	0	15

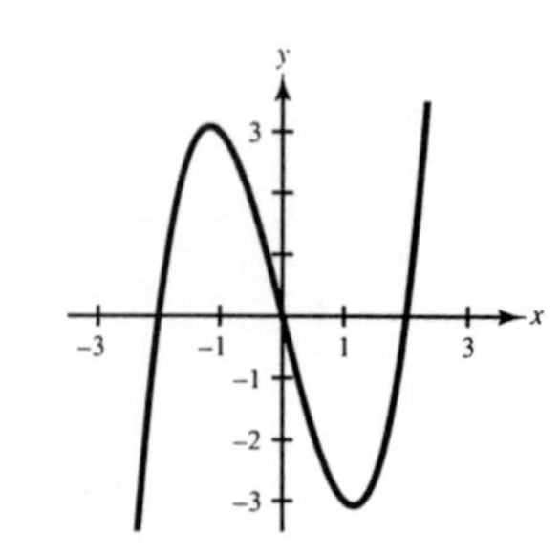

9. Center: $(3, \ -2)$

Radius: 4

$$(x-3)^2+(y+2)^2=4^2$$
$$x^2-6x+9+y^2+4y+4=16$$
$$x^2+y^2-6x+4y-3=0$$

10.
$$x^2+y^2-6x+2y+6=0$$
$$x^2-6x+\underline{9}+y^2+2y+\underline{1}=-6+9+1$$
$$(x-3)^2+(y-1)^2=4$$

Center: $(3, \ -1)$

Radius: 2

11. Parabola: Vertex $(0, \ -5)$

Intercepts: $(0, \ -5)$, $(\pm\sqrt{5}, \ 0)$

y-axis symmetry

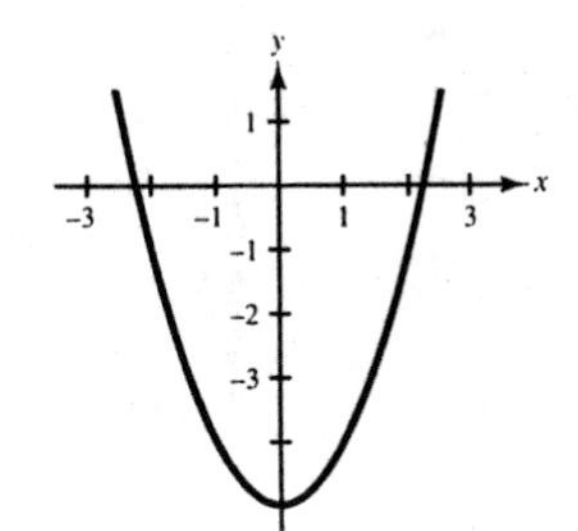

12. Intercepts: $(0, \ 3)$, $(-3, \ 0)$

x	0	1	-1	2	-2	-3	-4
y	3	4	2	5	1	0	1

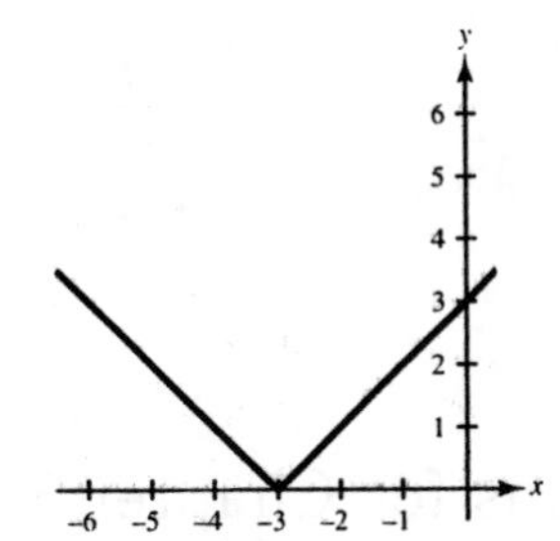

13. Intercepts: $(4, 0)$, $(0, 2)$

x	4	3	0	-5
y	0	1	2	3

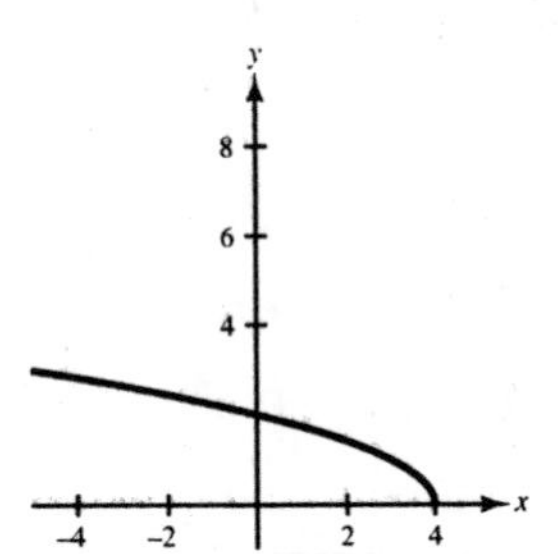

14.
$$m=\frac{-1-4}{3-2}=-5$$
$$y-4=-5(x-2)$$
$$y-4=-5x+10$$
$$y=-5x+14$$

15. $y=\frac{4}{3}x-3$

16. $2x + 3y = 0$

$$y = -\tfrac{2}{3}x$$

$$m_1 = -\tfrac{2}{3}$$

$$\perp m_2 = \tfrac{3}{2} \text{ through } (4, 1)$$

$$y - 1 = \tfrac{3}{2}(x - 4)$$

$$y - 1 = \tfrac{3}{2}x - 6$$

$$y = \tfrac{3}{2}x - 5$$

17. $x = -1$ is a vertical line, so the line through $(6, -5)$ is also vertical.
$x = 6$

18. $(5, 32)$ and $(9, 44)$

$$m = \frac{44 - 32}{9 - 5} = \frac{12}{4} = 3$$

$$y - 32 = 3(x - 5)$$

$$y - 32 = 3x - 15$$

$$y = 3x + 17$$

When $x = 20$, $y = 3(20) + 17$

$$y = \$77$$

19. $y = kx$ and $y = 30$ when $x = 5$

$$30 = k(5)$$

$$6 = k$$

$$y = 6x$$

20. $W = 2,100 + 0.08S$

Practice Test Solutions for Chapter 3

1. $f(x - 3) = (x - 3)^2 - 2(x - 3) + 1$

$$= x^2 - 6x + 9 - 2x + 6 + 1$$

$$= x^2 - 8x + 16$$

2. $f(3) = 12 - 11 = 1$

$$\frac{f(x) - f(3)}{x - 3} = \frac{(4x - 11) - 1}{x - 3}$$

$$= \frac{4x - 12}{x - 3}$$

$$= \frac{4(x - 3)}{x - 3}$$

$$= 4$$

3. $f(x) = \sqrt{36 - x^2} = \sqrt{(6 + x)(6 - x)}$

Domain: $[-6, 6]$

Range: $[0, 6]$

4. (a) $6x - 5y + 4 = 0$

$$y = \frac{6x + 4}{5} \quad \text{Function}$$

(b) $x^2 + y^2 = 9$

$$y = \pm\sqrt{9 - x^2} \quad \text{Not a function}$$

(c) $y^3 = x^2 + 6$

$$y = \sqrt[3]{x^2 + b} \quad \text{Function}$$

5.

x	0	1	2	3
y	1	3	5	7

x	-1	-2	0
y	2	6	0

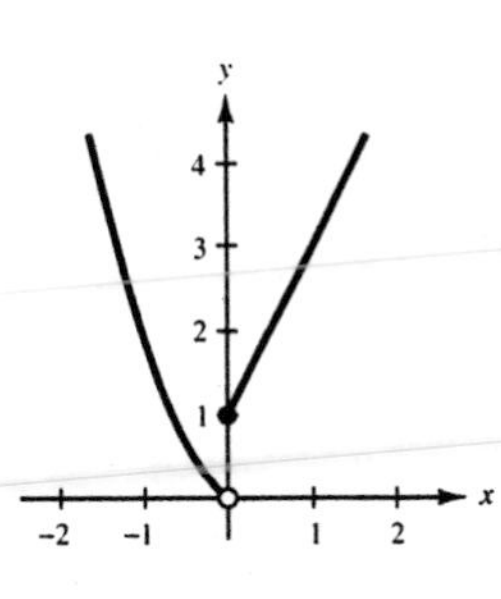

6.
$$\begin{aligned} f(g(x)) &= f(2x+3) \\ &= (2x+3)^2 - 2(2x+3) + 16 \\ &= 4x^2 + 12x + 9 - 4x - 6 + 16 \\ &= 4x^2 + 8x + 19 \end{aligned}$$

7.
$$\begin{aligned} f(x) &= x^3 + 7 \\ y &= x^3 + 7 \\ x &= \sqrt[3]{y-7} \\ f^{-1}(x) &= \sqrt[3]{x-7} \end{aligned}$$

8. x-intercepts: $(1, 0)$, $(5, 0)$

y-intercept: $(0, 5)$

Vertex: $(3, -4)$

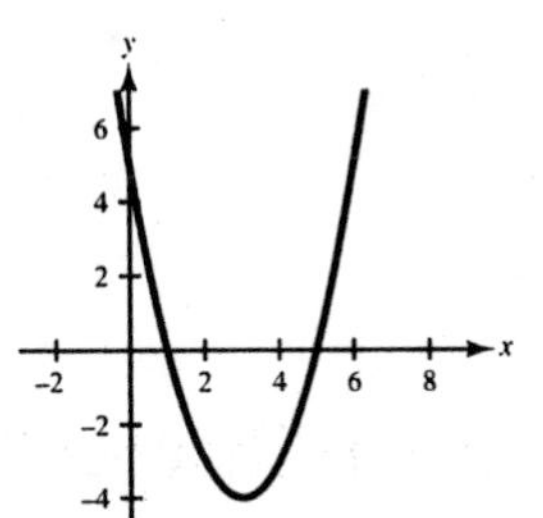

9. $a = 0.01,\ b = -90$

$$\frac{-b}{2a} = \frac{90}{2(.01)} = 4500 \text{ units}$$

10. Vertex $(1, 7)$ opening downward through $(2, 5)$

$y = a(x-1)^2 + 7$ Standard form

$5 = a(2-1)^2 + 7$

$5 = a + 7$

$a = -2$

$$\begin{aligned} y &= -2(x-1)^2 + 7 \\ &= -2(x^2 - 2x + 1) + 7 \\ &= -2x^2 + 4x + 5 \end{aligned}$$

11. $y = \pm(x-2)(3x-4)$

$y = \pm(3x^2 - 10x + 8)$

12. Leading coefficient: -3

Degree: 5

Moves down to the right and up to the left

13. $0 = x^5 - 5x^3 + 4x$

$= x(x^4 - 5x^2 + 4)$

$= x(x^2 - 1)(x^2 - 4)$

$= x(x + 1)(x - 1)(x + 2)(x - 2)$

$x = 0,\ x = \pm 1,\ x = \pm 2$

14. 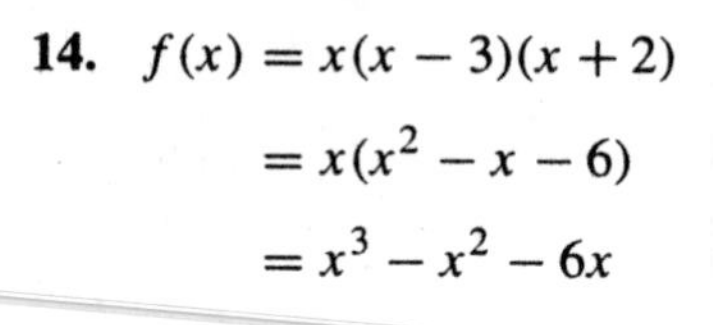

$f(x) = x(x - 3)(x + 2)$

$= x(x^2 - x - 6)$

$= x^3 - x^2 - 6x$

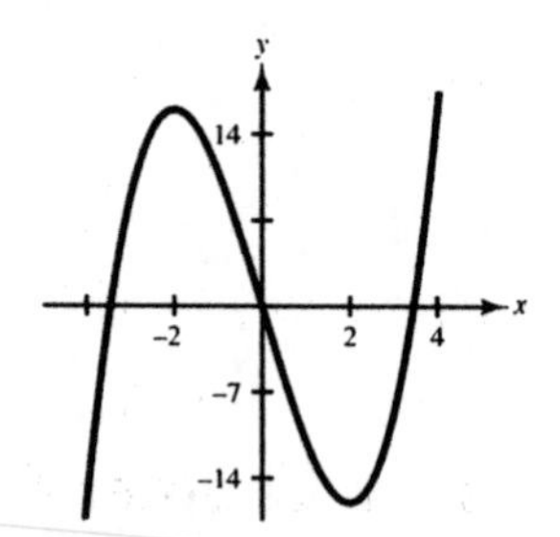

15. Intercepts: $(0,\ 0)$, $(0,\ \pm 2\sqrt{3})$

Origin symmetry

Moves up to the right.

Moves down to the left.

x	-2	-1	0	1	2
y	16	11	0	-11	-16

16. Vertical asymptote: $x = 0$

Horizontal asymptote: $y = \frac{1}{2}$

x-intercept: $(1, 0)$

17. Vertical asymptote: $x = 0$

Slant asymptote: $y = 3x$

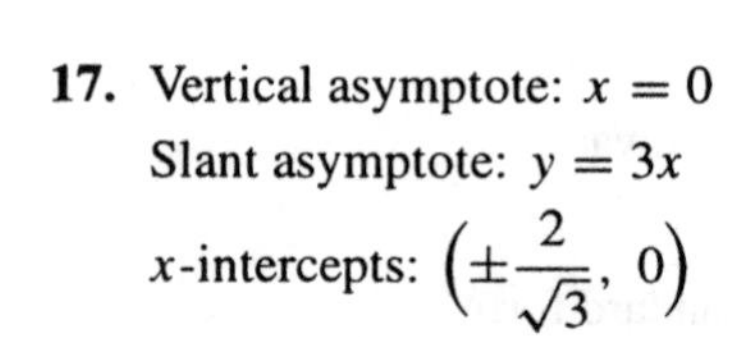

x-intercepts: $\left(\pm\frac{2}{\sqrt{3}},\ 0\right)$

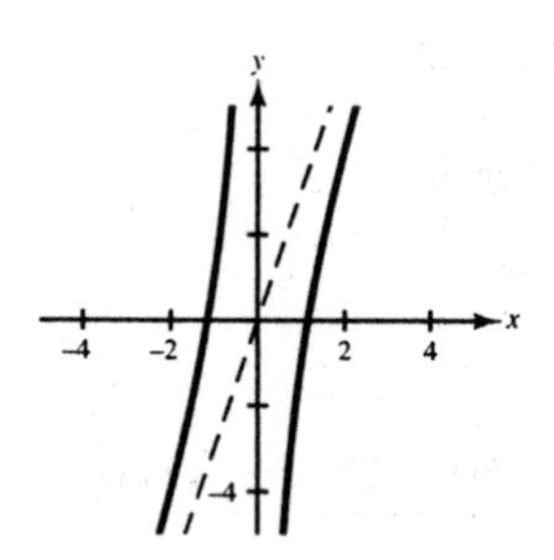

18. $y = \frac{8}{1} = 8$ is a horizontal asymptote. There are no vertical asymptotes.

19. $x = 1$ is a vertical asymptote. There are no horizontal asymptotes.

20. $f(x) = \dfrac{x-5}{(x-5)^2} = \dfrac{1}{x-5}$

Vertical asymptote: $x = 5$

Horizontal asymptote: $y = 0$

y-intercept: $\left(0, -\frac{1}{5}\right)$

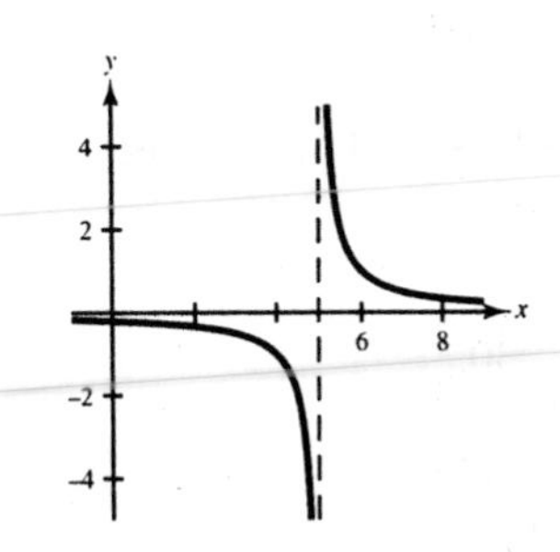

Practice Test Solutions for Chapter 4

1.

$$
\begin{array}{r}
3x^3 + 9x^2 + 20x + 62 + \dfrac{176}{x-3} \\
x-3\,\overline{)\,3x^4 + 0x^3 - 7x^2 + 2x - 10} \\
\underline{3x^4 - 9x^3} \qquad\qquad\qquad\quad \\
9x^3 - 7x^2 \qquad\qquad \\
\underline{9x^3 - 27x^2} \qquad\qquad \\
20x^2 + 2x \qquad \\
\underline{20x^2 - 60x} \qquad \\
62x - 10 \\
\underline{62x - 186} \\
176
\end{array}
$$

2.

$$
\begin{array}{r}
x - 2 + \dfrac{5x-13}{x^2+2x-1} \\
x^2+2x-1\,\overline{)\,x^3 + 0x^2 + 0x - 11} \\
\underline{x^3 + 2x^2 - x} \qquad \\
-2x^2 + x - 11 \\
\underline{-2x^2 - 4x + 2} \\
5x - 13
\end{array}
$$

3.

$$
\begin{array}{r|rrrrrr}
-5 & 3 & 13 & 0 & 0 & 12 & -1 \\
 & & -15 & 10 & -50 & 250 & -1310 \\
\hline
 & 3 & -2 & 10 & -50 & 262 & -1311
\end{array}
$$

$$\frac{3x^5 + 13x^4 + 12x - 1}{x+5} = 3x^4 - 2x^3 + 10x^2 - 50x + 262 - \frac{1311}{x+5}$$

4.

$$
\begin{array}{r|rrrr}
-6 & 7 & 40 & -12 & 15 \\
 & & -42 & 12 & 0 \\
\hline
 & 7 & -2 & 0 & 15
\end{array}
$$

$f(-6) = 15$

5. $0 = x^3 - 19x - 30$

Possible Rational Roots: $\pm 1, \pm 2, \pm 3, \pm 5, \pm 6, \pm 10, \pm 15, \pm 30$

$$\begin{array}{r|rrrr} -2 & 1 & 0 & -19 & -30 \\ & & -2 & 4 & 30 \\ \hline & 1 & -2 & -15 & 0 \end{array}$$

-2 is a zero.

$0 = (x + 2)(x^2 - 2x - 15)$

$0 = (x + 2)(x + 3)(x - 5)$

Zeros: $x = -2, \ x = -3, \ x = 5$

6. $0 = x^4 + x^3 - 8x^2 - 9x - 9$

Possible Rational Roots: $\pm 1, \pm 3, \pm 9$

$$\begin{array}{r|rrrrr} 3 & 1 & 1 & -8 & -9 & -9 \\ & & 3 & 12 & 12 & 9 \\ \hline & 1 & 4 & 4 & 3 & 0 \end{array}$$

$x = 3$ is a zero.

$0 = (x - 3)(x^3 + 4x^2 + 4x + 3)$

Possible Rational Roots of $x^3 + 4x^2 + 4x + 3 : \ \pm 1, \pm 3$

$$\begin{array}{r|rrrr} -3 & 1 & 4 & 4 & 3 \\ & & -3 & -3 & -3 \\ \hline & 1 & 1 & 1 & 0 \end{array}$$

$x = -3$ is a zero.

$0 = (x - 3)(x + 3)(x^2 + x + 1)$

There are no real zeros of $x^2 + x + 1$.

Zeros: $x = 3, \ x = -3$

7. $0 = 6x^3 - 5x^2 + 4x - 15$

Possible Rational Roots: $\pm 1, \pm 3, \pm 5, \pm 15, \pm \frac{1}{2}, \pm \frac{3}{2}, \pm \frac{5}{2}, \pm \frac{15}{2}, \pm \frac{1}{3}, \pm \frac{5}{3}, \pm \frac{1}{6}, \pm \frac{5}{6}$

8. $0 = x^3 - \frac{20}{3}x^2 + 9x - \frac{10}{3}$

$0 = 3x^3 - 20x^2 + 27x - 10$

Possible Rational Roots: $\pm 1, \pm 2, \pm 5, \pm 10, \pm\frac{1}{3}, \pm\frac{2}{3}, \pm\frac{5}{3}, \pm\frac{10}{3}$

$$\begin{array}{r|rrrr} 1 & 3 & -20 & 27 & -10 \\ & & 3 & -17 & 10 \\ \hline & 3 & -17 & 10 & 0 \end{array}$$

$0 = (x - 1)(3x^2 - 17x + 10)$

$0 = (x - 1)(3x - 2)(x - 5)$

Zeros: $x = 1,\ x = \frac{2}{3},\ x = 5$

9. Possible Rational Roots: $\pm 1, \pm 2, \pm 5, \pm 10$

$$\begin{array}{r|rrrrr} 1 & 1 & 1 & 3 & 5 & -10 \\ & & 1 & 2 & 5 & 10 \\ \hline & 1 & 2 & 5 & 10 & 0 \end{array}$$

$x = 1$ is a zero.

$$\begin{array}{r|rrrr} -2 & 1 & 2 & 5 & 10 \\ & & -2 & 0 & -10 \\ \hline & 1 & 0 & 5 & 0 \end{array}$$

$x = -2$ is a zero.

$f(x) = (x - 1)(x + 2)(x^2 + 5) = (x - 1)(x + 2)(x + 5i)(x - 5i)$

10. $x \approx 0.453$, Use your graphing utility with the TRACE and ZOOM keys.

11. $(6 + 5i) + (-2 + 4i) = (6 + (-2)) + (5i + 4i) = 4 + 9i$

12. $(11 - 5i) - (-2 + i) = 11 - 5i + 2 - i = 13 - 6i$

13. $(3 + 4i)(5 - 6i) = 15 - 18i + 20i - 24i^2 = 39 + 2i$

14. $(5 + \sqrt{-8})(7 - \sqrt{-8}) = (5 + 2\sqrt{2}i)(7 - 2\sqrt{2}i) = 35 - 10\sqrt{2}i + 14\sqrt{2}i - 8i^2$

$= 43 + 4\sqrt{2}i$

15. $(3 - 10i)^2 = 9 - 60i + 100i^2 = -91 - 60i$

16. $\dfrac{1 + 4i}{2 - 7i} \cdot \dfrac{2 + 7i}{2 + 7i} = \dfrac{2 + 7i + 8i + 28i^2}{4 + 49} = \dfrac{-26 + 15i}{53} = -\dfrac{26}{53} + \dfrac{15}{53}i$

17. $f(x) = (x - 2)[x - (3 + i)][x - (3 - i)]$

$= (x - 2)[x^2 - x(3 - i) - x(3 + i) + (3 + i)(3 - i)]$

$= (x - 2)[x^2 - 6x + 10] = x^3 - 8x^2 + 22x - 20$

18.

$3i$	1	4	9	36
		$3i$	$12i - 9$	-36
	1	$4 + 3i$	$12i$	0

19. Since $5i$ is a zero, as is $-5i$

$5i$	1	-8	25	-200
		$5i$	$-25 - 40i$	200
$-5i$	1	$-8 + 5i$	$-40i$	0
		$-5i$	$40i$	
	1	-8	0	

$f(x) = (x - 5i)(x + 5i)(x - 8)$

Zeros: $\pm 5i$, 8

20. Using the Quadratic Formula on each factor we have:

$$x = \frac{-6 \pm \sqrt{6^2 - 4(1)(-3)}}{2(1)} = \frac{-6 \pm \sqrt{48}}{2} = \frac{-6 \pm 4\sqrt{3}}{2} = -3 \pm 2\sqrt{3}$$

$$x = \frac{-1 \pm \sqrt{1^2 - 4(1)(11)}}{2(1)} = \frac{-1 \pm \sqrt{-43}}{2} = \frac{-1 \pm \sqrt{43}i}{2} = -\frac{1}{2} \pm \frac{\sqrt{43}}{2}i$$

Zeros: $-3 \pm 2\sqrt{3}$; $-\frac{1}{2} \pm \frac{\sqrt{43}}{2}i$

Practice Test Solutions for Chapter 5

1. $x^{3/5} = 8$

$x = 8^{5/3} = (\sqrt[3]{8})^5 = 2^5 = 32$

2. $3^{x-1} = \frac{1}{81}$

$3^{x-1} = 3^{-4}$

$x - 1 = -4$

$x = -3$

3. $f(x) = 2^{-x} = \left(\frac{1}{2}\right)^x$

x	-2	-1	0	1	2
$f(x)$	4	2	1	$\frac{1}{2}$	$\frac{1}{4}$

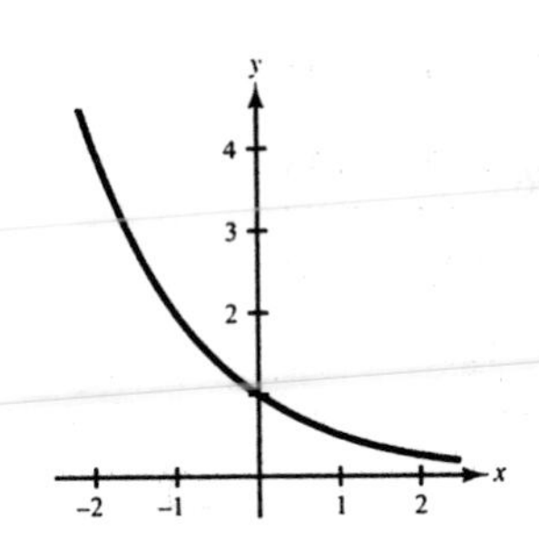

4. $g(x) = e^x + 1$

x	-2	-1	0	1	2
$g(x)$	1.14	1.37	2	3.72	8.39

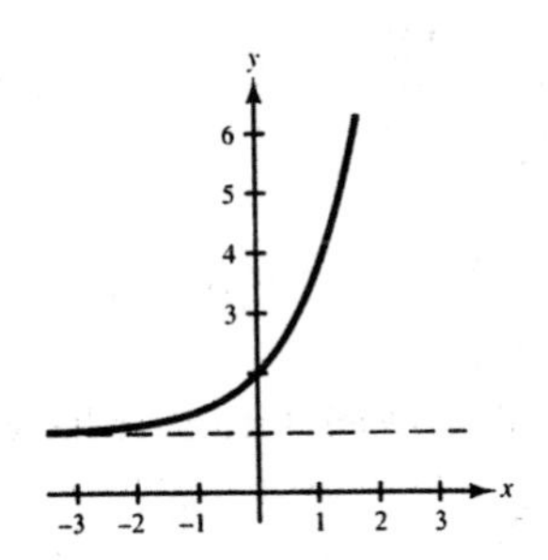

5. $A = P\left(1 + \frac{r}{n}\right)^{nt}$

(a) $A = 5000\left(1 + \frac{0.09}{12}\right)^{12(3)} \approx \$6{,}543.23$

(b) $A = 5000\left(1 + \frac{0.09}{4}\right)^{4(3)} \approx \$6{,}530.25$

(c) $A = 5000e^{(0.09)(3)} \approx \$6{,}549.82$

6. $7^{-2} = \frac{1}{49}$

$\log_7 \frac{1}{49} = -2$

7. $x - 4 = \log_2 \frac{1}{64}$

$2^{x-4} = \frac{1}{64}$

$2^{x-4} = 2^{-6}$

$x - 4 = -6$

$x = -2$

8. $\log_b \sqrt[4]{8/25} = \frac{1}{4}\log_b \frac{8}{25}$

$= \frac{1}{4}[\log_b 8 - \log_b 25]$

$= \frac{1}{4}[\log_b 2^3 - \log_b 5^2]$

$= \frac{1}{4}[3\log_b 2 - 2\log_b 5]$

$= \frac{1}{4}[3(0.3562) - 2(0.8271)]$

$= -0.1464$

9. $5\ln x - \frac{1}{2}\ln y + 6\ln z = \ln x^5 - \ln\sqrt{y} + \ln z^6 = \ln\left(\frac{x^5z^6}{\sqrt{y}}\right)$

10. $\log_9 28 = \frac{\log 28}{\log 9} \approx 1.5166$

11. $\log N = 0.6646$

$N = 10^{0.6646} \approx 4.62$

12.

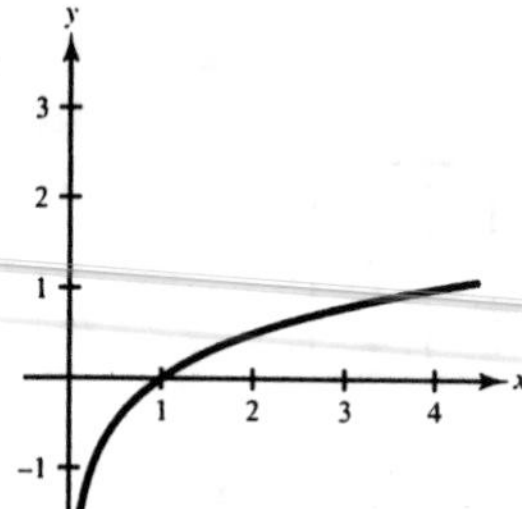

13. Domain: $x^2 - 9 > 0$

$(x + 3)(x - 3) > 0$

$x < -3 \text{ or } x > 3$

14.

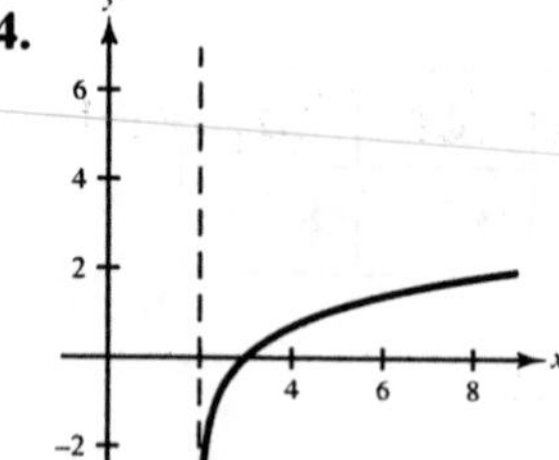

15. $\dfrac{\ln x}{\ln y} \neq \ln(x - y)$ since $\dfrac{\ln x}{\ln y} = \log_y x$

16. $5^x = 41$

$x = \log_5 41 = \dfrac{\ln 41}{\ln 5} \approx 2.3074$

17. $x - x^2 = \log_5 \frac{1}{25}$

$5^{x-x^2} = \frac{1}{25}$

$5^{x-x^2} = 5^{-2}$

$x - x^2 = -2$

$0 = x^2 - x - 2$

$0 = (x + 1)(x - 2)$

$x = -1 \text{ or } x = 2$

18. $\log_2 x + \log_2(x - 3) = 2$

$\log_2[x(x - 3)] = 2$

$x(x - 3) = 2^2$

$x^2 - 3x = 4$

$x^2 - 3x - 4 = 0$

$(x + 1)(x - 4) = 0$

$x = 4$

$x = -1$ (extraneous solution)

19. $$\frac{e^x + e^{-x}}{3} = 4$$
$$e^x(e^x + e^{-x}) = 12e^x$$
$$e^{2x} + 1 = 12e^x$$
$$e^{2x} - 12e^x + 1 = 0$$
$$e^x = \frac{12 \pm \sqrt{144 - 4}}{2}$$
$$e^x = 11.9161 \quad \text{or} \quad e^x = 0.0839$$
$$x = \ln 11.9161 \qquad x = \ln 0.0839$$
$$x \approx 2.4779 \qquad x \approx -2.4779$$

20. $$A = Pe^{rt}$$
$$12,000 = 6,000e^{0.13t}$$
$$2 = e^{0.13t}$$
$$0.13t = \ln 2$$
$$t = \frac{\ln 2}{0.13}$$
$$t \approx 5.3319 \text{ yrs or } 5 \text{ yrs } 4 \text{ months}$$

Practice Test Solutions for Chapter 6

1. $$x + y = 1$$
$$3x - y = 15 \Rightarrow y = 3x - 15$$
$$x + (3x - 15) = 1$$
$$4x = 16$$
$$x = 4$$
$$y = -3$$
Solution: $(4, -3)$

2. $$x - 3y = -3 \Rightarrow x = 3y - 3$$
$$x^2 + 6y = 5$$
$$(3y - 3)^2 + 6y = 5$$
$$9y^2 - 18y + 9 + 6y = 5$$
$$9y^2 - 12y + 4 = 0$$
$$(3y - 2)^2 = 0$$
$$y = \tfrac{2}{3}$$
$$x = -1$$
Solution: $\left(-1, \tfrac{2}{3}\right)$

3. $$x + y + z = 6 \Rightarrow z = 6 - x - y$$
$$2x - y + 3z = 0 \qquad 2x - y + 3(6 - x - y) = 0 \Rightarrow -x - 4y = -18$$
$$5x + 2y - z = -3 \qquad 5x + 2y - (6 - x - y) = -3 \Rightarrow 6x + 3y = 3$$
$$x = 18 - 4y$$
$$6(18 - 4y) + 3y = 3$$
$$-21y = -105$$
$$y = 5$$
$$x = 18 - 4y = -2$$
$$z = 6 - x - y = 3$$
Solution: $(-2, 5, 3)$

4. $x + y = 1$

$3x - y = 15 \Rightarrow y = 3x - 15$

$x + (3x - 15) = 1$

$4x = 16$

$x = 4$

$y = -3$

Solution: $(4, -3)$

5. $2x + 2y = 170 \Rightarrow y = \dfrac{170 - 2x}{2} = 85 - x$

$xy = 2800$

$x(85 - x) = 2800$

$0 = x^2 - 85x + 2800$

$0 = (x - 25)(x - 60)$

$x = 25$ or $x = 60$

$y = 60$ $\quad y = 25$

Dimensions: $60' \times 25'$

6.

$$\begin{aligned} 2x + 15y &= 4 &\Rightarrow\quad 2x + 15y &= 4 \\ x - 3y &= 23 &\Rightarrow\quad 5x - 15y &= 115 \\ & & \overline{7x \qquad\quad} &= 119 \end{aligned}$$

$x = 17$

$y = \dfrac{x - 23}{3} = -2$

Solution: $(17, -2)$

7.

$$\begin{aligned} x + y &= 2 &\Rightarrow\quad 19x + 19y &= 38 \\ 38x - 19y &= 7 &\Rightarrow\quad 38x - 19y &= 7 \\ & & \overline{57x \qquad\quad} &= 45 \end{aligned}$$

$x = \frac{45}{57} = \frac{15}{19}$

$y = 2 - x = \frac{38}{19} - \frac{15}{19} = \frac{23}{19}$

Solution: $\left(\frac{15}{19}, \frac{23}{19}\right)$

8.

$$\begin{aligned} 0.4x + 0.5y &= 0.112 &\Rightarrow\quad 0.28x + 0.35y &= 0.0784 \\ 0.3x - 0.7y &= -0.131 &\Rightarrow\quad 0.15x - 0.35y &= -0.0655 \\ & & \overline{0.43x \qquad\qquad} &= 0.0129 \end{aligned}$$

$x = \dfrac{0.0129}{0.43} = 0.03$

$y = \dfrac{0.112 - 0.4x}{0.5} = 0.20$

Solution: $(0.03, 0.20)$

9. Let x = amount in 11% fund and y = amount in 13% fund.

$x + y = 17000 \Rightarrow y = 17000 - x$

$0.11x + 0.13y = 2080$

$0.11x + 0.13(17000 - x) = 2080$

$-0.02x = -130$

$x = \$6500$ at 11%

$y = \$10,500$ at 13%

10. (4, 3), (1, 1), (−1, −2), (−2, −1)

$$n = 4,\ \sum_{i=1}^{4} x_i = 2,\ \sum_{i=1}^{4} y_i = 1,\ \sum_{i=1}^{4} x_i^2 = 22,\ \sum_{i=1}^{4} x_i y_i = 17$$

$$\begin{aligned} 4b + 2a &= 1 \quad \Rightarrow & 4b + 2a &= 1 \\ 2b + 22a &= 17 \quad \Rightarrow & -4b - 44a &= -34 \\ & & -42a &= -33 \end{aligned}$$

$$a = \tfrac{33}{42} = \tfrac{11}{14}$$

$$b = \tfrac{1}{4}\left[1 - 2\left(\tfrac{33}{42}\right)\right] = -\tfrac{1}{7}$$

$$y = ax + b = \tfrac{11}{14}x - \tfrac{1}{7}$$

11.

$$\begin{aligned} x + y &= -2 \quad \Rightarrow & -2x - 2y &= 4 & -9y + 3z &= 45 \\ 2x - y + z &= 11 & 2x - y + z &= 11 & 4y - 3z &= -20 \\ 4y - 3z &= -20 & -3y + z &= 15 & -5y &= 25 \\ & & & & y &= -5 \\ & & & & x &= 3 \\ & & & & z &= 0 \end{aligned}$$

Solution: $(3, -5, 0)$

12.

$$\begin{aligned} 4x - y + 5z &= 4 \quad \Rightarrow & 4x - y + 5z &= 4 \\ 2x + y - z &= 0 \quad \Rightarrow & -4x - 2y + 2z &= 0 \\ 2x + 4y + 8z &= 0 & -3y + 7z &= 4 \end{aligned}$$

$$\begin{aligned} 2x + 4y + 8z &= 0 \\ -2x - y + z &= 0 \\ 3y + 9z &= 0 \\ -3y + 7z &= 4 \\ 16z &= 4 \\ z &= \tfrac{1}{4} \\ y &= -\tfrac{3}{4} \\ x &= \tfrac{1}{2} \end{aligned}$$

Solution: $\left(\tfrac{1}{2}, -\tfrac{3}{4}, \tfrac{1}{4}\right)$

13. $3x + 2y - z = 5 \Rightarrow 6x + 4y - 2z = 10$

$6x - y + 5z = 2 \Rightarrow -6x + y - 5z = -2$

$$5y - 7z = 8$$

$$y = \frac{8 + 7z}{5}$$

$$3x + 2y - z = 5$$
$$12x - 2y + 10z = 4$$
$$15x + 9z = 9$$

$$x = \frac{9 - 9z}{15} = \frac{3 - 3z}{5}$$

Let $z = a$, then $x = \dfrac{3 - 3a}{5}$ and $y = \dfrac{8 + 7a}{5}$.

Solution: $\left(-\frac{3}{5}a + \frac{3}{5}, \frac{7}{5}a + \frac{8}{5}, a\right)$

14. $y = ax^2 + bx + c$ passes through $(0, -1)$, $(1, 4)$, and $(2, 13)$.

At $(0, -1)$, $-1 = a(0)^2 + b(0) + c \Rightarrow c = -1$

At $(1, 4)$, $4 = a(1)^2 + b(1) - 1 \Rightarrow 5 = a + b \Rightarrow 5 = a + b$

At $(2, 13)$, $13 = a(2)^2 + b(2) - 1 \Rightarrow 14 = 4a + 2b \Rightarrow -7 = -2a - b$

$$-2 = -a$$
$$a = 2$$
$$b = 3$$

Thus, $y = 2x^2 + 3x - 1$.

15. $s = \frac{1}{2}at^2 + v_0t + s_0$ passes through $(1, 12)$, $(2, 5)$, and $(3, 4)$.

At $(1, 12)$, $12 = \frac{1}{2}a + v_0 + s_0 \Rightarrow 24 = a + 2v_0 + 2s_0$

At $(2, 5)$, $5 = 2a + 2v_0 + s_0 \Rightarrow -5 = -2a - 2v_0 - s_0$

At $(3, 4)$, $4 = \frac{9}{2}a + 3v_0 + s_0$

$$19 = -a + s_0$$

$$15 = 6a + 6v_0 + 3s_0$$
$$-8 = -9a - 6v_0 - 2s_0$$
$$7 = -3a + s_0$$
$$-19 = a - s_0$$
$$-12 = -2a$$
$$a = 6$$
$$s_0 = 25$$
$$v_0 = -16$$

Thus, $s = \frac{1}{2}(6)t^2 - 16t + 25 = 3t^2 - 16t + 25$.

16. $x^2 + y^2 \geq 9$

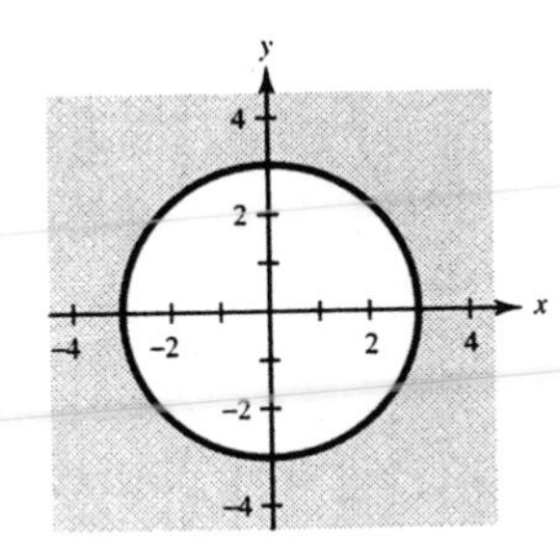

17. $x + y \leq 6$

$x \geq 2$

$y \geq 0$

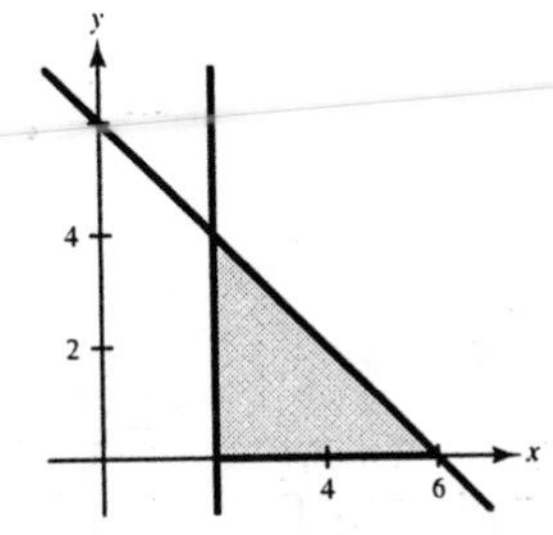

18. Line through (0, 0) and (0, 7): $x = 0$

Line through (0, 0) and (2, 3):

$y = \frac{3}{2}x$ or $3x - 2y = 0$

Line through (0, 7) and (2, 3):

$y = -2x + 7$ or $2x + y = 7$

Inequalities: $x \geq 0$

$3x - 2y \leq 0$

$2x + y \leq 7$

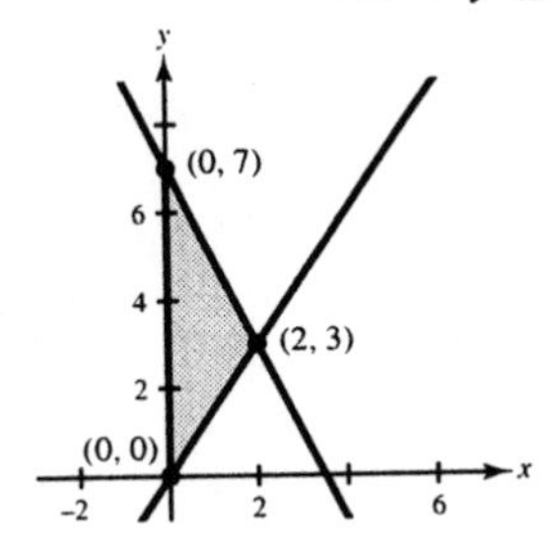

19. Vertices: (0, 0), (0, 7), (6, 0), (3, 5)

$C = 30x + 26y$

At (0, 0), $C = 0$

At (0, 7), $C = 182$

At (6, 0), $C = 180$

At (3, 5), $C = 220$

The maximum value of C is 220.

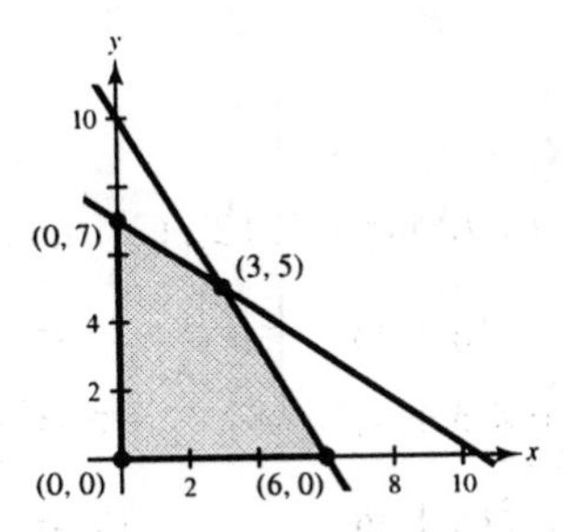

20. $x^2 + y^2 \leq 4$

$(x - 2)^2 + y^2 \geq 4$

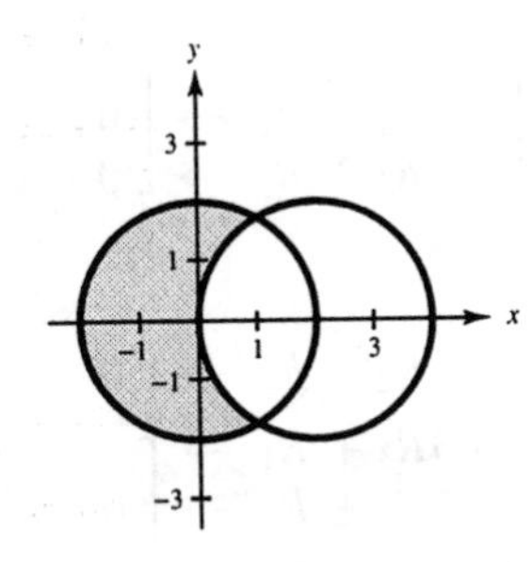

Practice Test Solutions for Chapter 7

1. $\begin{bmatrix} 1 & -2 & 4 \\ 3 & -5 & 9 \end{bmatrix} \; -3R_1 + R_2 \rightarrow \begin{bmatrix} 1 & -2 & 4 \\ 0 & 1 & -3 \end{bmatrix} \; 2R_2 + R_1 \rightarrow \begin{bmatrix} 1 & 0 & -2 \\ 0 & 1 & -3 \end{bmatrix}$

2. $3x + 5y = 3$

$2x - y = -11$

$$\left[\begin{array}{cc:c} 3 & 5 & 3 \\ 2 & -1 & -11 \end{array}\right] \; -R_2 + R_1 \rightarrow \left[\begin{array}{cc:c} 1 & 6 & 14 \\ 2 & -1 & -11 \end{array}\right]$$

$$-2R_1 + R_2 \rightarrow \left[\begin{array}{cc:c} 1 & 6 & 14 \\ 0 & -13 & -39 \end{array}\right]$$

$$-\tfrac{1}{13}R_2 \rightarrow \left[\begin{array}{cc:c} 1 & 6 & 14 \\ 0 & 1 & 3 \end{array}\right]$$

$$-6R_2 + R_1 \rightarrow \left[\begin{array}{cc:c} 1 & 0 & -4 \\ 0 & 1 & 3 \end{array}\right]$$

Answer: $x = -4, \; y = 3$

3. $2x + 3y = -3$

$3x + 2y = 8$

$x + y = 1$

$$\left[\begin{array}{cc:c} 2 & 3 & -3 \\ 3 & 2 & 8 \\ 1 & 1 & 1 \end{array}\right] \quad \begin{matrix} R_3 \\ \\ R_1 \end{matrix} \quad \left[\begin{array}{cc:c} 1 & 1 & 1 \\ 3 & 2 & 8 \\ 2 & 3 & -3 \end{array}\right]$$

$$\begin{matrix} \\ -3R_1 + R_2 \rightarrow \\ 2R_1 + R_3 \rightarrow \end{matrix} \left[\begin{array}{cc:c} 1 & 1 & 1 \\ 0 & -1 & 5 \\ 0 & -1 & 5 \end{array}\right]$$

$$\begin{matrix} R_2 + R_1 \rightarrow \\ -R_2 \rightarrow \\ -R_2 + R_3 \rightarrow \end{matrix} \left[\begin{array}{cc:c} 1 & 0 & 6 \\ 0 & 1 & -5 \\ 0 & 0 & 0 \end{array}\right]$$

Answer: $x = 6, \; y = -5$

4. $x + 3z = -5$

$2x + y = 0$

$3x + y - z = 3$

$$\left[\begin{array}{ccc:c} 1 & 0 & 3 & -5 \\ 2 & 1 & 0 & 0 \\ 3 & 1 & -1 & 3 \end{array}\right] \begin{matrix} \\ -2R_1 + R_2 \rightarrow \\ -3R_1 + R_3 \rightarrow \end{matrix} \left[\begin{array}{ccc:c} 1 & 0 & 3 & -5 \\ 0 & 1 & -6 & 10 \\ 0 & 1 & -10 & 18 \end{array}\right]$$

$$\begin{matrix} \\ \\ -R_2 + R_3 \rightarrow \end{matrix} \left[\begin{array}{ccc:c} 1 & 0 & 3 & -5 \\ 0 & 1 & -6 & 10 \\ 0 & 0 & -4 & 8 \end{array}\right]$$

$$\begin{matrix} -3R_3 + R_1 \rightarrow \\ 6R_3 + R_2 \rightarrow \\ -\tfrac{1}{4}R_4 \rightarrow \end{matrix} \left[\begin{array}{ccc:c} 1 & 0 & 0 & 1 \\ 0 & 1 & 0 & -2 \\ 0 & 0 & 1 & -2 \end{array}\right]$$

Answer: $x = 1, \; y = -2, \; z = -2$

5. $\begin{bmatrix} 1 & 4 & 5 \\ 2 & 0 & -3 \end{bmatrix}\begin{bmatrix} 1 & 6 \\ 0 & -7 \\ -1 & 2 \end{bmatrix} = \begin{bmatrix} -4 & -12 \\ 5 & 6 \end{bmatrix}$

6. $3A - 5B = 3\begin{bmatrix} 9 & 1 \\ -4 & 8 \end{bmatrix} - 5\begin{bmatrix} 6 & -2 \\ 3 & 5 \end{bmatrix}$

$= \begin{bmatrix} 27 & 3 \\ -12 & 24 \end{bmatrix} - \begin{bmatrix} 30 & -10 \\ 15 & 25 \end{bmatrix} = \begin{bmatrix} -3 & 13 \\ -27 & -1 \end{bmatrix}$

7. $AB = \begin{bmatrix} 9 & 1 \\ -4 & 8 \end{bmatrix}\begin{bmatrix} 6 & -2 \\ 3 & 5 \end{bmatrix} = \begin{bmatrix} 57 & -13 \\ 0 & 48 \end{bmatrix}$

$BA = \begin{bmatrix} 6 & -2 \\ 3 & 5 \end{bmatrix}\begin{bmatrix} 9 & 1 \\ -4 & 8 \end{bmatrix} = \begin{bmatrix} 62 & -10 \\ 7 & 43 \end{bmatrix}$

8. False since

$(A + B)(A + 3B) = A(A + 3B) + B(A + 3B) = A^2 + 3AB + BA + 3B^2$

9. $\left[\begin{array}{cc:cc} 1 & 2 & 1 & 0 \\ 3 & 5 & 0 & 1 \end{array}\right] \begin{array}{r} \\ -3R_1 + R_2 \rightarrow \end{array} \left[\begin{array}{cc:cc} 1 & 2 & 1 & 0 \\ 0 & -1 & -3 & 1 \end{array}\right]$

$\begin{array}{r} 2R_2 + R_1 \rightarrow \\ -R_2 \rightarrow \end{array} \left[\begin{array}{cc:cc} 1 & 0 & -5 & 2 \\ 0 & 1 & 3 & -1 \end{array}\right]$

$A^{-1} = \begin{bmatrix} -5 & 2 \\ 3 & -1 \end{bmatrix}$

10. $\left[\begin{array}{ccc:ccc} 1 & 1 & 1 & 1 & 0 & 0 \\ 3 & 6 & 5 & 0 & 1 & 0 \\ 6 & 10 & 8 & 0 & 0 & 1 \end{array}\right] \begin{array}{r} \\ -3R_1 + R_2 \rightarrow \\ -6R_1 + R_3 \rightarrow \end{array} \left[\begin{array}{ccc:ccc} 1 & 1 & 1 & 1 & 0 & 0 \\ 0 & 3 & 2 & -3 & 1 & 0 \\ 0 & 4 & 2 & -6 & 0 & 1 \end{array}\right]$

$\begin{array}{r} -R_2 + R_1 \rightarrow \\ \frac{1}{3}R_2 \rightarrow \\ -4R_2 + R_3 \rightarrow \end{array} \left[\begin{array}{ccc:ccc} 1 & 0 & \frac{1}{3} & 2 & -\frac{1}{3} & 0 \\ 0 & 1 & \frac{2}{3} & -1 & \frac{1}{3} & 0 \\ 0 & 0 & -\frac{2}{3} & -2 & -\frac{4}{3} & 1 \end{array}\right]$

$\begin{array}{r} \frac{1}{2}R_3 + R_1 \rightarrow \\ R_3 + R_2 \rightarrow \\ -\frac{3}{2}R_3 \rightarrow \end{array} \left[\begin{array}{ccc:ccc} 1 & 0 & 0 & 1 & -1 & \frac{1}{2} \\ 0 & 1 & 0 & -3 & -1 & 1 \\ 0 & 0 & 1 & 3 & 2 & -\frac{3}{2} \end{array}\right]$

$A^{-1} = \begin{bmatrix} 1 & -1 & \frac{1}{2} \\ -3 & -1 & 1 \\ 3 & 2 & -\frac{3}{2} \end{bmatrix}$

11. (a) $x + 2y = 4$
$3x + 5y = 1$

$$\left[\begin{array}{cc:cc} 1 & 2 & 1 & 0 \\ 3 & 5 & 0 & 1 \end{array}\right] \begin{array}{c} \\ -3R_1 + R_2 \to \end{array} \left[\begin{array}{cc:cc} 1 & 2 & 1 & 0 \\ 0 & -1 & -3 & 1 \end{array}\right]$$

$$\begin{array}{r} -2R_2 + R_1 \to \\ -R_2 \to \end{array} \left[\begin{array}{cc:cc} 1 & 0 & -5 & 2 \\ 0 & 1 & 3 & -1 \end{array}\right]$$

$$X = A^{-1}B = \begin{bmatrix} -5 & 2 \\ 3 & -1 \end{bmatrix}\begin{bmatrix} 4 \\ 1 \end{bmatrix} = \begin{bmatrix} -18 \\ 11 \end{bmatrix}$$

$x = -18,\ y = 11$

(b) $x + 2y = 3$
$3x + 5y = -2$

$$X = A^{-1}B = \begin{bmatrix} -5 & 2 \\ 3 & -1 \end{bmatrix}\begin{bmatrix} 3 \\ -2 \end{bmatrix} = \begin{bmatrix} -19 \\ 11 \end{bmatrix}$$

$x = -19,\ y = 11$

12. $\begin{vmatrix} 6 & -1 \\ 3 & 4 \end{vmatrix} = 24 - (-3) = 27$

13. $\begin{vmatrix} 1 & 3 & -1 \\ 5 & 9 & 0 \\ 6 & 2 & -5 \end{vmatrix} \begin{matrix} 1 & 3 \\ 5 & 9 \\ 6 & 2 \end{matrix} = (-45 + 0 - 10) - (-54 + 0 - 75) = 74$

14. $\begin{vmatrix} 1 & 4 & 2 & 3 \\ 0 & 1 & -2 & 0 \\ 3 & 5 & -1 & 1 \\ 2 & 0 & 6 & 1 \end{vmatrix} = \begin{vmatrix} 1 & 2 & 3 \\ 3 & -1 & 1 \\ 2 & 6 & 1 \end{vmatrix} + 2\begin{vmatrix} 1 & 4 & 3 \\ 3 & 5 & 1 \\ 2 & 0 & 1 \end{vmatrix} = 51 + 2(-29) = -7$

Expansion along Row 2.

15. $\begin{vmatrix} 3 & 0 & 0 \\ 0 & 3 & 0 \\ 0 & 0 & 3 \end{vmatrix} = 3(3)(3)\begin{vmatrix} 1 & 0 & 0 \\ 0 & 1 & 0 \\ 0 & 0 & 1 \end{vmatrix} = -3^3\begin{vmatrix} 1 & 0 & 0 \\ 0 & 0 & 1 \\ 0 & 1 & 0 \end{vmatrix}$

True

16. $\begin{vmatrix} 6 & 4 & 3 & 0 & 6 \\ 0 & 5 & 1 & 4 & 8 \\ 0 & 0 & 2 & 7 & 3 \\ 0 & 0 & 0 & 9 & 2 \\ 0 & 0 & 0 & 0 & 1 \end{vmatrix} = 6(5)(2)(9)(1) = 540$

17. Area $= \pm\frac{1}{2}\begin{vmatrix} 6 & 3 & 1 \\ -1 & 4 & 1 \\ 2 & -5 & 1 \end{vmatrix} = \pm\frac{1}{2}(60) = 30$ square units.

18. $\begin{vmatrix} 0 & -7 & 1 \\ 3 & 2 & 1 \\ -4 & -19 & 1 \end{vmatrix} = 0$ Therefore, the points are collinear.

19. $\begin{vmatrix} x & y & 1 \\ 4 & -3 & 1 \\ \frac{1}{2} & 7 & 1 \end{vmatrix} = 0$

$-10x - \frac{7}{2}y + \frac{59}{2} = 0$

$20x + 7y - 59 = 0$

20. $\begin{matrix} \text{S T} \\ \text{A N} \\ \text{D} \\ \text{F I} \\ \text{R M} \end{matrix} \begin{bmatrix} 19 & 20 \\ 1 & 14 \\ 4 & 0 \\ 6 & 9 \\ 18 & 13 \end{bmatrix} \begin{bmatrix} 3 & -1 \\ 7 & 5 \end{bmatrix} = \begin{bmatrix} 197 & 81 \\ 101 & 69 \\ 12 & -4 \\ 81 & 39 \\ 145 & 47 \end{bmatrix}$

Cryptogram: 197 81 101 69 12 −4 81 39 145 47

Practice Test Solutions for Chapter 8

1. $a_n = \dfrac{2n}{(n+2)!}$

$a_1 = \dfrac{2(1)}{3!} = \dfrac{2}{6} = \dfrac{1}{3}$

$a_2 = \dfrac{2(2)}{4!} = \dfrac{4}{24} = \dfrac{1}{6}$

$a_3 = \dfrac{2(3)}{5!} = \dfrac{6}{120} = \dfrac{1}{20}$

$a_4 = \dfrac{2(4)}{6!} = \dfrac{8}{720} = \dfrac{1}{90}$

$a_5 = \dfrac{2(5)}{7!} = \dfrac{10}{5040} = \dfrac{1}{504}$

$\left\{\dfrac{1}{3}, \dfrac{1}{6}, \dfrac{1}{20}, \dfrac{1}{90}, \dfrac{1}{504}, \ldots\right\}$

2. $a_n = \dfrac{n+3}{3^n}$

3. $\displaystyle\sum_{i=1}^{6}(2i-1) = 1+3+5+7+9+11 = 36$

4. $a_1 = 23,\ d = -2$

$a_2 = a_1 + d = 21$

$a_3 = a_2 + d = 19$

$a_4 = a_3 + d = 17$

$a_5 = a_4 + d = 15$

$\{23, 21, 19, 17, 15, \ldots\}$

5. $a_1 = 12,\ d = 3,\ n = 50$

$a_n = a_1 + (n-1)d$

$a_{50} = 12 + (50-1)3 = 159$

6. $a_1 = 1$

$a_{200} = 200$

$S_n = \dfrac{n}{2}(a_1 + a_n)$

$S_{200} = \dfrac{200}{2}(1 + 200) = 20,100$

7. $a_1 = 7,\ r = 2$

$a_2 = a_1 r = 14$

$a_3 = a_2 r = 28$

$a_4 = a_3 r = 56$

$a_5 = a_4 r = 112$

$\{7,\ 14,\ 28,\ 56,\ 112,\ \ldots\}$

8. $\displaystyle\sum_{n=0}^{9} 6\left(\frac{2}{3}\right)^n,\ a_1 = 6,\ r = \frac{2}{3},\ n = 9$

$$S_n = \frac{a_1(1 - r^n)}{1 - r}$$

$$= \frac{6[1 - (2/3)^9]}{1 - (2/3)} \approx 17.5318$$

9. $\displaystyle\sum_{n=0}^{\infty} (0.03)^n,\ a_1 = 1,\ r = 0.03$

$$S_n = \frac{a_1}{1 - r} = \frac{1}{1 - 0.03} = \frac{1}{0.97} = \frac{100}{97} \approx 1.0309$$

10. $\displaystyle\frac{(n+3)!}{n!} = \frac{(n+3)(n+2)(n+1)n!}{n!} = (n+3)(n+2)(n+1)$

11. $\displaystyle A = \sum_{i=1}^{240} 50\left(1 + \frac{0.07}{12}^i\right) \approx \$26,198.27$

12. $\displaystyle {}_{13}C_4 = \frac{13!}{(13-4)!4!} = 715$

13. $(x+3)^5 = x^5 + 5x^4(3) + 10x^3(3)^2 + 10x^2(3)^3 + 5x(3)^4 + (3)^5$

$= x^5 + 15x^4 + 90x^3 + 270x^2 + 405x + 243$

14. ${}_{12}C_5 x^7(-2)^5 = -25,344x^7$

15. $\displaystyle {}_{30}P_4 = \frac{30!}{(30-4)!} = 657,720$

16. $6! = 720$ ways

17. ${}_{12}P_3 = 1320$

18. $P(2) + P(3) + P(4) = \frac{1}{36} + \frac{2}{36} + \frac{3}{36}$

$= \frac{6}{36} = \frac{1}{6}$

19. $P(K,\ B10) = \frac{4}{52} \cdot \frac{2}{51} = \frac{2}{663}$

20. Let A = probability of no faulty units.

$$P(A) = \left(\frac{997}{1000}\right)^{50} \approx 0.8605$$

$$P(A') = 1 - P(A) \approx 0.1395$$